MODELLING AND COMPUTATION IN ENGINEERING

PROCEEDINGS OF THE INTERNATIONAL CONFERENCE ON MODELLING AND COMPUTATION IN ENGINEERING, CMCE 2010, HONG KONG, 6–7 NOVEMBER 2010

Modelling and Computation in Engineering

Editor

Jinrong Zhu
North China Electric Power University, Beijing, China

CRC Press
Taylor & Francis Group
Boca Raton London New York Leiden

CRC Press is an imprint of the
Taylor & Francis Group, an **informa** business
A BALKEMA BOOK

Sponsored by Society for Resources, Environment and Engineering

CRC Press/Balkema is an imprint of the Taylor & Francis Group, an informa business

Typeset by MPS Limited, a Macmillan Company, Chennai, India

Published by: CRC Press/Balkema
P.O. Box 447, 2300 AK Leiden, The Netherlands
e-mail: Pub.NL@taylorandfrancis.com
www.crcpress.com – www.taylorandfrancis.co.uk – www.balkema.nl

ISBN: 978-0-415-61516-7 (Hbk)
ISBN: 978-0-203-82985-1 (eBook)

Modelling and Computation in Engineering – Zhu (ed)
© 2011 Taylor & Francis Group, London, ISBN 978-0-415-61516-7

Table of Contents

Modelling and Computation in Engineering – Zhu (ed)
© 2011 Taylor & Francis Group, London, ISBN 978-0-415-61516-7

Preface

In recent years the theory and technology of modelling and computation in engineering has expanded rapidly, and has been widely applied in various kinds of engineering projects. The 2010 International Conference on Modeling and Computation in Engineering (CMCE Hong Kong, November 6–7, 2010), sponsored and hosted by the Society for Resources, Environment and Engineering, provided a forum for discussion on this highly topical subject. The 37 contributions selected from 130 submissions cover the state-of-the-art on a broad range of topics, including:

- Tunnelling
- Seismic reduction technologies
- Wind-induced vibration control
- Asphalt-rubber concrete
- Open boundary field problems

A special word of thanks goes to all committee members and the editors from Taylor & Francis/ CRC Press / Balkema for their excellent work. We hope that the book will attract much interest from academics, leading engineers, industry researchers and scholar students in engineering and engineering-related disciplines.

Jinrong Zhu

Modelling and Computation in Engineering – Zhu (ed)
© 2011 Taylor & Francis Group, London, ISBN 978-0-415-61516-7

Influence analysis of core rock construction in super large section and span tunnel

Dingheng Zhou
Technical University of Munch, Munich, Germany

Liqiao Cao
China Railway Siyuan Survey and Design Group Co., Ltd, Hubei, China

Yongfeng Ma
China Petroleum East China Design Institute, Shandong, China

Zhan Shi
Tongji University, Shanghai, China

ABSTRACT: Different from normal tunnel, the mechanical behaviors of super large section tunnel with two bores and eight lanes are more complex and the construction methods are also diversified. Based on the largest four-lane super large section and span highway tunnel in China, the influence of core rock excavation and temporary support dismantling in super large section and span tunnel is analyzed in detail with the numerical analysis method. Some conclusions have been drawn. (1) During the construction of super large section and span tunnel, the excavation of the core rock has a great effect on the deformation and stress state of the surrounding rock and tunnel support. (2) When excavating the core rock and dismantling the temporary support, the stress change of the zone above the inverted arch is larger than that of the inverted arch zone. (3) The effect on the support stress in the left and right heading caused by the dismantling of temporary support is different and the stress change of the left heading larger than that of the right heading. The experience and data presented in the paper can be refereed in the design, construction and research of similar tunnels.

1 INTRODUCTION

Recently, many four-lane super large section tunnels have been built all over China to suit for multiply-lane highway construction. There are some developments about the stress, deformation and stability research on rockmass and lining in four-lane super large section tunnels. As to theory analysis aspect, Qu Haifeng proposed a new load mode for large cross-section and low flat-ratio tunnel. Some researchers such as Sun Xiang, Huang Lunhai, Wu Mengjun, Xu Chongbang and LI Zhigang have done some work on numerical analysis and model test. Those researches are mainly about model test and numerical analysis of dynamic construction process and the mechanical characteristics. Except that, Yuan Yong proposed a pre-built and prestressed innovative support method for four-lane super large section tunnel and made a validation about the pre-built and prestressed structure. Huang Chengzao proposed cross anchor to control the deformation of core rock in four-lane tunnel. As for in-situ test on four-lane tunnel, Chen Gengye analyzed the stress characteristics of four-lane tunnel based on stress monitoring. And Zhou Dingheng made a detailed analysis of rock deformation based on in-situ test of super large section and span tunnel. Based on one four-lane tunnel, Liu Heng finished some research about on the durability of tunnel structure. At present, the design and construction standard of super large section and span tunnels has not been formed and the related research is limited. Therefore, it is essential to do further research about four-lane tunnels. Except that, due to the characteristics of large excavation span and low flat-ratio, the construction sequences conversion is complicated and multiply blast has multiply disturbance to the surrounding rock. Especially, the core rock stress is very complicated. Thus, there tends to be many stability problems such as large section and tunnel collapse. Therefore, further research on super large section and span tunnel is an important content of rock tunnels.

Based on the largest four-lane super large section and span rock tunnel in China, the effect on the preliminary lining safety caused by core rock excavation and temporary support dismantling is analyzed with site monitoring and numerical simulation methods.

Table 1. Physical-mechanical parameters.

Material	γ/ kN·m^{-1}	E/GPa	μ	φ/(°)	c/MPa
Gravelly clayey	16.6	0.14	0.38	24.5	40
Completely decomposed granite	19	0.2	0.35	100	21
Highly decomposed granite	21	0.65	0.3	30	130
Weakly decomposed granite	26.1	2.5	0.21	1055	47.6
Backfill	25	1.0	0.2	1500	50

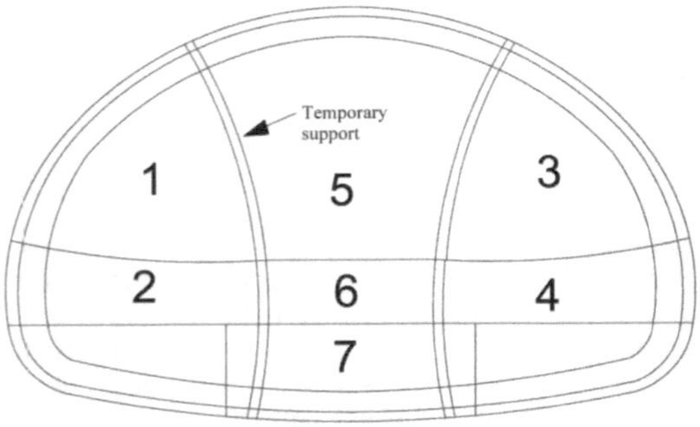

Figure 1. Design construction sequence.

2 BACKGROUNDS

The construction of this two bores and eight lanes tunnel began from January, 2005 to May, 2006, with the normal speed 100 km/h. The length of the left and right line is 1010 m and 1006 m with the distance of two tunnels 25 m. The rock includes second to fifth class rock with the maximum depth 98 m. The large excavation width and height are 20.7 m and 13.58 m. According to the engineering geological exploration, the rock layer is simple. The bedrock is granite intrusive mass with the rock type monzonitic granite. The rock parameters are listed in table 1.

The design construction sequence in soft rock is shown in figure 1. The firstly-excavated left heading and the secondly-excavated right heading can be divided into two parts. The excavation of the core rock has three stages. The structure is composite lining with preliminary supports such as anchor bolt and steel fiber reinforced concrete. Except that, other support measures such the steel support and pre-grouting with small ducts are also available.

3 NUMERICAL SIMULATION

3.1 *Numerical model*

The K5 + 870 section has been chosen as the simulation section. According to the incidence of tunnel

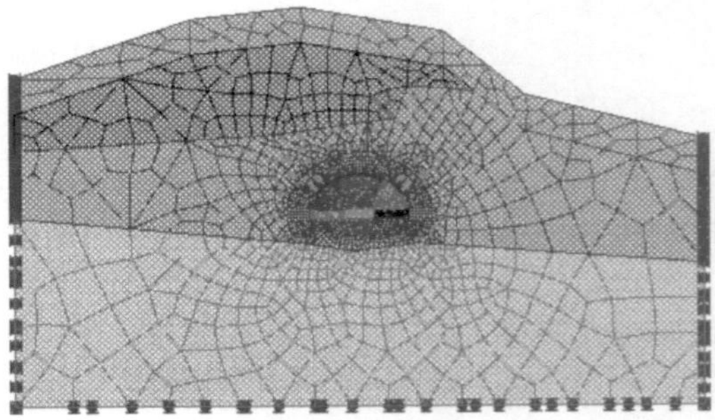

Figure 2. Mesh division of model.

Table 2. Supporting structures parameters.

Support structure	E/GPa	γ/ (kN·m^{-1})	A/m^2	I/m^4
System bolt	210	77	0.000491	/
Temporary bolt	210	77	0.000491	/
Preliminary lining	27.8	22	0.27	0.00164
Secondary lining	31	26	0.6	0.018

excavation, the model size is described as following. The distance from the left or right boundary to the tunnel is 3 times of the span and the distance from the bottom boundary to the tunnel is also 3 times of tunnel span. The upper boundary is the ground surface. The physical-mechanical parameters are listed in table 1.

Because of the bad geotechnical condition and complicated rock stress field, the failure modes of rock have diversity, such as tension rupture, brittle shear failure or plastic failure. And those failure types are regarded as plastic failure. Therefore, the ideal elasto-plastic model and Drucker-Prager yield criterion are chosen in FEM analysis.

Boundary-restraint condition: 2-dimenssion plane strain model is used in the numerical simulation with the left and right boundary horizontal restraint and the bottom boundary vertical restraint. And the top is free with restraint. The mesh division is shown in figure 2.

The construction process is achieved by the elements provide by some commands such as "Excavation", "Backfill" and "Copy of linear material". And the rock and backfill are simulated by 4-point solid elements. The anchor bolts, preliminary lining and spray concrete layer of core rock is simulated with 2-D bar element. As to the simulation of the secondary lining, steel arch and temporary steel arch and horizontal support of core rock, the 2-D beam element is adopted. The supporting structures parameters are listed in table 2.

3.2 *Dynamic simulation of construction scheme*

The dynamic simulation process of tunnel construction is: (1) upper-stage of left stage; (2) preliminary

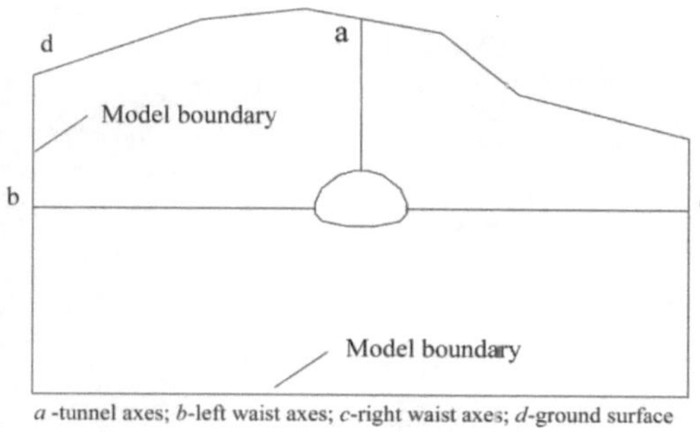

Figure 3. Schematic diagram of displacement acquisition.

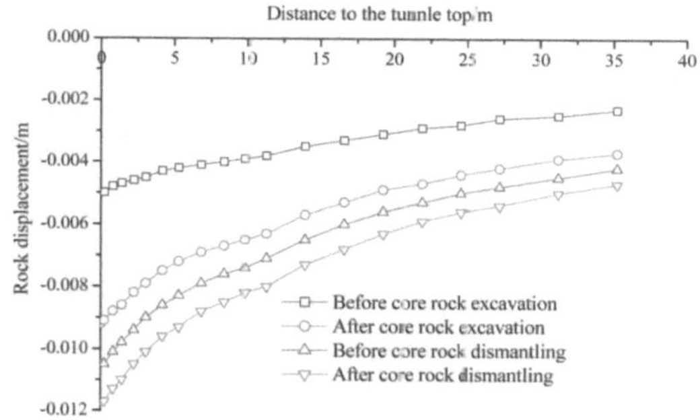

Figure 4. Comparison of rock displacement near arch crown.

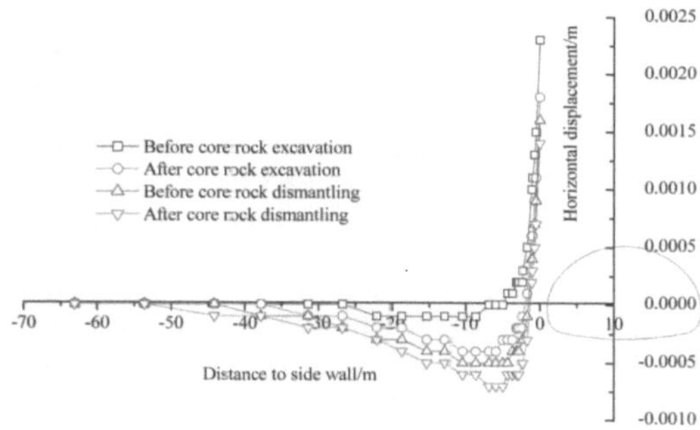

Figure 5. Comparison of horizontal displacement of left side.

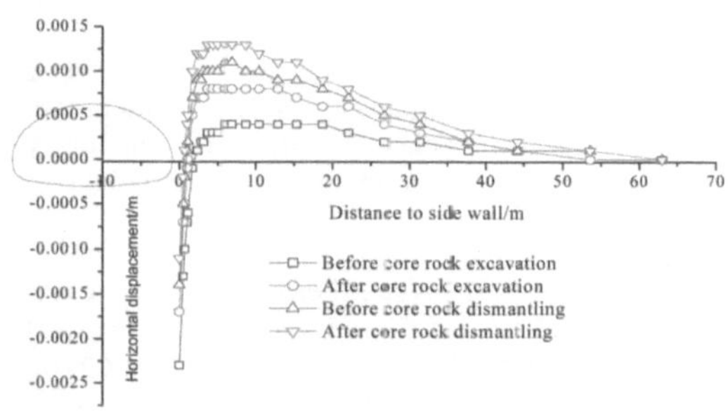

Figure 6. Comparison of horizontal displacement of right side.

lining and temporary support; (3) upper-stage of right stage; (4) preliminary lining and temporary support; (5) bottom-stage of right stage; (6) preliminary lining and temporary support; (7) secondary lining in invert arch; (8) backfill; (9) bottom-stage of left stage; (10) preliminary lining and temporary support; (11) secondary lining in invert arch; (12) backfill; (13) upper-part in core rock; (14) preliminary lining; (15) middle-part in core rock; (16) temporary support dismantling of core rock; (17) bottom-part in core rock; (18) preliminary lining; (19) secondary lining; (20) backfill; (21) secondary lining of other parts.

4 NUMERICAL RESULTS ANALYSIS

4.1 *Displacement analysis*

The effect of excavation and dismantling of core rock can be reflected with the surrounding rock displacement. The displacement of the feature points is collected in numerical simulation. And the sign of surrounding rock is accorded with that of the model coordinate.

4.1.1 *Rock displacement above the tunnel top*

The crown settlement is an important index for surrounding rock stability and safety evaluation of construction. The rock displacement above the tunnel top before and after the core rock excavation and temporary support dismantling is shown in figure 4. It can be seen from figure 4 that the core rock excavation causes a great change of the rock displacement. The displacement change at the tunnel top is about 4 mm, which is 90% of the former displacement. The displacement decreases with the distance to the tunnel top the rock. Compared with excavation of core rock, the rock displacement caused by temporary support dismantling is smaller. And the difference for different distance to the tunnel top is not apparent.

4.1.2 *Horizontal displacement*

The horizontal displacement of line b and c in figure 3 is shown in figure 5 and 6. Based on figure 5 and 6, it can be drawn that the core rock excavation and temporary support dismantling can reduce the horizontal displacement and control the horizontal convergence. Thus, the tunnel construction is benefit to the deformation control of side walls.

4.1.3 *Ground subsidence*

The ground subsidence before and after the core rock excavation and temporary support dismantling is shown in figure 7. The ground subsidence change of the midline is the largest. The subsidence change caused by excavation of core rock is about 2 mm. The ground subsidence at the both side become small with the distance from that point to the midline. As same as rock displacement above the tunnel top, the ground

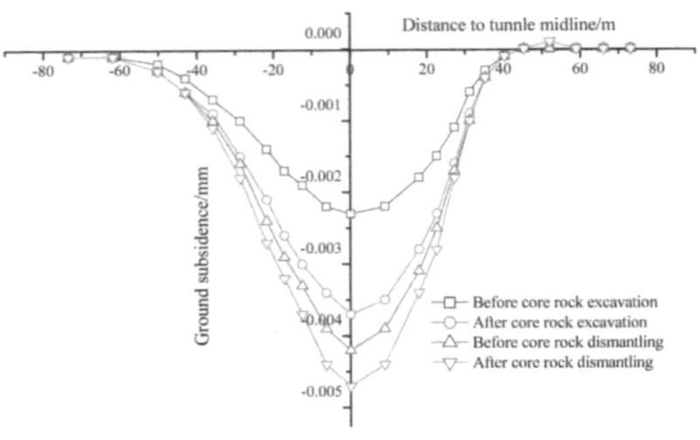

Figure 7. Comparison of surface settlement.

subsidence caused by the dismantling of temporary support is small.

Based on the numerical analysis of rock displacement above the top, rock displacement in side wall direction and ground subsidence results, it can be drawn that the excavation of core rock causes a great change of surrounding rock displacement. Except that, the displacement change above the tunnel top is larger than the displacement change in the side wall direction. Thus, the rock deformation should be controlled during the core rock construction in super large section and span tunnel and the key point to deformation control is the top deformation.

4.2 *Stress analysis*

The effect of different construction procedures can be reflected by the rock stress distribution, the force of the preliminary lining and secondary lining. Therefore, the state of the rock stress distribution, the force of the preliminary lining and secondary lining before and after the core rock excavation and temporary support dismantling are compared in the following.

4.2.1 *Rock stress*

The change of rock stress state before and after excavation of core rock and dismantling of temporary support is shown in figure 8. Comparing figure 8(a) and 8(b), the core rock excavation caused a larger change of rock stress. Before core rock excavation, stress concentration appears at the top of the left and right hole and the backfill zone of two holes is in extremely small tension stress condition. After core rock excavation, the rock stress state change greatly. The stress concentration at the top of the left and right hole has been controlled with the concentration zone smaller. But the rock stress level is improved and tension stress appears at the top of the left hole. After core rock excavation, the anchor bolts are construction and the rock stress are better controlled with no apparent change (figure 8(c)). Comparing figure 8(c) and 8(d), the rock stress changes with a certain extent. While compared with excavation procedure, the change is smaller. The apparent change is that the tension zone at the top of left hole reduces. Except that, the stress state of the down-stage of core rock changes a littlie.

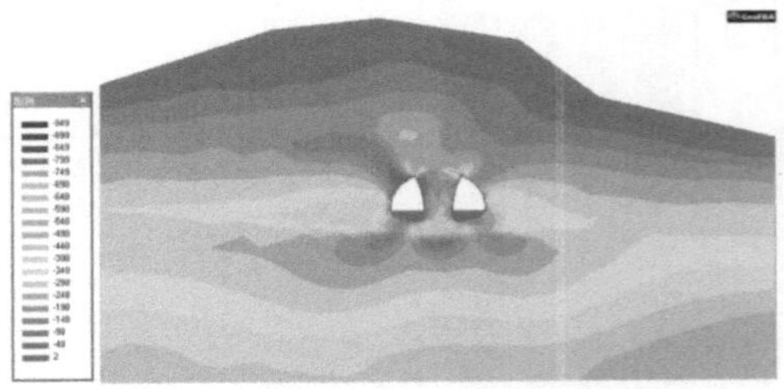

(a) Before excavation of core rock

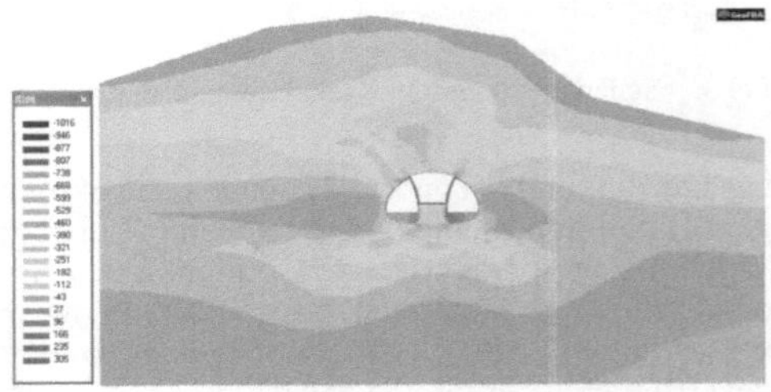

(b) After excavation of core rock

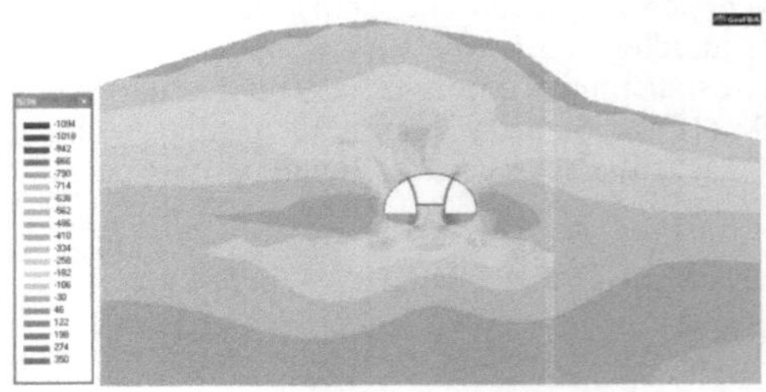

(c) Before dismantling of temporary support

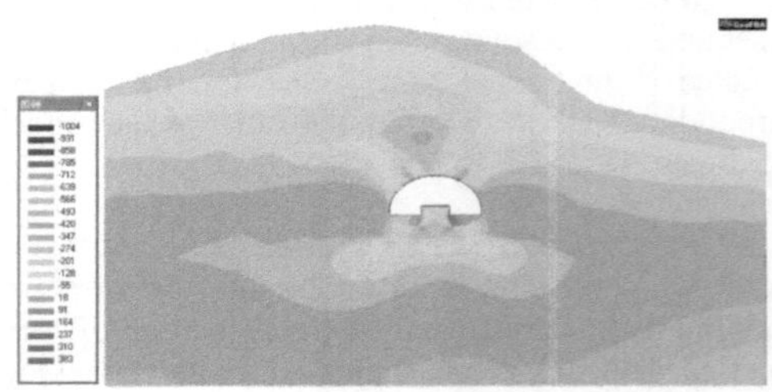

(d) After dismantling of temporary support

Figure 8. Stress of core rock excavation and support dismantling.

4.2.2 *Stress of support structure*

The support stress of the feature points is collected in numerical simulation (figure 9). With the axial force as the subject, the effect on support structure stress caused by core rock excavation and temporary support dismantling is analyzed. The axial force of the preliminary lining before and after core rock excavation and temporary support dismantling is listed in table 3. It can be seen from table 3 that the support structure stress change greatly and the change caused by excavation procedure is larger than that of dismantling procedure. Except that, the effect on the support

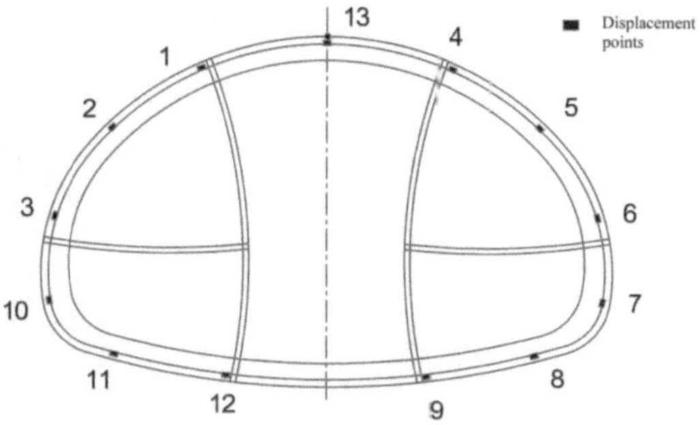

Figure 9. Schematic diagram of stress acquisition.

Table 3. Comparison of internal force in lining (kN)

Stress No	Stage-1	Stage-2	Stage-3	Stage-4
1	259	381	491	602
2	310	638	779	934
3	168	534	670	828
4	349	486	556	571
5	510	864	1020	1136
6	511	942	1101	1267
7	388	789	930	1074
8	119	150	170	172
9	91	34	−56	−71
10	224	558	676	816
11	133	183	198	192
12	133	−24	9	−16
13	/	/	113	211

structure stress caused by core rock excavation is different with different positions. The stress change above the inverted arch is larger than that of inverted arch zone. The change difference of support structure stress of the left and right hole is not apparent. It means the excavation effect on two holes is not apparently different. As same as excavation procedure, the stress change above the inverted arch is larger than that of inverted arch zone. But the dismantling procedure has different influence on the support stress between the left and right hole.

5 CONCLUSIONS

Based on the displacement and stress simulation before and after core rock excavation and temporary support dismantling, the experience and data presented in the paper can be refereed in the design, construction and research of super large section and span tunnels.

(1) During the construction of super large section and span tunnel, based on the displacement and stress results, the core rock excavation has a great influence on the deformation and stress of rock and lining. That procedure is the key point of tunnel construction control. Thus, the deformation and stress monitoring should be strengthened to ensure the tunnel construction safety.

(2) When doing core rock excavation and temporary support dismantling, the rock displacement at the tunnel top is large and the horizontal displacement of side wall is small. Thus, the crown settlement monitoring should be strengthened and the crown settlement control should be paid attention to.

(3) During the construction of super large section and span tunnel, the stress change above the inverted arch caused by excavation and dismantling is greatly larger than that of inverted arch zone. The support dismantling has different influence on left and right hole with the construction effect on the support stress of the left hole larger than that of the right hole. Therefore, the control of support stress in the upper stage zone should be strengthened, especially the stress change during temporary support dismantling to ensure the preliminary lining bears load effectively .

REFERENCES

Chengzao H, Zongxue Y & Xiaorong Z (2007). Cross anchor and its application in the fracture zone in highway tunnel. *Chinese Journal of Underground Space and Engineering* 3(5): 923–927.

Chongbang X, Caichu X & Hehua Z (2009). Optimum analysis of construction scheme of multi-arch tunnel with eight traffic lane. *Chinese Journal of Rock Mechanics and Engineering* 28(1): 66–73.

Dingheng Z, Haifeng Q & Yongchang C et al. (2009). In-situ test on surrounding rock deformation of super large section and span tunnel. *Chinese Journal of Rock Mechanics and Engineering* 8(9):1773–1782.

Gengye C, Bin L & Wan Mingfu et al. (2005). Analysis of stress monitoring of a large-span highway tunnel in Hanjiangling mountain. *Chinese Journal of Rock Mechanics and Engineering* 24(Supplement 2):5509–5515.

Lunhai H, Wei L & Mengjun W (2007). Model test on single four-lane highway tunnel excavation. *Highway Tunnel* (4): 10–15.

Mengjun W & Lunhai H (2006). Research on dynamic construction mechanics of four-lane highway tunnel. *Chinese Journal of Rock Mechanics and Engineering* 25 (Supplement 1): 3057–3062.

Mingfu W, Zhe H & Jianping L et al. (2007). Analysis of stability about excavation and support of super-span road tunnel. *Journal of Liaoning Technical University (Natural Science* 26(1):71–73.

Mingfu W, Hong H & Bin L (2007). Investigation on excavating a 4-lane road tunnel through indoor model test, *Journal of Northeastern University (Natural Science)* 28(2):266–269.

Yong Y & Shenghui W (2008). Numerical modeling of prebuilt and prestressed innovative support system for super cross-section highway tunnel with lower flat-ratio. *Rock and Soil Mechanics* 29(1): 240–244.

Zhigang L, Wenqi D & Zhongcun Y (2007). Simulation and analysis of core rock in flat and large span highway tunnel. *Chinese Journal of Underground Space and Engineering* 3(4): 627–632.

Modelling and Computation in Engineering – Zhu (ed)
© 2011 Taylor & Francis Group, London, ISBN 978-0-415-61516-7

Research on the application of BVP members in the seismic reduction technology for core-outrigger structures

Zhongliang Deng
Building Department, College of Civil engineering, Tongji University, Shanghai China

Feifei Sun & Guoqiang Li
Building Department, College of Civil engineering, Tongji University, Shanghai China
State Key Laboratory for Disaster Reduction in Civil Engineering, Tongji University, Shanghai, China

ABSTRACT: The core-outrigger structure is an excellent and efficient structure style, especially for the super high-rise building, however, the arrangement of outriggers causes the stiffness break inevitably, which will be the potential hazard in the earthquake. The mature and wide application of energy dissipation technology provides a reasonable solution for the defect of core-outrigger structures. This paper proposes a new seismic reduction technology (BVP) for the core-outrigger structure, which lays the buckling-restrained brace (BRB) with the viscous damper as the buckling-restrained column (BRC) in parallel, this specific structural assembly will be arranged below the outrigger, which connects with exterior columns in series. The stiffness break alleviation mechanism of BVP installed in the core-outrigger structure is also analyzed. One practical projects applied with three seismic reduction plan are compared, which contains the one with the parallel dissipation structural assembly of BRB and the viscous damper (BVPS), the one with viscous dampers located below the outrigger near the exterior column (VDS) and the original common core-outrigger structure (CCOS). By comparing the different reaction of the three models by using IDA method (Incremental dynamic analysis method), the different seismic reduction effect could be achieved in different intensity region.

1 INTRODUCTION

The core-outrigger structure (COS) is an efficient lateral resistant system, which had been widely used in the super high buildings, such as: global financial center in Shanghai (492 m), Jinmao building (421 m), and abuilding Shanghai Center (680 m), etc. However, the arrangement of outriggers causes the stiffness break, which may be the potential danger in the rare earthquake. In addition, the setting of multi-defensive line for COS is rather difficult (Li Guoqiang, et al.). Energy dissipation technology is kind of passive control measures for structures, which uses some specific members to dissipate energy inputted by the earthquake, consequently, the main structure is protected (T.T Song).

2 SEISMIC REDUCTION STRATEGY FOR COS

The author proposes a new seismic reduction method for COS by combining the BRB and viscous damper. Since the stiffness of outriggers is rather larger, the deformation of COS under lateral loads mainly concentrated on the exterior column, as illustrated in the figure 1. Consequently, the position between outriggers and exterior columns is the most optimal location of energy dissipation members.

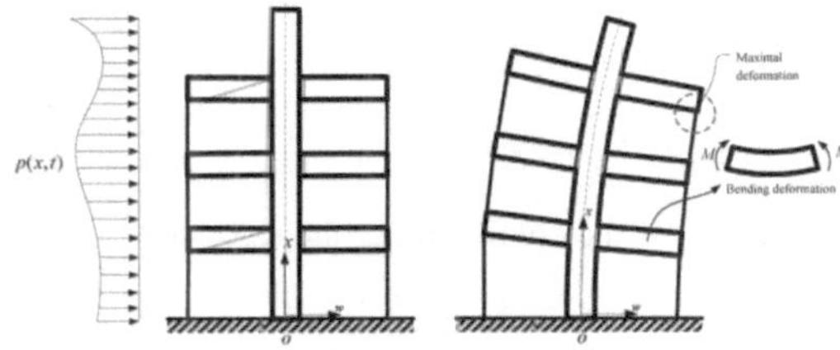

Figure 1. Schematic diagram: Deformation of COS under lateral load.

3 MECHNANICAL PROPERTY OF BVP

BVP consists of BRB and VD in parallel, as in figure 2, the basic property of BRB is simulated by using double-linear model, in which α is the post stiffness ratio, $\alpha = 0.01\sim0.02$, $F_{b,y}$ is yielding capacity of BRB, $F_{b,\max}$ is the ultimate capacity, $u_{b,y}$, $u_{b,\max}$ is yielding and ultimate displacement of BRB respectively, k_b is elastic stiffness, k_b' is the secondary stiffness, k_b^* is the equivalent linear stiffness. k_d'' is the lost stiffness of VD.

According to the figure 3, the mechanical property of BVP can be presented:

$$F_D = \begin{cases} k_b u + C_c \dot{u} & u \le u_y \\ k_b u_y + \alpha k_b (u - u_y) + C_d \dot{u} & u > u_y \end{cases} \tag{1}$$

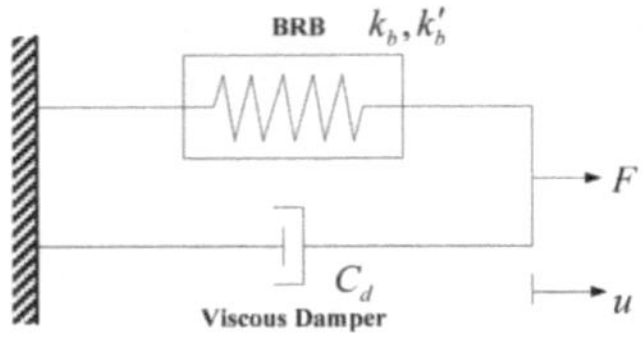

Figure 2. The BVP model.

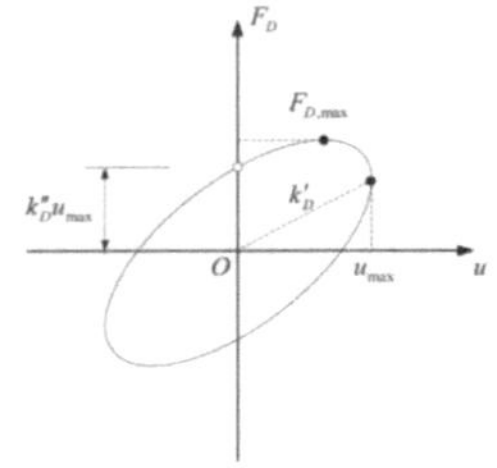

Figure 3. Hysteresis loop of BVP.

In which, $\mu_D = \mu_b = \frac{u_{b,\max}}{u_y}$ is the ductility factor.

For simplification, the equation above can be disposed:

$$F_D = k_b^* u + C_d \dot{u} \tag{2}$$

$$k_b^* = \frac{k_b[1+\alpha(\mu-1)]}{\mu} \tag{3}$$

Energy dissipated in one cycle E_D:

$$E_D = 4k_b(1-\alpha)(\mu-1)u_{b,y}^2 + \pi C_d \omega u_0^2 \tag{4}$$

Maximal force:

$$F_{D,\max} = \sqrt{k_b^{*2} + (C_d\omega)^2}\, u_0 \tag{5}$$

In which, C_d is the coefficient of the damper.

4 IMPROVEMENT MECHANISM OF THE STIFFNESS BREAK IN THE COS

Owing to the setting of outriggers changes the vertical stiffness along the height, the stiffness break is inevitable. As can be seen in the figure 4, the horizontal shearing force equilibrium can be obtained:

$$T^l + C^r - V_T^u - V^T = 0 \tag{6}$$

The couple equilibrium caused by the outrigger can be got:

$$T^l + C^r = \frac{M_T^u + M_T^d}{H_o - h} \tag{7}$$

Then,

$$V^T = \frac{M_T^u + M_T^d}{H_o - h} - V_T^u = \frac{\sum M}{H_o - h} - V_T^u \tag{8}$$

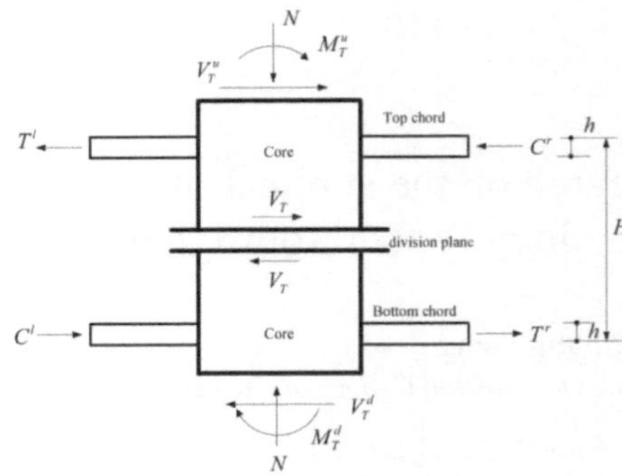

Figure 4. Simplified analysis diagram for stiffness break in the COS.

Equation (6) indicates that the magnitude of shearing break depends on the stiffness and the height of the outrigger. BRB in the BVP can buckled in some kind of earthquake, which decreases stiffness, which means $\sum M$ becomes smaller, then the shearing break and subsequent moment will be reduced, then the deformation is also reduce.

5 NUMERICAL EXAMPLE

According to the deformation style of COS, setting energy dissipation members under the outrigger is an efficient way to improve the seismic property of COS by applying passive energy dissipation technology. The following analysis will be based on a numerical example, in which two finite model of BVP and VD will be analyzed and compared with the original COS model. For simplification, the name of three models are as follows:

(1) BVP is set under the outrigger (BVPS).
(2) Viscous damper is set under the outrigger (VDS).
(3) Common core-outrigger structures (CCOS).

5.1 *Project overview*

The core-outrigger structure consists of 39 floors, 148 meters; the core is composed with central brace frame. Classification of design earthquake is the first group, the site classification is III, site intensity is 7degree (0.1 g), two outriggers are located on the 19th and 35th floor respectively. The distance between grids of central brace frame is 7.5 meters; the span of frame beam is 9 meters. The component size of three models above is listed in the table 1.

5.2 *Model establishment and related parameters*

The model is established in the SAP2000 V14. The material of the whole structure is Q345. The BVP and VD are designed according to the equivalent energy, which means the energy dissipated in one cycle of BVP and VD is same. The damping coefficient of VD is 3000 N · s/mm, the damping coefficient of VD in the BVP is the same, BRB of BVP is design

Table 1. Components size.

Name of components	Location	Size (mm)
Exterior columns	Up	1200 × 1200 × 200
	mid	1500 × 1500 × 200
	bottom	1500 × 1500 × 250
Frame beam		I400 × 300 × 12 × 8
Column	Central	I800 × 700 × 35 × 30
Beam	brace frame	I600 × 500 × 22 × 18
Brace		I400 × 300 × 12 × 8
Beam	Top	I600 × 500 × 22 × 18
Brace	outrigger	I450 × 350 × 16 × 12
Beam	Mid	I600 × 500 × 22 × 18
Brace	outrigger	I450 × 350 × 16 × 12

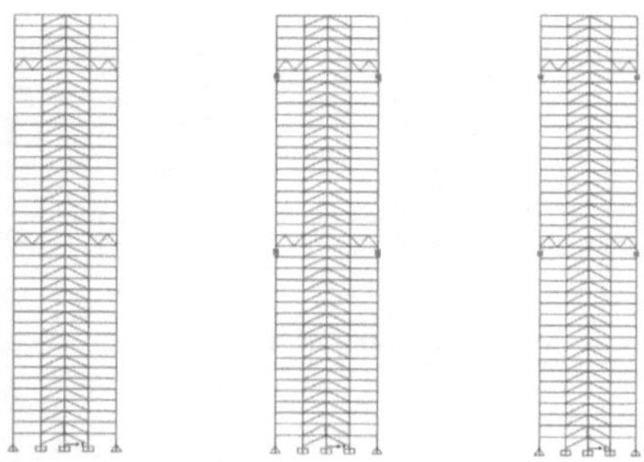

Figure 5. Analysis model.

Table 2. Period comparison.

Period(s)	1	2	3	4
CCOS	2.647	0.823	0.420	0.271
BVPS	2.891	0.897	0.437	0.282
VDS	4.277	1.176	0.590	0.525

as being 'elastic in the mid-earthquake, plastic in the rare earthquake', hence, the section of BRB is I800 × 800 × 30 × 25. The VD and BRB are simulated by the damper and plastic-wen model, in which $\alpha = 0.02$, the index of Bouc-Wen differential equation is 2.

5.3 *Model analysis*

The vibration model of the 3 models is similar, whose periods are listed in the table 2.

5.4 *Elasto-plastic time history and IDA analysis*

The elasto-plastic property of the frame is simulated by applying fiber hinges, for different sections of columns and beams, the distribution of fiber hinges can be

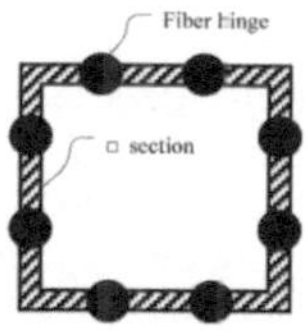

Figure 6. Distribution of fiber hinges of □ section.

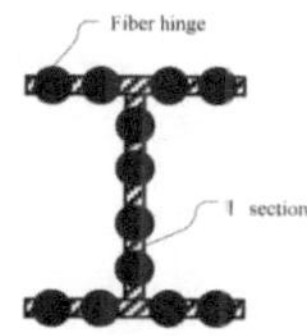

Figure 7. Distribution of fiber hinges of I section.

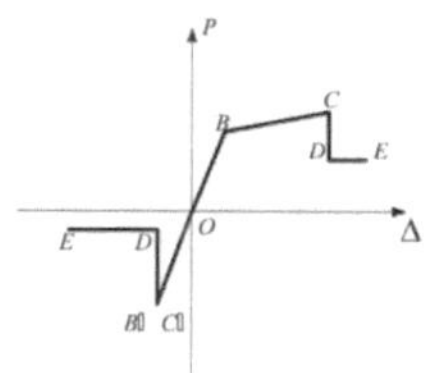

Figure 8. Axial force hinges.

referred to figure 6 and figure 7. The length of plastic zone on the beam and column is assumed as 8% of the length of components. The plastic property of brace is simulated by P hinge, whose hysteresis rule is assumed as Takeda, and the skeleton curve can be seen in the figure 8. This assumption is only to simulate the pinch phenomenon of brace approximately. Although this assumption may not be exact to simulate the post-buckling behavior, however, it could provide approximate effect of brace in the earthquake.

Incremental Dynamic Analysis method (IDA) [16] is a sort of parameter analysis method which was developed recently to evaluate the structural performance under seismic ground motion.

For COS, the lateral deformation is control by moment and shear, the time consistence doesn't exist like the frame structure, the former seismic index for COS is selected and the relationship between maximal story drift angle and PGA is also considered. The PGA in this paper for IDA analysis is as follows: 220 gal, 400 gal, 620 gal, 800 gal, 900 gal.

Owing to paper limit, only results of PGA = 900 gal at the 20th s is chosen. Since the fiber hinge can not appear directly in the SAP2000, the P hinge can be got intuitively; hence, the structural damage could be estimated by the combined results of P hinge and fiber hinge. The P hinge of three models can be seen from figure 9. ○ indicates the yielding point, Δ indicates the hinge has lost its ultimate bearing capacity, × indicates the hinge had lost its load capacity completely.

5.5 *Deformation analysis*

From figure 10~12, the story drift angle of CCOS, BVPS and VDS could be found. The maximal story drift angle appears at the 18th floor, where the damper was located. In the figure 12, the floors near the outrigger are sensitive to be damaged. For BVPS, the deformation appears the proportional as the PGA increases. Figure 13 is the story drift angle ratio of the maximal value above the mid-outrigger and the one below the mid-outrigger (DF), which characterize the deformation magnitude of the COS. For BVPS, the deformation is rather uniform, DF = 1. The deformation of VDS concentrates on the lateral drift of floors above the mid-outrigger, hence, DF < 1. Detailed data of the deformation of three models could be found in the table 3.

Plastic hinges

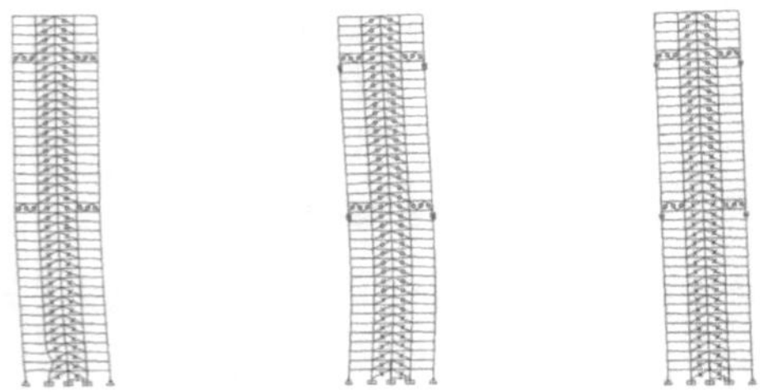

(a) CCOS (b) BVPS (c) VDS

Figure 9. The occurrence of plastic hinges of CCOS, BVPS, VDS (PGA = 900 gal, t = 20 s).

Table 3. Summary of IDA analysis-deformation.

	PGA(gal)	Story drift angle: Peak value	Story drift angle: Occurring floor	DF	Top drift (mm)
CCOS	220	1/412	10	1.44	241.2
	400	1/252	7	1.53	410.2
	620	1/92	5	2.69	812.8
	800	1/81	5	2.36	1023
	900	1/79	2~5	2.27	1132
BVPS	220	1/386	18	1.40	323.9
	400	1/303	18	1.03	574.1
	620	1/185	18	1.03	899.9
	800	1/134	25	9.63	1196
	900	1/111	3	1.03	1400
VDS	220	1/224	20~33	0.80	565.9
	400	1/124	20~33	0.79	1012
	620	1/84	20~33	0.77	1473
	800	1/56	18	0.65	1903

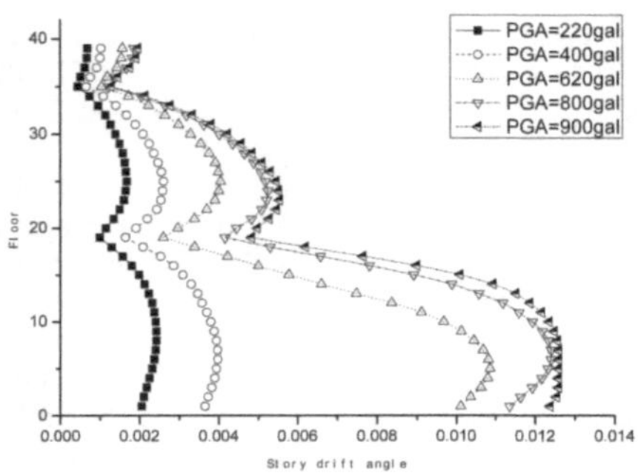

Figure 10. Story drift angles of CCOS.

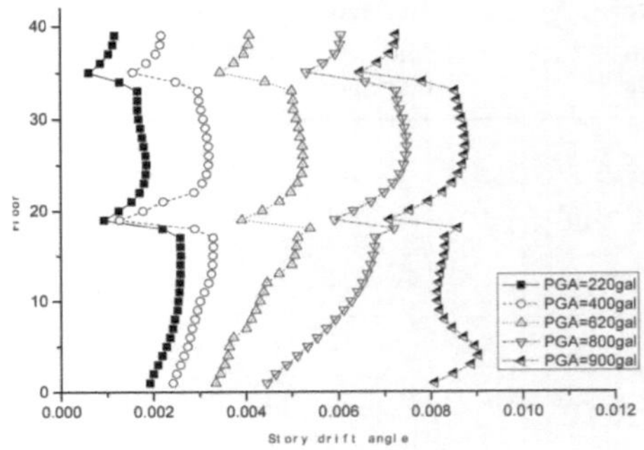

Figure 11. Story drift angles of BVPS.

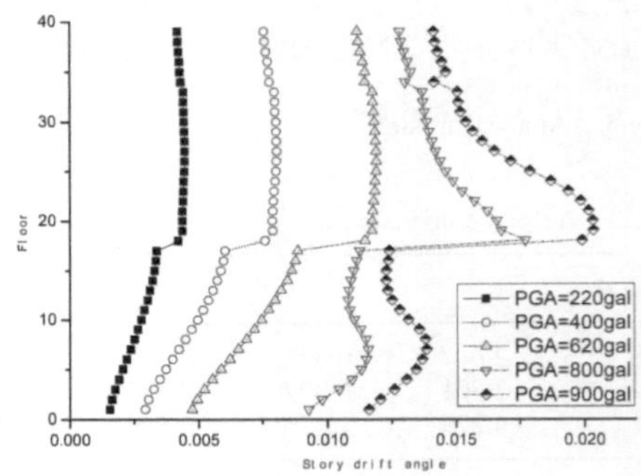

Figure 12. Story drift angles of VDS.

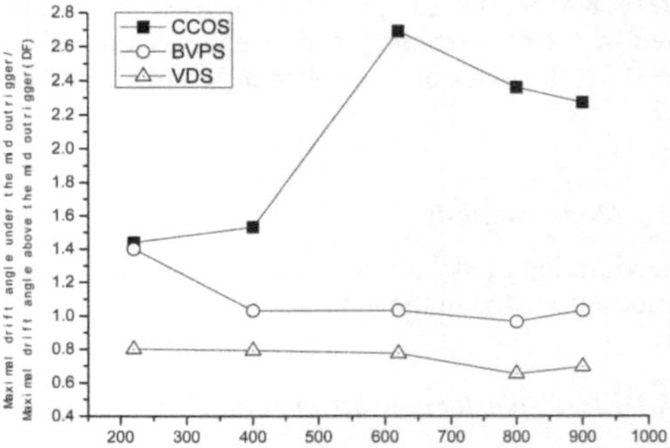

Figure 13. PGA-story drift angle ratios.

5.6 *Mechanical analysis*

Figure 15 is the IDA curve: Maximal shear force and maximal top drift, which illustrates that the relative deformation of stories in VDS is nearly linear with the PGA increases, and the maximal deformation occurs at the floor where the VD is located. CCOS's curve turns at the PGA = 400 gal, while the maximal story drift angle of BVPS equals to CCOS at PGA = 400 gal, which is less than the CCOS and VDS when PGA > 620 gal. Before PGA = 400 gal, the main structure of CCOS remains elastic. For fiber hinges of bottom beams and beams near the outrigger, where the element becomes partial plastic, all the plastic hinge concentrates on the brace of the core.

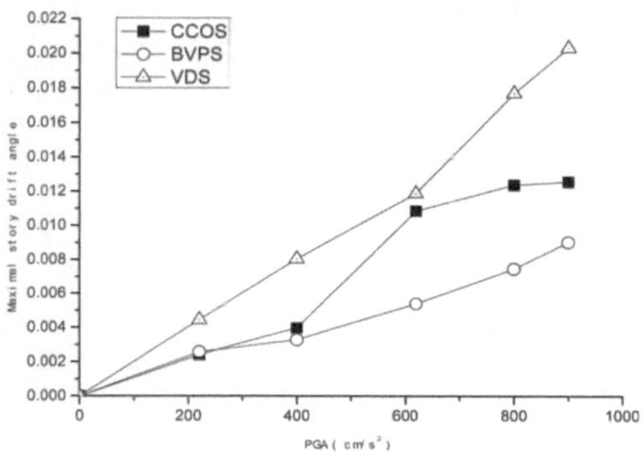

Figure 14. PGA-story drift angles.

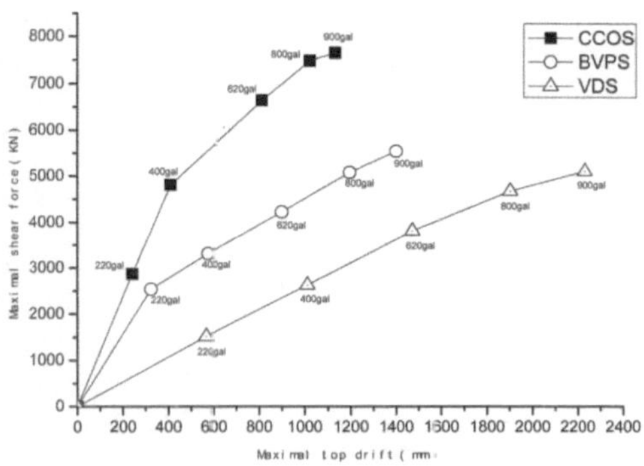

Figure 15. IDA curve: Maximal shear force and maximal top drift.

For further investigation of the mechanical property of the models, the ratio of participation in the anti-bending (RPAB) is defined as follows:

RPAB=Maximal moment provided by the bottom columns/Total overturning moment.

From the table 4, owing to paper limit, only results of PGA > 620 are chose, the base shearing forces and overturning moment accretes as the PGA increases. The incremental amplitude of mechanical behavior is bigger than BVPS and VDS. The base shearing force of BVPS is slightly larger than the one of VDS, when PGA > 400 gal, the discrepancy is not observably. For RPAB comparison, CCOS nearly is 57.6%, BVPS is about 40%, VDS is only 20%. Because of buckling of BRB, the anticipation ability of exterior columns decreases nearly 20%.

The overturning moment and the shearing force of BVPS is not significantly different, owing the anticipation of exterior columns in BVPS, the moment of interior columns in BVPS reduces approximately 50% compared with CCOS and VDS, the BVP could protect the core from earthquake and reduces the degree of damage of the core.

5.7 *Hystersis loops*

For comparing the hysteresis property of BVP and VD, maximal energy dissipated in the hysteresis loop could be found in the table 5. In the same seismic ground motion, the BVP could provide much larger lost stiffness and storage stiffness. In addition, the maximal energy dissipated in the hysteresis loop of BVP and VD are approximately consistent. For PGA = 400 gal, the hysteresis loop of BVP and VD could be seen in the figure 16.

6 CONCLUSIONS

The mechanical property of BVP and stiffness reduction mechanism for COS by applying BVP as well as the a new seismic reduction technology is proposed by the author, an actual engineering sample is compared by CCOS, BVPS and VDS, then some available conclusions could be obtained as follows:

(1) The seismic property of BVPS is the most excellent compared with CCOS and VDS, whose overall and local deformation is controllable.

Table 4. Summary of IDA analysis-mechanical analysis.

	PGA (gal)	Maximal base shearing force (KN)	Total overturning moment ($\times 10^8$N $\cdot$ m)	RPAB (%)	Maximal axial force of interior columns (KN)	Maxima moment of interior columns (N $\cdot$ m)
CCOS	800	7479	5.56	58.85	15517.63	10560000.00
	900	7647	5.83	59.55	17421.47	11630000.00
BVPS	800	5083	4.43	37.72	17370.20	4073401.90
	900	5544	4.87	37.63	17648.89	5788251.06
VDS	800	4677	4.26	19.90	23564.91	8200821.49
	900	5110	4.29	23.02	22800.66	11410000.00

The deformation style is uniform along the building, no matter the arrangement of outriggers.

(2) BVP could make the exterior column participate overall anti-bending, and buckled timely in the rare earthquake, which protects the core and relieves the harmful stiffness break effect to avoid the occurrence of weakness story, and elevate the seismic property.

(3) In conclusion, for existed projects, the seismic plan of merely applying viscous dampers under the outrigger in the core-outrigger structure should pay more attention to excessive deformation near the stories of VD and the protection of the core.

REFERENCES

Chen, F.S & Qiu, G.H & Fan. Z. 2004. *Building Structures Design (2nd Edition)*. China Building Industry Press.

Li, G.Q. 2004. *Design of high-rise steel buildings*. China Architecture & Building Press.

Lu, X.Z & Ye, L.P et al. 2009. *Seismic elastic-plastic analysis-principles, model and practice on the ABAQUS, MSC.MRC and SAP2000*. China Architecture & Building Pres.

Song, T.T & Dargush, G.F. 1997. *Passive energy dissipation systems in structural engineering*. John Wiley & Sons.

Smith, R & Willford, M. 2007. *Damped outriggers for tall buildings*. The Arup Journal.

Smith, R.J & Willford, M. R. 2007. *The damped outrigger concept for tall buildings*. The structural design of tall and special buildings, 16: 501–517.

Seismic Design of Building (GB50011-2001). 2008. China Architecture & Building Press.

Vamvatsikos, D & Comel A. 2002. *Incremental dynamic analysis*. Earthquake Engineering and Structural Dynamics, 31(3): 491–514.

Modelling and Computation in Engineering – Zhu (ed)
© 2011 Taylor & Francis Group, London, ISBN 978-0-415-61516-7

Relaxation modulus prediction of asphalt-rubber concrete based on micromechanics

N.S. Guo & Y.Q. Tan
School of Transportation Science and Engineering, Harbin Institute of Technology, Harbin, China

Z.C. Wang & Y.H. Zhao
Institute of Road and Bridge Engineering, Dalian Maritime University, Dalian, China

ABSTRACT: Crumb rubber modified asphalt concrete (CRMAC) has been found to be an effective material comparing to the common asphalt concrete for its favorable engineering performance. Relaxation modulus of asphalt concretes is one of the fundamental engineering properties. Although laboratory tests provide the way to obtain the value of modulus of CRMAC, a predictive model based on the microstructure of CRMAC is more desirable considering saving of energy source. A modified Mori-Tanaka's theory for the effective moduli of composite materials is used to establish the effective relaxation modulus of CRMAC. "Four-unit, five-parameter" and Burgers models are employed to express the viscoelastic properties of asphalt concrete and crumb rubber in the model, respectively. The relaxation modulus of CRMAC is then predicted with the newly developed viscoelastic micromechanics model. Laboratory test results show that a discrepancy exists between the predicted and measured relaxation modulus. The reasons for the discrepancy between the measured and predicted results may be attributed to interface bonding, interaction of inclusions and matrix. Thus the studies on mechanical properties of crumb rubber and microstructure characteristics of CRMAC are essential to be carried out to obtain the satisfactory viscoelastic micromechanics model.

1 INTRODUCTION

Over 300 million vehicle tires have been generated until 2010 in China. Of these, approximately 200 million are scrap tires and added to stockpiles, landfills, or illegal dumps (Ren et al. 2009). The alternative use of crumb rubber as an additive in asphalt concrete has been discussed and researched for the last 30 years. The crumb rubber modified asphalt concrete (CRMAC) can be broken down into wet and dry processes. The wet process means the crumb rubber as an additive used in the asphalt binder, mixing of the crumb rubber with the binder and allowing time for their reaction prior to mixing with the aggregate. The dry process means rubber as an aggregate, simultaneous mixing of the crumb rubber, the binder, and the aggregate. Several researchers have proved that CRMAC have favorable performance with respect to resistance to moisture damage, high temperature stability and resistance to cracking at low temperature in laboratory tests (Zhang et al. 2005).

It is well known that asphalt concrete is viscoelastic material. The relaxation modulus is a fundamental engineering property that represents the rheological behavior of viscoelastic materials and the capacity of asphalt pavement to resistance to cracking under low temperature. The interrelation between relaxation modulus and creep compliance are available in the viscoelasticity models (Burgers model, "Four-unit, five-parameter" model (Xu 1991) are employed to depict the creep behavior of the asphalt concrete). Then the relaxation modulus can be determined from a simple laboratory test. However, current research mainly focuses on the creep property by means of tests rather than investigate the relaxation property of CRMAC. There is much less viscoelastic theory analysis, especially relaxation property of CRMAC is necessary to know.

Micromechanical models have been long used to analyze and predict the mechanical property of the composite. Based on Mori-Tanaka's theory, the effective elastic moduli of a particle-reinforced composite with spherical inclusions are constituted by Weng (1984). In addition, the effective elastic moduli of ribbon-reinforced composites are derived by Zhao & Weng (1990). Guo et al. (2006) have presented a model to predict equivalent stiffness modulus of fiber-reinforced asphalt concrete according to the model of ribbon-reinforced composites. A developed micromechanical model is proposed to investigate the viscoelastic property of fiber reinforced concrete. In the model, the fiber reinforced asphalt concrete is considered as a two phases composite of asphalt concrete as the viscoelastic matrix and fiber as the elastic inclusion, and the "Four-unit, five-parameter" viscoelastic model is employed to describe creep properties of fiber reinforced asphalt concrete (Guo et al. 2008).

Many micromechanical models are available to study the mechanical property of the composite, whereas the inclusions are all considered as linear elastic material rather than account for nonlinear material in these models. The CRMAC can be regarded as a composite with asphalt concrete and crumb rubber, and the two constituents of CRMAC are viscoelastic material, thus it is more desirable to take into account the material properties of the two phases in a micromechanical model for analyzing the relaxation property of CRMAC. It can also help us better to understand the viscoelastic behavior of CRMAC, and analyze the effect of the two constituents.

The primary objective of this study is to develop a micromechanical model to investigate the relaxation property of CRMAC. The newly developed model has the capability of taking into account the viscoelastic effect of the two constituents. Laboratory experiments are used to verify the proposed model. The secondary objective is to investigate the effect of crumb rubber considered as viscoelastic material on the relaxation property of CRMAC.

2 VISCOELASTICITY THEORY

2.1 *Constitutive equations*

The general form of the linear viscoelastic stress-strain relations are given by

$$\sigma(t) = \int_{-\infty}^{t} G(t-\tau)\frac{d\varepsilon}{d\tau}d\tau, \varepsilon(t) = \int_{-\infty}^{t} J(t-\tau)\frac{d\sigma}{d\tau}d\tau \quad (1)$$

Where $G(t)$ = the relaxation function, $J(t)$ = the creep function.

To establish the relationship between them, it is convenient to employ the Laplace transform. Taking the Laplace transforms of Equation 1, using the convolution can be expressed as

$$\sigma(s) = sG(s)\varepsilon(s), \quad \varepsilon(s) = sJ(s)\sigma(s) \quad (2)$$

Where s = the transform variable, the Equation 2 can be represented as follows

$$J(s) = 1/(s^2G(s)) \quad (3)$$

2.2 *Viscoelasticity model*

Burgers model may describe viscoelastic properties of asphalt concrete and rubber (Zhou et al. 2001; Lin et al. 2007). The Equation 4 presents the relationship between time and Burgers model's parameters as

$$\varepsilon(t) = \sigma_0[1/E_M + t/\eta_M + (1-\exp(-E_V t/\eta_V))/E_V] \quad (4)$$

Where σ_0 = applied stress, t = time of loading, E_M = Young's modulus of immediate elasticity, E_V = modulus of delayed elasticity, η_M = coefficient of viscosity, η_V = coefficient of elastic delay viscosity.

A modified Burgers model named "Four-unit and five-parameter" is presented to describe viscoelastic property of asphalt concrete. The modified model makes up the defect of Burgers model and indicates effectively deformation characteristic of asphalt concrete. The experimental data show that the viscoelastic property of CRMAC is similar to plain asphalt concrete. Thus, the "Four-unit and five-parameter" model is employed to describe the viscoelastic property of CRMAC. Lin et al. (2007) have proved that Burgers model can depict accurately the viscoelastic property of rubber. Therefore Burgers model is used to describe the viscoelastic property of crumb rubber.

3 EFFECTIVE MODULI OF THE PARTICLE EINFORCED COMPOSITE

Based on Eshelby's (1957) equivalence principle and Mori-Tanaka's (1973) concept of average stress. The inclusion phase will be referred to as phase 1 and the matrix as phase 0 in the two-phase system. The bulk and shear moduli of the *rth* phase are denoted by κ_r and μ_r, and volume fraction of the inclusion by c. When the two phases are assumed homogeneous and isotropic and the inclusions are spherical, the equations for the effective bulk and shear moduli of the composite are proposed by Weng (1984)

$$\frac{\kappa}{\kappa_0} = 1 + \frac{c(\kappa_1 - \kappa_0)}{(1-c)(\kappa_1 - \kappa_0)\alpha + \kappa_0}$$
$$\frac{\mu}{\mu_0} = 1 + \frac{c(\mu_1 - \mu_0)}{(1-c)(\mu_1 - \mu_0)\beta + \mu_0} \quad (5)$$

Where $\alpha = \frac{1+\nu_0}{3(1-\nu_0)}$, $\beta = \frac{2(4-5\nu_0)}{15(1-\nu_0)}$, ν_0 = Poisson's ratio of matrix.

While the two phases are viscoelastic or viscoplastic material, the bulk and shear moduli of the constituents in Equation 5 are secant moduli. The two phases (asphalt concrete as phase 0 and crumb rubber as phase 1) are both considered as viscoelastic material in this study.

4 EFFECTIVE RELAXATION MODULI OF CRMAC

To analyze the viscoelasticity of CRMAC and establish the relationship between effective moduli and time, in the transformed domain Equation 5can be expressed as follows

$$\frac{\kappa^{TD}}{\kappa_0^{TD}} = 1 + \frac{c(\kappa_1^{TD} - \kappa_0^{TD})}{(1-c)(\kappa_1^{TD} - \kappa_0^{TD})\alpha^{TD} + \kappa_0^{TD}}$$
$$\frac{\mu^{TD}}{\mu_0^{TD}} = 1 + \frac{c(\mu_1^{TD} - \mu_0^{TD})}{(1-c)(\mu_1^{TD} - \mu_0^{TD})\beta^{TD} + \mu_0^{TD}} \quad (6)$$

Where κ_r^{TD} = bulk modulus, μ_r^{TD} = shear modulus of the *rth* phase in the transformed domain, respectively, phase 0 as asphalt concrete, phase 1 as crumb rubber, c is volume fraction of crumb rubber, and, ν_0^{TD} is Poisson's ratio of asphalt concrete.

k^{TD} and μ^{TD} are given by

$$\kappa^{TD} = E^{TD}/(3(1-2\nu^{TD})), \quad \mu^{TD} = E^{TD}/(2(1+\nu^{TD})) \qquad (7)$$

Then the Young's modulus in the transformed domain can be expressed as

$$E^{TD} = \frac{9\kappa^{TD}\mu^{TD}}{3\kappa^{TD} + \mu^{TD}} \qquad (8)$$

With the Burgers model, the Young's modulus in the transformed domain E^{TD} is given by (Li & Weng 1995)

$$E^{TD} = \frac{E_M \eta_M (E_V + \eta_V s)s}{E_M E_V + [\eta_M \eta_V - E_M(\eta_M + \eta_V)]s + \eta_M \eta_V s^2} \qquad (9)$$

With the four-unit and five-parameter model, the creep compliance $J(t)$ is given in Equation 9 as

$$J(t) = 1/E_1 + (1-\exp(-Bt))/AB + (1-\exp(-E_2 t/\eta_2))/E_2 \qquad (10)$$

Thus, $J(s)$ can be obtained through Laplace transforms, substituting it into the Equation 10, then

$$E^{TD} = \frac{ABE_1(E_2 + \eta_2 s)(B+s)}{(E_2 + \eta_2 s)[E_1 B + AB(B+s)] + ABE_1(B+s)} \qquad (11)$$

Burgers model and four-unit and five-parameter model are employed to describe viscoelastic properties of rubber and asphalt concrete, respectively. First, substituting Equation 9 into Equation 7, we can obtain bulk modulus κ_1^{TD} and shear modulus μ_1^{TD} of crumb rubber. For asphalt concrete, substituting Equation 11 into Equation 7, we have bulk modulus κ_0^{TD} and shear modulus μ_0^{TD} of asphalt concrete. Also, substituting the equations obtained into Equation 6, 8, we can obtain the Young's modulus of CRMAC E^{TD}. Finally, substituting E^{TD} into Equation 3, the effective relaxation modulus in the transformed domain $G(s)$ can be obtained. However, it is difficult to obtain the final results especially by using of inverse Laplace transform to derive the effective relaxation modulus of CRMAC $G(t)$. The test data show that Poisson's ratio of CRMAC is similar to asphalt concrete matrix. Thus, it is assumed that Poisson's ratio of CRMAC ν, asphalt concrete matrix ν_0, and asphalt concrete matrix in the transformed domain ν_0^{TD} are approximately equal, as $\nu = \nu_0 = \nu_0^{TD}$. Substituting κ^{TD} into Equation 6, the Young's modulus of CRMAC in the transformed domain can be expressed as follows

$$E^{TD} = 3\kappa_0^{TD}(1-2\nu_0)\left(1 + \frac{c(\kappa_1^{TD} - \kappa_0^{TD})}{(1-c)(\kappa_1^{TD} - \kappa_0^{TD})\alpha^{TD} + \kappa_0^{TD}}\right) \qquad (12)$$

Then substituting Equation 12 into Equation 3, taking the inverse Laplace transforms of the equations derived above, the effective relaxation modulus $G(t)$ of CRMAC can be obtained.

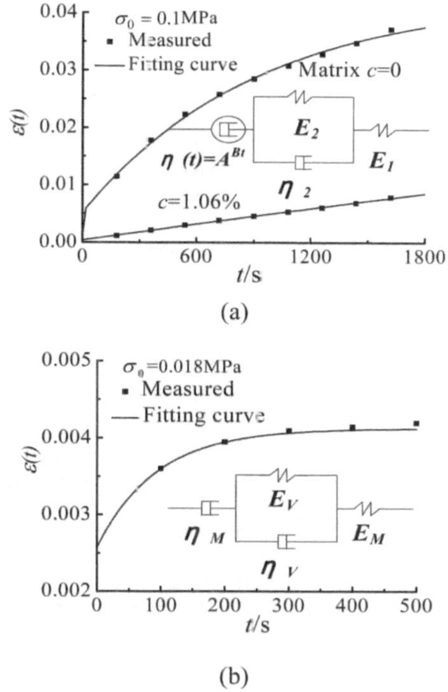

Figure 1. Variation of creep strain with loading time.

5 COMPARISONS BETWEEN PREDICTED AND MEASURED RESULTS

Zeng (2006) presents an experimental study on the creep behavior of CRMAC with different volume fractions of crumb rubber. Figure 1 (a) shows the experimental results and the fitting curve of CRMAC with the loading time from 0 to 1800s at 50°. Lin et al. (2007) provide test results for creep behavior of butyl rubber with different test temperatures (25°, 50° and 100°). The test results and fitting curve of rubber with loading time from 0 to 500 s at 50° are shown in Figure 1 (b). The experimental data of CRMAC with crumb rubber mesh 40, mass fraction $c_m = 10\%$ (by mass of asphalt; volume fraction $c = c_m \times 0.106$) are chosen to compare with the predicted results. The Poisson's ratio of asphalt concrete ν_0 and crumb rubber ν_1 are assumed as 0.35 and 0.49, respectively.

The parameters presented in Equation 4 and Equation 10 are achieved by means of mathematic iterative methods. The non-linear fit function is used to carry out the complicated computing process. The fitting results are shown in Table 1–2.

However, Burgers, Four-unit and five-parameter viscoelastic model are only suitable to some certain range. The fitting results can not be expressed accurately by the viscoelastic models provided that beyond that range. For instance, it is meaningless that the fitting result of the Young's modulus of delayed elasticity E_2 is negative. Therefore the experimental data for

Table 1. Viscoelastic parameters of CRMAC.

Crumb rubber %	E_1 MPa	E_2 MPa	η_2 MPa·s	A MPa	B MPa
0	87.30	2.64	2443.21	15.18	1.65
1.06	288.80	1.20	21063.95	249.24	11.52

Table 2. Viscoelastic parameters of rubber.

E_M MPa	E_V MPa	η_M 10^{10}MPa · s	η_V MPa · s
6.99	11.57	1.7585	1083.66

CRMAC with mass fraction 20% and 30% (by mass of asphalt) of crumb rubber can not be used to analyze the viscoelastic model.

It is expedient for comparison to employ the logarithmic time scale. Figure 2 shows the comparison of the measured (by viscoelastic model) and predicted (by micromechanical model) effective relaxation modulus $G(t)$ at 50° of CRMAC with 1.06 % volume fraction of crumb rubber.

It can be seen that the predicted results have good consistence with the measured ones. The effective relaxation modulus $G(t)$ decrease drastically from 0 to 400 s, and tend to stable value after 400 s, which means that the effect of time on $G(t)$ is limited after 400 s. The predicted curve intersects with the measured at 90 s, and the corresponding $G(t)$ is 9.5 MPa. The predicted results are much lower than the measured ones from 0 to 90 s and always bigger than them after 90 s. The discrepancy between the predicted and measured may be explored as follows:

Based on Mori-Tanaka's theory, it is assumed that the interface of the two phases of composite united absolutely. However, the interface of crumb rubber and asphalt concrete may be loose partly in CRMAC.

The viscoelastic parameters and elastic modulus are derived from the test of butyl rubber and crown of tires, respectively. The discrepancy exists between the mechanical property of this agglomerate rubber and crumb rubber. Thus, to better predict the effective relaxation modulus of CRMAC, the viscoelastic parameters and Young's modulus of crumb rubber need to be determined.

The inclusions are viscoelastic material in this study. Thus, it is necessary to analyze the effect of the viscoelasticity of crumb rubber inclusions on the predicted effective relaxation modulus of CRMAC. It is assumed that crumb rubber is an elastic material, and the Young's modulus of crumb rubber is equivalent to that of the crown of tires E_c according to the presented study (Liu and Wang 2002). Therefore the Young's modulus in Equation 7 can be written as $E^{TD} = E_c = 9.6$ MPa.

From Figure 3, it can be seen that the predicted results of crumb rubber considered as elastic material

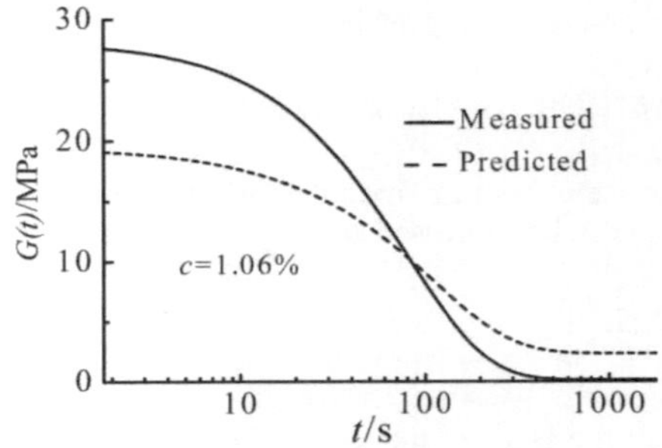

Figure 2. Comparison of the measured and predicted results.

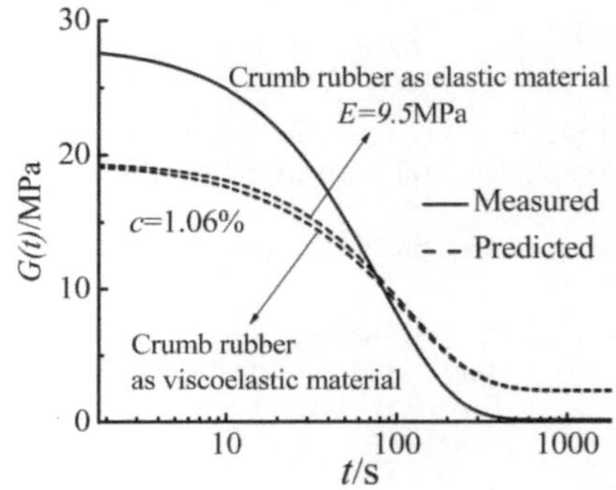

Figure 3. Effect of viscoelasticity of crumb rubber on $G(t)$.

are very close to that of crumb rubber considered as viscoelastic material. This indicates that the effect of the viscoelasticity of crumb rubber on the predicted $G(t)$ is limited. It can also be seen that the predicted results of crumb rubber considered as elastic material are closer to the measured values than that of crumb rubber considered as viscoelastic material from 0 to 90 s. However, the predicted error is slightly larger when crumb rubber is considered as elastic material after 90 s. The reason may be attributed to the effect of loading time on the mechanical property of inclusions. During 90 s of this study, crumb rubber behaves more like an elastic material. Therefore the discrepancy between the predicted and measured is smaller before 90 s. Crumb rubber develops more viscoelastic property and the predicted effective modulus of crumb rubber considered as viscoelastic material are closer to the measured values after 90 s. Thus, it is more desirable to neglect the viscoelasticity of crumb rubber before 90 s. To better predict the long-term effective relaxation modulus of CRMAC, the viscoelasticity of crumb rubber are necessary to be considered for the further research.

6 SUMMARY AND CONCLUSIONS

A micromechanical model is proposed for analyzing the relaxation modulus of CRMAC in this study. The microstructure is represented by two-phase systems:

the viscoelastic crumb rubber as the spherical inclusion and the viscoelastic asphalt concrete as the matrix. With Burgers model of crumb rubber and four-unit and five-parameter model of CRMAC, the effective relaxation moduli is used to describe the viscoelasticity of CRMAC by applying a modified Mori-Tanaka's theory for the effective moduli of composite. The measured results are also compared with the predicted ones from the proposed model. The following conclusions can be drawn.

The developed micromechanical model can predict the effective relaxation modulus of CRMAC, and has the capability of considering the viscoelastic effect on the predicted results.

The laboratory test is limited to one type of volume fraction of crumb rubber, thus the verification of the developed model is preliminary. More tests on different volume fraction of crumb rubber are needed to further verify this micromechanical viscoelastic model.

The tests of mechanical properties of crumb rubber are essential to be carried out for further analyzing the viscoelastic properties of CRMAC.

ACKNOWLEDGMENT

This work is supported by the Doctoral Foundation Ministry of Education of The People's Republic of China under Grant No. 20092125120005, and the Fundamental Research Funds for the Central Universities under Grant No.2009QN052. The writers gratefully acknowledge this support.

REFERENCES

Eshelby, J.D. 1957. The determination of the elastic field of an ellipsoidal inclusion and related problems. *Proceedings of Royal Society. London* 240: 367–396.

Guo, N.S. et al. 2008. Relaxation property of fiber reinforced asphalt concrete. *Journal of Building Material* 11(1): 28–32.

Guo, N.S. et al. 2006. Equivalent stiffness moduli of fiber reinforced asphalt concrete. *Journal of Highway and Transportation Research and Development* 23(9): 23–26.

Itoh, K. et al. 2003. Centrifugal simulation of vibration reduction generated by high-speed trains using rubber-modified asphalt foundation and EPS barrier. *International Journal of Physics Model* 3(2): 1–10.

Li, J. & Weng, G.J. 1995. Effect of a viscoelastic interphase on the creep and stress/strain behavior of fiber-reinforced polymer matrix composites. *Composites part B* 27: 89–598.

Lin, S. et al. 2007. Nonlinear creep behavior of viscoelastic material butyl rubber. *Materials for Mechanical Engineering*31(7): 35–41.

Liu, Y. & Wang, W.M. 2002. Experimental study on mechanical characteristics of tire compounds. *China Rubber Industry* 49(6): 325–329.

Mori, T. & Tanaka, K. 1973. Average Stress in the Matrix and Average Elastic Energy of Materials with Misfitting Inclusions. *Acta Metallurgic* 21: 571–574.

Ren, Z.W. et al. 2009. Recycling situation and prospects of scrap tires in China, *China Resources Comprehensive Utilization* 27(6): 12–14.

Weng, G.J. 1984. Some elastic properties of reinforced solids with special reference to isotropic ones containing spherical inclusions. *International Journal of Engineering Science* 22: 845–856.

Xu, S.F. 1991. A rheological model representing the deformation behavior of asphalt mixtures. *Journal of Beijing Institute of Civil Engineering and Architecture*1: 57–65.

Zeng, W. 2006. Experiment studies on creep of the rubber powder pitch compound with dry-blending method. *Rubber Asphalt in Pavement Engineering Applied Technology Seminar* 72–77.

Zhang, L.P. et al. 2005. Laboratory investigation of performance of discarded tire rubber modified asphalt mixes. *Journal of Shenyang jianzhu University*21(4): 293–296.

Zhao, Y.H. & Weng, G.J. 1990. Effective elastic moduli of ribbon-reinforced composites. *Transactions of the ASME* 57: 158–167.

Zhou, Z.G. et al. 2001. Research on the method of testing viscoelastic parameters of bituminous mixtures. *Journal of Changsha Communications University* 21(4): 23–28.

Modelling and Computation in Engineering – Zhu (ed)
© 2011 Taylor & Francis Group, London, ISBN 978-0-415-61516-7

Effect of forward swept leading edge on a transonic axial compressor rotor stall margin

YaFeng Shi, Hu Wu, Mi Li & Kai Mao
School of Power and Energy, Northwestern Polytechnical University, Xi'an city, ShaanXi province, China

LiXin Chen
Electric power design institute, Zhengzhou city, HeNan province, China

ABSTRACT: The flow field of a transonic axial flow compressor rotor (NASA Rotor 37) has been numerically simulated, and the detailed experimental measurements of this rotor prove the correctness of numerical computation. On the basis of rotor 37 original geometry, the forward swept rotor is produced by moving the tip span section forwards by 75% tip chord length, the 75% span section forwards by 20% tip chord length, the mid span section backwards by 3% tip chord length, the 25% span section backwards by 1% tip chord length. The computational results of forward swept rotor show that the stall margin is largely improved.

1 INTRODUCTION

Compressor performance at the peak of the pressure rise characteristic is limited by aerodynamic instabilities, which lead to rotating stall and surge. Rotating stall has harassed the researchers from the onset of axial compressor industry, and it can deteriorate the compressor performance and lead to excessive blade vibration which possibly makes blade broken, so improving the compressor stability is the aim of many researchers in this field. To avoid the rotating stall and improve the stability, it is necessary to firstly reveal the instability mechanism.

Casing treatment can delay the rotating stall in Axial compressor, which shows the relationship of rotating stall with blade tip region flow. Cumpsty (1989) discovered the endwall boundary layer at rotor inlet has been quickly thicken before stall. Hah (1999) investigated the flow field of a low speed axial compressor and indicated the stall inception with the tip clearance leakage vortex. Furukawa (1999) pointed out the tip clearance vortex breakdown as an indicator near stall. The work by Mailach (2000) and Maerz (2002) indicated the rotating instabilities were caused by the tip clearance vortex. Adamczyk (1993) studied the effect of variations in tip clearance on the performance of a transonic rotor, and showed that the shock/vortex interaction plays a major role in determining the compressor flow range. The work by Hah (2001) showed that a transonic rotor begins to stall with the tip clearance spillage emerging near the adjacent blade pressure side at blade leading edge. From the work by these researchers, it can be concluded that the tip clearance leakage flow plays an important role on the rotating stall.

Sweep is the inclination of the leading edge against the flow direction. If the outer span of a profile is positioned upstream relative to mid-span, then it is called "forward sweep". On forward-swept blades, the endwall regions can be relieved, as wadia (1998) showed in his investigations. The goal of this paper is to examine the flow phenomena induced by forward sweep and to assess the influence of forward sweep on the tip leakage vortex and rotating stall in a transonic axial compressor rotor.

2 ROTOR 37 AERODYNAMIC DESIGN AND GEOMETRY

NASA Rotor 37 is selected as the original rotor for its experimental investigation which has been provided by Suder (1997). The rotor design pressure ratio is 2.106 at a mass flow of 20.19 kg/s, and its peak efficiency is 0.889. The rotor design rotation speed is 17188r/min, and its choked mass flow is 20.93 kg/s. The inlet relative Mach number is 1.13 at the hub and 1.48 at the tip at the design speed of 454 m/s. The rotor aspect ratio is 1.19 and the hub/tip radius ratio is 0.70. The rotor has multiple circular arc (MCA) blade shapes and 36 blades, and it has a tip clearance of 0.356 mm.

3 COMPUTATIONAL PROCEDURE

A block-structured H-O-H grid is used with a total of 506193 nodes. The stagnation pressure of 101325Pa, stagnation temperature of 288 K and 0° flow angle are given at the inlet. The characteristic line of rotor

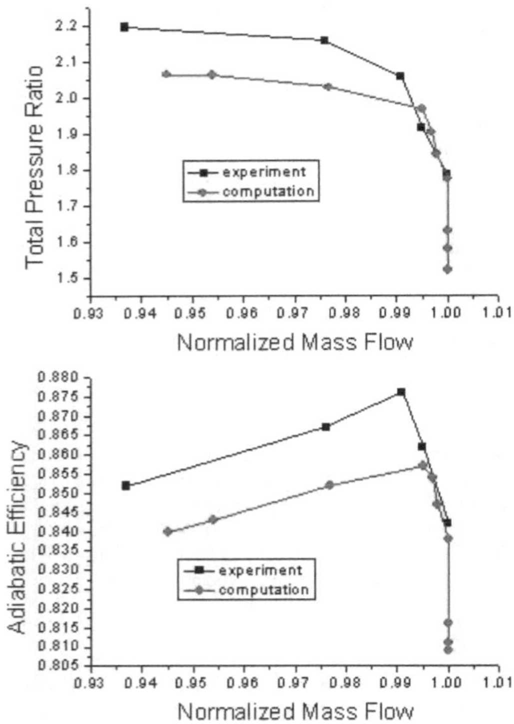

Figure 1. The experimental and computational performance under 100% design rotation speed.

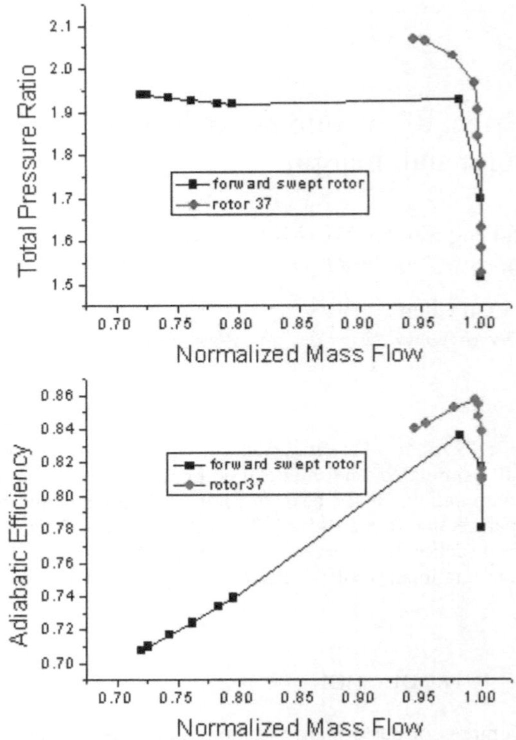

Figure 2. The forward swept rotor and rotor 37 performance under 100% design rotation speed.

performance is acquired by changing the exit imposed static pressure.

An explicit cell-centered 2nd-order-accurate finite volume scheme is used to solve the Reynolds-averaged Navier-Stokes equations. To speed up convergence to steady state, local time-stepping, residual smoothing and successive mesh refinement are applied.

The stalling point of the rotor is taken to be the point at which the calculation fails to converge as the exit static pressure is increased.

4 COMPARISION WITH EXPERIMENTAL RESULTS

The experimental and computational stagnation pressure ratio and adiabatic efficiency of rotor37 are plotted against normalized mass flow (the ratio of mass flow to choked flow) in Figure 1.

From Figure 1, it can be seen that the computational results is basically distributed in accordance with the experimental results with minor difference, which proves the correctness of computation method.

5 FORWARD SWEPT ROTOR

In order to study the effects of sweep the datum rotor 37 is modified by sweeping the blade sections along their local chord line. In doing this the blade sections are not changed and no attempt is made to optimize the design for the new flow environment. Hence, the modified designs do not necessarily represent the best that could be produced for a specified amount of sweep. However, the objectives are to obtain a physical understanding of the effect of sweep upon the flow behavior and the designs are considered suitable for this.

The forward swept rotor is produced by moving the tip span section forwards by 75% tip chord length, the 75% span section forwards by 20% tip chord length, the mid span section backwards by 3% tip chord length, the 25% span section backwards by 1% tip chord length. It is important to note that the entire casing is moved with the tip section so that the tip speed remains the same.

The computational procedure to forward swept rotor is the same with the rotor 37.

The computational stagnation pressure ratio and adiabatic efficiency of forward swept rotor and rotor 37 are plotted against normalized mass flow (the ratio of mass flow to choked flow) in Figure 2.

From Figure 2, it can be seen the stable mass flow range in the forward swept rotor is improved three

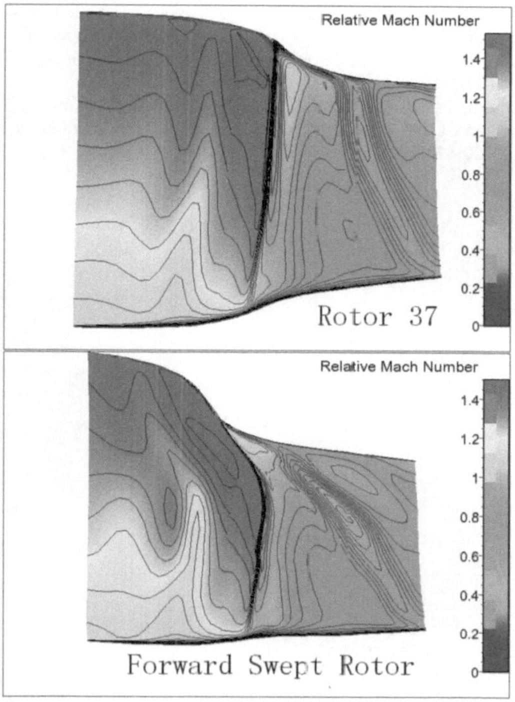

Figure 3. Contours of relative mach number at mid pitch surface under 100% design rotation speed.

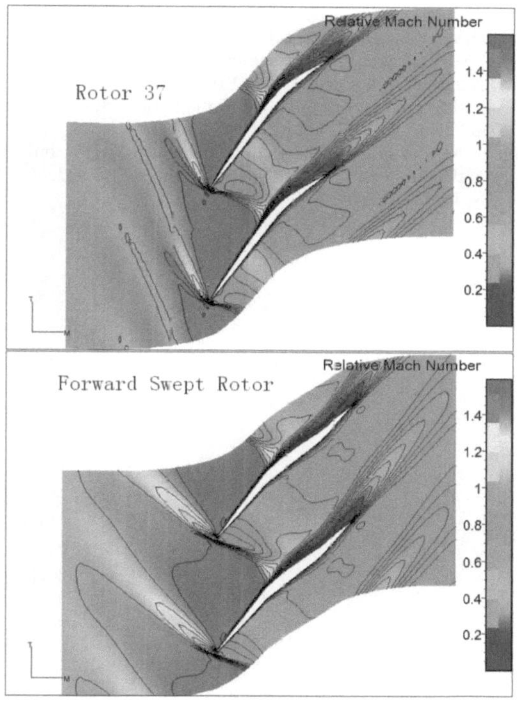

Figure 4. Contours of relative mach number at 70% blade span under 100% design rotation speed.

times compared with the datum rotor 37, however the forward swept rotor has lower total pressure ratio and adiabatic efficiency which may be because it is produced by simply displacing the datum sections with no attempt to optimize their blade sections.

6 THE EFFECTS OF FORWARD SWEEP TO THE SHOCK STRUCTURE AND STALL MARGIN

Figure 3 shows the contour of relative mach number near peak efficiency at mid pitch surface under design rotation speed.

It can be seen that the passage shock structure of forward swept rotor is different from the rotor 37. Near the tip, the shock in forward swept rotor is weaker than in rotor 37, and the passage flow velocity is accordingly higher, which decreases the accumulating of low energy tip vortex and deters the endwall blockage. This accounts for the good stall margin of forward swept rotor.

Figure 4 gives the Contours of relative mach number at 70% blade span. It can be seen that the shock in rotor 37 is attached to the leading edge, however the Forward swept rotor moves the shock away from the leading edge, which also improves the flow stability and lowers the stall mass flow.

7 CONCLUSIONS

Forward swept rotor can largely improve the stall margin, and it can change the passage shock structure, which apparently improves the flow field near rotor tip. Using forward swept rotor as a measure to control rotating stall should be considered during the designing process of transonic rotor.

REFERENCES

Adamczyk, J. J. & Celestina, M. L. 1993. The Role of Tip Clearance in High-Speed Fan Stall. ASME Journal of Turbomachinery, 115: 28–39.

Cumpsty, N. A. 1989. Part-circumference Casing Treatment and the Effect on Compressor Stall, ASME Paper, 89-GT -312.

Furukawa, M. & Inoue, M. 1999. The Role of Tip Leakage Vortex Breakdown in Compressor Rotor Aerodynamics. ASME Journal of Turbomachinery, 121(3): 469–480.

Furukawa, M. & Saiki, K. 2000.Unsteady Flow Behavior Due to Breakdown of Tip Leakage Vortex in an Axial Compressor Rotor at Near-Stall Condition. ASME Paper, 2000-GT-666.

Hah, C. & Rabe, D C. 2001. Role of Tip Clearance Flows on Flow Instability in Axial Flow Compressors. ISABE Paper, 2001–1223.

Hah, C. & Schulze, R. 1999. Numerical and Experimental Study for Short Wavelength Stall Inception in a Low-Speed Axial Compressor. ISABE Paper, 99–7033.

Maerz, J. & Hah, C. 2002. An Experimental and Numerical Investigation into the Mechanism of Rotating Instability. ASME Journal of Turbomachinery, 124: 367–374.

Mailach, R. & Lehmann, I. 2000. Rotating Instabilities in an Axial Compressor Originating from the fluctuating Blade Tip Vortex. ASME Paper, 2000-GT-506.

Suder, K. L. 1998. Blockage Development in a Transonic, Axial Compressor Rotor. ASME Journal of Turbomachinery, 120(3): 465–476.

Suder, K. L. & Celestina, M. L. 1997. Experiment and Computational Investigation of the Tip Clearance Flow in a Transonic Axial Compressor Rotor. ASME Journal of turbomachinery, 118: 218–229.

Wadia, A. R. & Szucs P. N. 1998. Inner workings of aerodynamic sweep, ASME Journal of Turbomachinery, 120:671–682.

Modelling and Computation in Engineering – Zhu (ed)
© 2011 Taylor & Francis Group, London, ISBN 978-0-415-61516-7

Research on inherent characteristics of the wind turbine tower based on field testing

R.L. Ma, Y.Q. Ma & H.Q. Liu
Department of Building Engineering, University of Tongji, Shanghai, China

ABSTRACT: Based on the theory of random vibration and system identification, the ambient vibration tests of the three wind turbine towers in wind-power station of the Inner Mongolia Wulanyiligeng were carried out. The method of coupling overall modeling of blade, hub, nacelle and tower was put forward, and the numerical stimulation and tests results show that the wind turbine towers can effectively avoid resonance, and meet the standard design requirements of Germanischer Lloyd. The vibrational forms of wind turbine tower mainly are lateral bending vibration, forth-and-back bending vibration and torsional vibration; the translational damping ratio in the first phrase is about 1.75%, and the torsional damping ratio in the first phrase about 0.6%. The overall modeling shows excellent consistency with the tests results, which can benefit the wind-induced dynamic response analysis and the vibration control research on the wind turbine tower system.

1 INTRODUCTION

With the fast development of the wind turbine orienting to the large-scaled and flexible and in order to ensure the system working reliably, we should keep it in good inherent characteristics. During rotor's rotating, if the incentive frequency or passing frequency produced by the rotation approaches the inherent frequency of the tower, it will cause the tower in great vibration or even in resonance. Vibration can make the tower causing greater dynamic stress, structure fatigue so that it will also shorten the service life of the machine (Burton et al., 2001). Therefore, it has profound significance of the research on the inherent characteristics of the wind turbine tower.

Currently, the research on the inherent characteristics of the wind turbine tower mostly adopts the methods of analytic mechanics (Miller et al., 1978; Sheu, 1978), multi-body dynamics (Wright et al., 1999; Lee, 2002) and the theory of finite element (Lobtiz, 1984; Murtagh et al., 2005). However, the report of the inherent characteristics based on the field measurement is hardly to be available. Since the wind turbine tower is always situated in severely changeable environment with the special structure that at the top of the tower there is an installed rotor, therefore, the field measurement is an comparatively reliable way to obtain the inherent characteristics of the wind turbine tower, as well as an way to provide the forceful verification for the theoretical analysis.

The paper deals with the inherent characteristics of the three wind turbine towers based on the ambient vibration testing in wind-power station of the Inner Mongolia Wulanyiligeng, we obtain the valuable first-hand information of the tower's inherent frequency and damping ratio. Meanwhile, on the basis of the characteristics of the wind turbine main components and the previous models (Bazeos et al., 2002; Lavassas et al., 2003), this paper puts forward an method of the overall model considering coupling of the blade, the hub, the nacelle and the tower, conducted modal analysis, and collected the inherent frequencies and formation of the tower from the system. Finally, by comparison with the measured value, we verify the correctness of the overall modeling and pave the way for the wind-induced dynamic response analysis and the vibration control research on the wind turbine tower system.

2 PROJECTING BACKGROUND AND MEASURING STRATEGY

2.1 *Projecting background*

The tested wind turbines in wind-power station of the Inner Mongolia Wulanyiligeng are situated in the average altitude for 1380 m with the southwest prevailing wind direction. There are 200 sets wind turbines installed there in the type of 1500 KW/77. the hub height is 65 m, diameter of the rotor is 77 m, and rotating speed is 17.3 rmp. However, we choose model N53,N65 and N73, three of them as the tested wind turbines.

2.2 *Measuring-point arrangement and testing methods*

We use piezoelectric acceleration transducers to measure acceleration time curve of the towers and use SVSA software for data collection, the sampling frequency for 50 Hz, the sampling length for 9216. Suppose X is along the head, Y perpendicular to the head and Z vertical to the head. There are eight measuring-points in all during the test. Lay out point 1, 5 and 7 along the X direction, point 2, 6 and 8 along the Y direction and point 3 and 4 parallel to the X direction. The measuring-points are positioned respectively on the platform 2, 3 and 4, in Figure 1 to Figure 3. The sensor absorbs are facilitated inside of the tower by its own magnet.

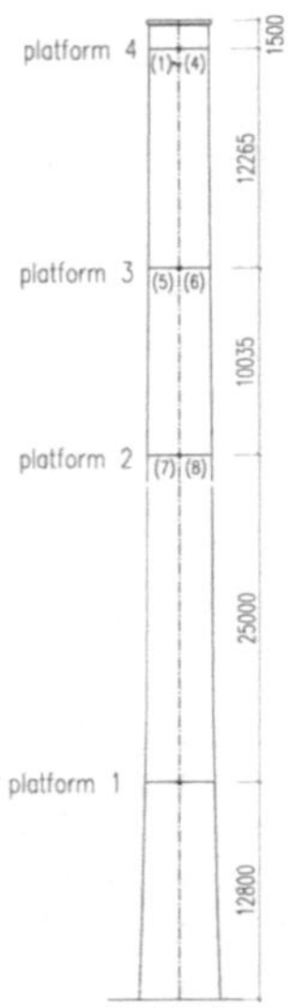

Figure 1. Measuring-point elevation layout.

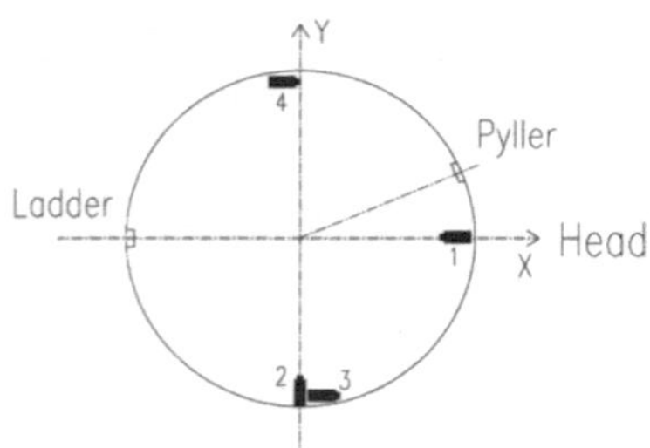

Figure 2. Measuring-point plan layout.

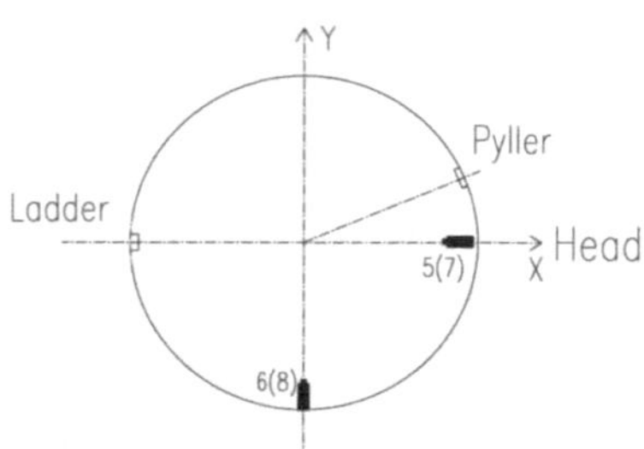

Figure 3. Measuring-point plan layout.

3 TEST DATA ANALYSIS AND PROCESSING

3.1 *Analysis method*

Under the case of the input signal unknown, we directly utilize the pulsation responded signal of the wind turbine tower to discern the modal parameter of structure. Through the peaks of power spectrum curve we get the inherent frequency of the overall vibration in the system (He et al., 2001; Yan et al., 2010). Since the partial resonant may exist, we need a comprehensive analysis and judgment from the power spectrum diagram of the measuring-points in the same direction and then make a decision. Use the half-power point method to determine the damping ratio. For translation and torsion coupled vibration, we can identify them with " + " and " – " of the acceleration of symmetry points. When signals add together, it can offset torsional signals and gains double translational signal; when signals subtract each other, it can offset translational signals and gains double torsional signals.

3.2 *Test results*

We get the acceleration time curve of each measuring-point through the test. We get the power spectrum curve through the frequency spectrum analysis of the point 1, point 2, point 5, point 6, point 7 and point 8. Figure 4 and Figure 5 are the power spectrum curves of point 5 and 6 on the N65 wind turbine tower. We get the tower's translation frequency and the damping ratio. The subtraction of the accelerated speeds of point3 and point4 divided by the distance of them, and through the torsional power spectrum analysis of the tower, we get the tower's torsional frequency and the damping ratio. Figure 6 is the torsional power spectrum curve on the N65 wind turbine tower. N53, N73 and N65 almost keep in the same testing results. Table 1 is the testing results of the inherent frequencies and damping ratio of the previous five phrases of the three towers.

Table 1 shows that the inherent frequencies and damping ratio of the previous five phrases of the three towers is basically the same values. Averaging them,

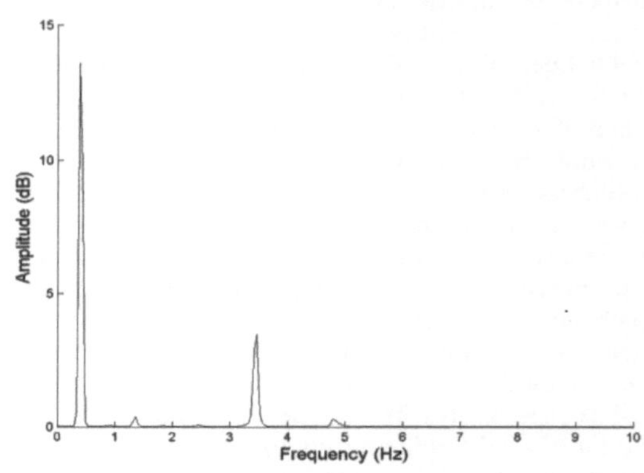

Figure 4. Power spectrum curves of point 5 on the N65 wind turbine tower.

we get that the inherent frequencies of the previous five phrases of the towers are 0.4028 Hz, 0.4150 Hz, 1.1963 Hz, 3.4415 Hz and 3.5052 Hz which meet the design requirements of Germanischer Lloyd guideline, and the wind turbine towers can effectively avoid the resonance. Translational damping ratio of the first phrases is 1.75% and torsion damping ratio is 0.56%.

At present recommended value of damping ratio is not given in Germanischer Lloyd guideline and other related wind turbine guidelines, and we often use damping ratio (1%) of high-rise steel structures. We can see that the translational damping ratio of the wind turbine tower is larger, so we recommend the parameters in the dynamic analysis.

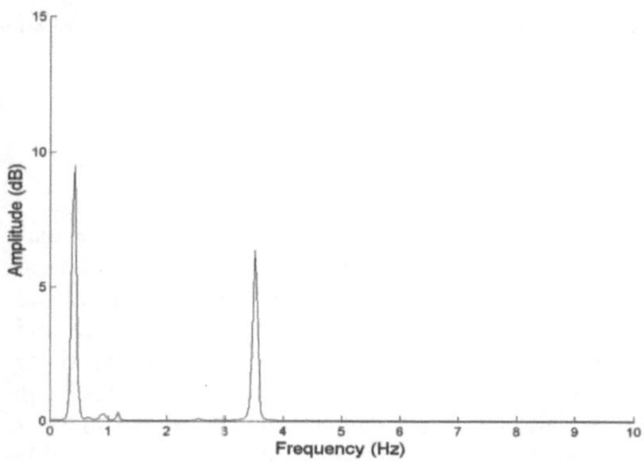

Figure 5. Power spectrum curves of point 6 on the N65 wind turbine tower.

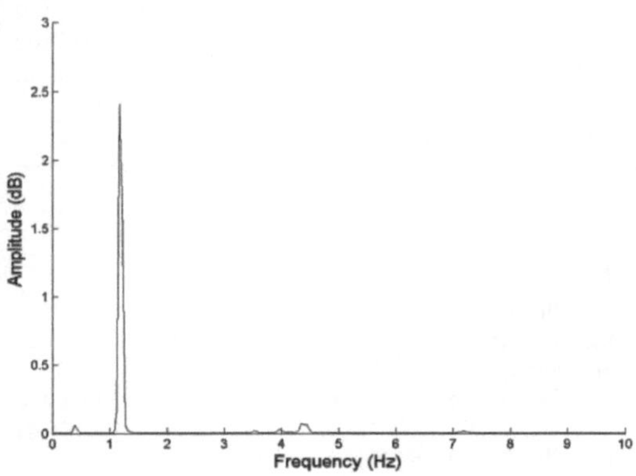

Figure 6. Torsional power spectrum curve on the N65 wind turbine tower.

4 FINITE ELEMENT ANALYSIS

A wind turbine consists of three blades and a hub connected to a nacelle, which in turn is rigidly connected to the top of a wind turbine tower. Coupling among them have a great influence for natural characteristics of the tower. this paper puts forward an method of the overall model considering coupling of the blade, the hub, the nacelle and the tower, conducted modal analysis, and collected the inherent frequencies and formation of the tower from the system.

4.1 *Overall model of wind turbine system*

The test wind turbine consists of three tower sections and its hub height is 65 m. The section diameter and thickness at base are 4 m and 26 mm and Its section diameter and thickness at top are 2.57 m and 12 mm. the elastic modulus and density of the tower were taken as $2.1 \times 10^{11} N/m^2$ and $7850 kg/m^3$. According to the geometric characteristics and major fore characteristics of the tower, this paper uses eight-node shell element—shell 181.

According to the principle of rigidity equivalent, the three blades are assumed to be hollow rectangular cross-section cantilever beam and simulated by beam 189 element. Each blade is 37.3 m in length with a hub radius of 3 m. the section of blade root is $3 m \times 0.8 m \times 0.01 m$ and the apex section is $0.3 m \times 0.08 m \times 0.01 m$. The blades are made of GRP and regard as the orthogonal anisotropic material. the density of the tower were taken as $2100 kg/m^3$. The axial elastic modulus is $6.25 \times 10^{10} N/m^2$, the radial elastic modulus is $1.65 \times 1010 N/m^2$, and the tangential elastic modulus is $5.5 \times 10^9 N/m^2$. The Poisson ratio is 0.22.

As the nacelle and the hub structure is complicated and the internal detail features are not required attention in the overall analysis, the nacelle and the hub can be simplified to the quality of points and simulated by mass 21 element. This paper sets element parameters to consider its quality, moment of inertia and eccentric position, and then the quality of points, top of the tower and the blades root are connected rigidly through CERIG order of ANSYS. As the base is relatively

Table 1. Inherent frequencies and damping ratio of the previous five phrases of the three towers.

Setting	Number of wind turbine	1st lateral bending	1st forth-and-back bending	1st twisting	2nd lateral bending	2nd forth-and-back bending
Measured Frequency (Hz)	N53	0.4028	0.4150	1.1963	3.4422	3.5522
	N65	0.4028	0.4150	1.1963	3.4546	3.5278
	N73	0.4028	0.4150	1.1963	3.4478	3.5411
Measured Damping Ratio (%)	N53	1.77	1.73	0.54	0.28	0.27
	N65	1.76	1.74	0.58	0.26	0.26
	N73	1.78	1.72	0.56	0.30	0.28

large stiffness of the upper structure, the bottom of the tower is considered as full consolidation. Through the above method, the integrated finite element model was established. The model was illustrated in Figure 7.

4.2 *Modal analysis of wind turbine tower*

This paper uses the method of BLOCK LANCZOS to conduct modal analysis. Table 2 shows the inherent frequencies and description of modes of the previous ten phrases of the wind turbines.

Table 2 shows that the first mode and second mode are lateral bending vibration and forth-and-back bending vibration in the first phrase respectively, and the frequency respectively are 0.4357 Hz and 0.4375 Hz. The fifth mode appears torsional vibration in the first phrase and the frequency is 1.2322 Hz. The ninth and tenth modes appear lateral bending vibration and forth-and-back bending vibration in the second phrase respectively, and the frequency respectively are 3.4572 Hz and 3.4769 Hz. By comparison with the measured value, the theoretical value is a little higher. But they are consistent basically on the whole. This is mainly because the theoretical value is calculated in certain hypothesis foundation which have a difference with the actual structure, and in addition pulsation testing itself has certain error. The primary purpose for the testing was to understand the modes of the wind turbine and to verify the correctness of the overall modeling for wind power tower and pave the way for the wind-induced dynamic response analysis and the vibration control research on the wind turbine tower system.

Figure 7. Overall model of the wind turbine system.

5 CONCLUSIONS

Based on the three wind turbine towers, this paper carried out the ambient vibration testing of the inherent characteristics and obtained the valuable first-hand information of the tower's inherent frequency and damping ratio. It provides the basis of appraisal for the existing theoretical model and also provides basic date for the wind-induced dynamic response analysis and the vibration control research on the wind turbine tower system. At present Germanischer Lloyd guideline and other related wind turbine guidelines required dynamic analysis for the whole wind turbine, but recommended value of damping ratio was not given. On the basis of the characteristics of the wind turbine and the previous models, this paper puts forward an

Table 2. The inherent frequencies and Description of Modes of the previous ten phrases.

Mode number	Analytical Frequency (Hz)	Description of modes
1	0.4357	1st lateral bending vibration of the tower.
2	0.4375	1st forth-and-back bending vibration of the tower.
3	1.0524	1st out-of-plane vibration of the blades: the upper blade remain unchanged and the bottom two blades vibrate in the opposite direction.
4	1.1762	1st out-of-plane vibration of the blades the upper blade and the bottom two blades vibrate in the opposite direction.
5	1.2322	1st out-of-plane vibration of the blades: the three blades vibrate in the opposite direction. 1st torsional vibration of the tower
6	1.4025	2nd out-of-plane vibration of the blades: the upper blade remain unchanged and the bottom two blades vibrate in the opposite direction.
7	2.3398	1st in-plane vibration of the blades: the three blades vibrate in the opposite direction.
8	3.202	2nd out-of-plane vibration of the blades the upper blade remain unchanged and the bottom two blades vibrate in the opposite direction. 1st in-plane vibration of the blades: the upper blade remain unchanged and the bottom two blades vibrate in the opposite direction.
9	3.4572	1st in-plane vibration of the blades: the upper blade and the bottom two blades vibrate in the opposite direction. 2nd lateral bending vibration of the tower.
10	3.4769	2nd out-of-plane vibration of the blades: the upper blade and the bottom two blades vibrate in the opposite direction. 1st in-plane vibration of the blades: the upper blade remain unchanged and the bottom two blades vibrate in the opposite direction. 2nd forth-and-back bending vibration of the tower.

method of overall model, whose correctness is verified compared with the measured value.

REFERENCES

Bazeos, N., Hatzigeorgiou, G.D., Hondros, I.D. et al., 2002. Static, seismic and stability analysis of a prototype wind turbine stell tower. Engineering Structures, 24(8): 1015–1025.

Burton, T., Sharpe, D., Jenkins, N., 2001. Wind energy handbook. New York: John Wiley & Sons.

He, M.J., Ma, R.L., Huang, Z., 2001. Measurements and analysis of model parameter of broadcast and television steel tower. Journal of Tongji University, 29(4): 862–866.

Lee, D., Hodges, D.H., Patil, M.J., 2002. Multi-flexible-body dynamic analysis of horizontal axis wind turbines. Wind Energy, 25(5): 281–300.

Lobtiz, D.W., 1984. A Nanstran-based computer program of structural dynamic analysis of horizontal axis wind turbine. Proceedings of the Horizontal Axis Wind Turbine Technology Workshop, Department of Energy and NASA-Lewis, Cleveland.

Miller, R.H., Dugundji, J., Wendell, J., 1978. Dynamic of horizontal axis wind turbine. Wind Energy Conversion. MIT, ASRL-IR-184-9, DOE COO-4131-T1.

Murtagh, P.J., Basu, B., Broderick, B.M., 2005. Along-wind response of a wind turbine tower with blade coupling subjected to rotationally sampled wind loading. Engineering Structures, 27(8): 1209–1219.

Sheu, D.L., 1978. Effects of tower motion on dynamic response of windmill rotor. Wind Energy Conversion, VO1 VII, ASRL-IR-184-13, DOE COO-4131-T1.

Thomas, G.C., 1983. Modal testing of a rotating wind turbine. Sandia National Laboratories, SAND82-0631.

Wright, A.D., Kelley, N.D., Osgood, R.M., 1999. Validation of a model for a two-bladed flexible rotor system: progress to date. Proceedings of the 37th AIAA Aerospace Sciences Meeting and Exhibit: 293–307.

Yan X.M., He M.J., Ma R.L., 2010. Full-scale measurements of ambient vibration around two adjacent transmission towers. Journal of Vibration and Shock, 29(3): 77–78.

Modelling and Computation in Engineering – Zhu (ed)
© 2011 Taylor & Francis Group, London, ISBN 978-0-415-61516-7

Map-matching algorithm based on the index mechanism

Jing Zhai & Hantao Zhao
School of Vehicle Engineering, Harbin Institute of Technology

Hongyan Mao
Library, Harbin Institute of Technology

Wencai Sun
School of Traffic&Transportation, Jilin University

ABSTRACT: The emergency vehicle equipped with GPS plays an important role in the highway rapid response behavior. In order to remove the display error produced by the GPS locating data and the E-map, the matching algorithm has been put forward based on the map index and the shortest distance algorithm. According to the difference between highways and urban roads, the proposed map-matching algorithm links the vehicle location information with the road information of E-map database, achieving the effective and rapid search of the road, thus, the vehicle's position relative to E-map can be determined. In order to validate the map-matching algorithm, the simulation program is developed using VB and MapObject module. The results of test show that the algorithm can greatly decrease the influence of the locating error and reduce the ratio of mismatch.

1 INTRODUCTION

In the terminal monitoring software system of highway Emergency Command Center, real-time access to the location of patrol vehicle is a prerequisite for dynamically monitoring. GPS vehicle location data and electronic map data exit inevitably of a variety of errors, which tend to bring the deviation from the vehicles trajectory of the actual road. At present, there are mainly two kinds of treatment methods to solve this problem (Sun Wencai 2006). One is to take action to improve GPS positioning accuracy and precision of electronic map, the other is the use of map-matching method. The current study of map-matching algorithm takes advantage of a variety of theories. Commonly used algorithm includes the shortest distance algorithm, semi-deterministic algorithm, probability and statistics algorithm, map matching algorithm based on fuzzy logic or based on the cost function and so on (Chen Jiayu 2004). These algorithms have their own conditions of application, and most of them are applicable to intensive urban roads. But for the scattered highways, the above matching algorithm can not be fully applicable. In view of this, one matching algorithm based on the map index is put forward which suits for the topographical features of highway road.

2 ESTABLISHMENT OF MAP INDEX

Based on GPS location information and driving directions, the so-called map index mechanism takes the dispersed stake number of VL (Virtual Line, which is composed of discrete points and used to calculate) as the identification point, searches highway layer data stored in the space database, and then quickly determines the matching path (Zhang Xiaoguo 2003). After the matching road has been found, the matching point of GPS location can be accessed by the shortest distance algorithm.

Table 1. Spatial data structure of VL.

NodeID	RoadID	DiretionID	X	Y
Ni	001	01	…	…
Nj	001	02	…	…
Nk	002	02	…	…
Nl	003	02	…	…

The basic source data used in highway electronic map is a series of SN (Stake Number of corresponding point on highway) with a certain distance interval, these data are measured on the basis of highway centerline. However, the road needs to be distinguished in different directions of traffic.

In this paper, the structure of highway spatial data is adjusted. The centerlines spatial data of bilateral lane carriageway is added to the original data. As shown in the Table 1, field NodeID is the main keywords of node table, which means the various marking points of lane carriageway centerlines. Field RoadID expresses the name codes of the highways. Field DirectionID expresses the name of the different directions of the same highway. Field X and Field Y express the coordinates of the corresponding marking points. After this structure adjustment, the map index mechanism for road searching is established.

3 ALGORITHM DESIGN

3.1 *GPS data acquisition and preprocessing*

The GPS positioning data includes longitude, latitude, speed, direction, time, etc. The data should be stored to prevent loss, and then extracted by application software to change data format.

3.2 *Determination of the data effectiveness*

Because of the block of the tall buildings, the GPS data drift phenomenon maybe happen in data receiving process, which often generates invalid data. So, these abnormal positioning data should be filtered out before map-matching. The criteria of data characteristics, velocity and acceleration need to be added in the algorithm to determine the data validity.

3.3 *Road index*

The road number of highway network is limited. Therefore, the entire map database of the highway can be searched as a candidate object. In order to build up search circle, two-dimensional coordinates (x,y) is taken for the center, r means the radius, J means the node of VL, G represents the GPS data point. Then the following formula can be used to determine whether the nodes are located at the search circle.

$$(x_G - x_J)^2 + (y_G - y_J)^2 \leq r^2 \tag{1}$$

If the map object in the circle, which is the composition identification point of VL, belongs to the same road in one direction, then, this object is the matching road. If the map object in the circle belongs to the same road in different direction, the matching road should be further screened based on the direction of vehicles. If the angle between road direction and vehicle direction is less than 180 degrees, this road direction is judged to be the vehicle direction. If there is no map object in the circle, the vehicle is considered to enter the square, gas stations, service areas and other open spaces, vehicle positioning does not require matching.

3.4 *Amendments to the shortest distance algorithm*

According to the road attribute, the shortest distance algorithm set up a match threshold in advance. The matching point can be determined by calculating the shortest distance path between the GPS point and the corresponding road. In this paper, the vertical point algorithm is used to improve the shortest path algorithm. Matching points can be the projector from GPS point to the road. Assuming that the number of identification points located at the search circle is N, as shown in Figure 1.

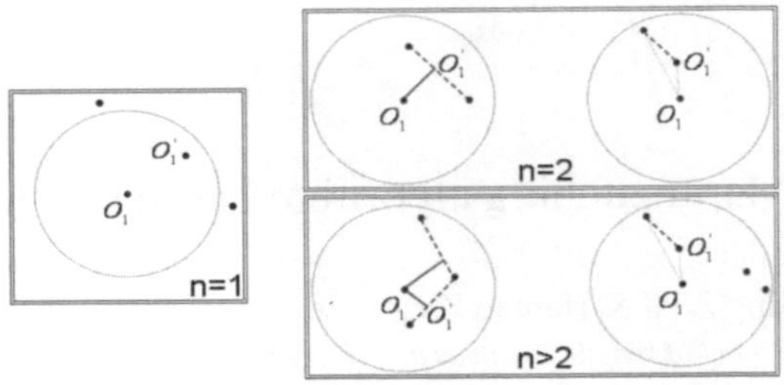

Figure 1. Determination of the matching points.

Then,

(1) If $N = 1$, the identification point is the match point.
(2) If $N = 2$, vertical line is made from vehicle location point to the line segment connected by the two identification points. If the pedal point is located at the segment, the pedal point is the matching point. If else, the identification point close to the center is the matching point.
(3) If $N > 2$, vertical lines are made from vehicle location point to the line segments connected by the adjacent identification points. The pedal point nearest to the center is the matching point. If the pedal points are located out of the segments, the identification point nearest to the center is the matching point.

4 KEY TECHNOLOGIES

4.1 *VL*

As shown in Figure 2, the VL is determined by the translational of the SN. It is worth noting that the VL is in fact composed of discrete points. Therefore, the discrete points of VL can be gotten by the coordinates of the SN.

The VL is not simple level translational, vertical translational, or translational of a certain angle of the SN. For the highway with relatively large curvature, such as a larger arc or ring roads, simple translational will cause coordinates offset. The geometric principle is used to calculate the coordinates of VL, as shown in Figure 3. The symbols of k1 to k5 are five continuous SN, (x, y) are their corresponding geodetic coordinates. Through making vertical line from k2 to the line segment connected with k1 and k3, the geodetic coordinates of k2′ and k2″ can be calculated based on pavement width, coordinate (x1, y1) and coordinate (x3, y3). So, all the coordinates of discrete points of VL can be figured out. The coordinates of destination and starting point need to take translational approach.

4.2 *Radius of search circle*

How to determine the scope of the neighboring regions of the GPS locating points is the key constraint factor in algorithm performance, that is, the determination of the radius of search circle. Too large radius will

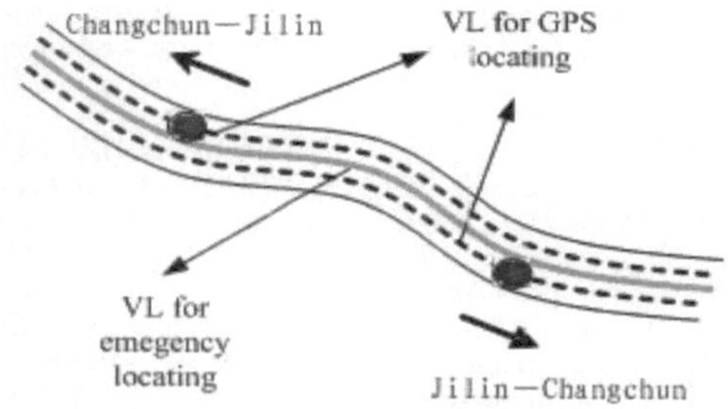

Figure 2. Difference between VL for GPS locating and VL for emergency locating.

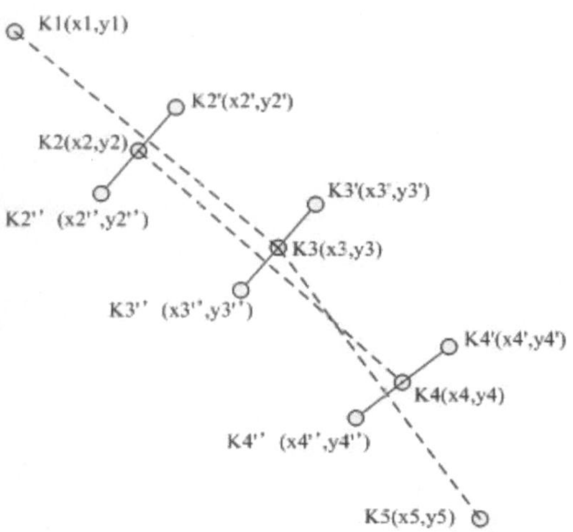

Figure 3. Coordinates determination of discrete points of VL for GPS locating.

increase the volume of the calculation and lead to miscarriage of justice. Too small radius will lead to road object can not be found in the neighborhood of the GPS locating points.

The errors in the process of vehicle positioning are mainly produced from the following sections (Su Jie, 2001): GPS positioning errors (ΔE_1), electronic map error (ΔE_2) and coordinate projection transformation error (ΔE_3).

Under the normal state of VL composed of consecutive points, the error range of GPS locating point should be one round-domain. The center is the GPS locating point, the radius is the sum of these three errors. That is,

$$r_{\max} = \Delta E_1 + \Delta E_2 + \Delta E_3 \quad (2)$$

In addition, because the VL is linked by marking points with distance less than or equal to 50 meters, the maximum radius of search circle can be calculated as shown in Figure 4 and formula (3).

$$r'_{\max} = \sqrt{r_{\max}^2 + 25^2} \quad (3)$$

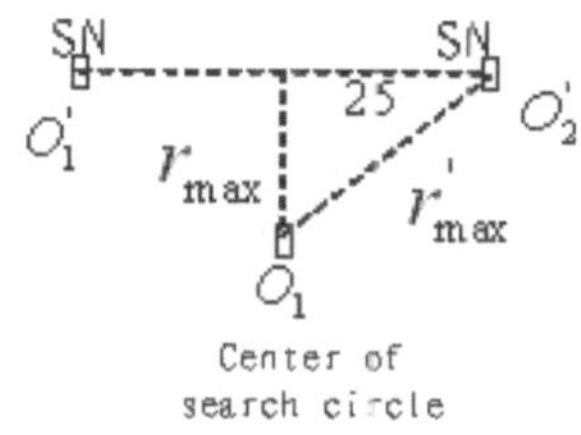

Figure 4. The maximum radius value of search circle.

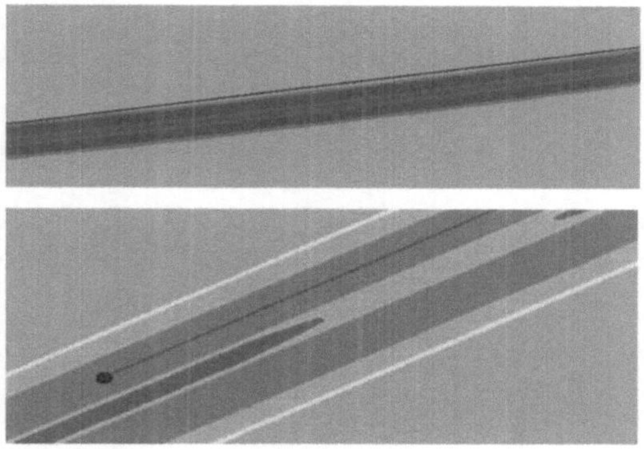

Figure 5. Comparison of vehicle location before and after the use of map-matching algorithm.

5 ALGORITHM REALIZATION

The positioning function is achieved by the control of MapObject. MapObject contains a dynamic tracking control layer known as TrackingLayer, which is always situated at the top of the map window, and its greatest feature is the faster refresh rate. So, the dynamic tracking control layer provides interface for setting up GPS positioning system, realizes real-time tracking in the vehicle monitoring and dispatching system. With vehicle moving, the position can be dynamic shown. Combined with TrackingLayer, the map-matching algorithm can be realized by VB language.

In order to verify the validity of map-matching algorithm, the effect of vehicle location before and after the use of algorithm for mobile vehicles is contrasted. The International Airport – Changchun eastern section of the Changchun – Jilin expressway is selected to test. The vehicle location is amended by the map-matching algorithms based on indexing mechanism, the effect of positioning is shown in Figure 5. The results show that the road can be quickly searched and the vehicle can be accurately located.

6 CONCLUSIONS

In accordance with highway characteristics, the concept of the VL is put forward based on SN. Then the matching algorithm is proposed based on the map index and the shortest distance algorithm, and the key technologies of the algorithm are studied. The principle of this arithmetic is simple and its operation is convenient. This arithmetic can reduce the ratio of mismatch and improve the performance of the dynamic monitoring and controlling system. However, in the course of the road index, the direction of vehicles is chosen to match the map object belonging to the same road in different direction, which will lead to GPS data drift and should be given further research.

REFERENCES

Chen Jiayu (2004), "Research and Realization of Vehicle Monitor and Control System Based on GPS/GPRS/GIS", Master Degree thesis, Fuzhou: Fuzhou University.

Su Jie, Zhou Dongfang, Yue Chunsheng (2001), "Real-time Map-matching Algorithm in GPS Navigation System for Vehicles". Acta Geodaetica et Cartographica Sinica, Vol.30, No.3, 252–256.

Sun Wencai(2006), "Study on dynamic monitoring and controlling technology for vehicle with GPS on freeway", Master Degree thesis, Changchun: Jilin University. 22–24.

Zhang Xiaoguo, Wang Qing (2003), "The Relationship among Vehicle Positioning Performance, Map Quality, and Sensitivities and Feasibilities of Map-matching Algorithms". Proc IEEE Intelligent Vehicles Symposium 2003, 468–473.

Modelling and Computation in Engineering – Zhu (ed)
© 2011 Taylor & Francis Group, London, ISBN 978-0-415-61516-7

Laboratory model experiments on dynamic behavior of road structures under repeated traffic loads

Zheng Lu, Hailin Yao, Jie Liu & Mengling Hu
State Key Laboratory of Geomechanics and Geotechnical Engineering, Institute of Rock and Soil Mechanics, Chinese Academy of Sciences, Wuhan, China

ABSTRACT: A laboratory full-size model experiment is presented in this paper in order to better understand the dynamic behaviour of road structures under repeated traffic loads. Different transducers measuring vertical pressures, elastic deformation and cumulative plastic deformation of road structure layers are installed. The distributions of dynamic stress and elastic deformation of road structures are obtained, and the variation laws of irreversible deformation with the number of loading cycles are also derived from the experiments. Furthermore, a three dimensional multi-layered computing model is employed to compare the experimental results with theoretical ones. The conclusions developed herein may be used as a preliminary reference in road pavement and subgrade engineering design.

1 INTRODUCTION

The study of dynamic behaviour of road structures under vehicle loads is one of the important researches in the domain of transport. With the rapid development of economy and mechanical technology, heavy loadings and the traffic volume on the highway become much greater than ever before. Thus, the subject of the dynamic response of road structures to repeated traffic loads is of great significance in the field of geotechnical and transportation-related engineering.

The dynamic characteristics of road structures under traffic loads were early investigated by using analytical method. Senddon (1952) firstly discussed the vibration of the elastic ground under a moving load. He considered the two-dimensional problem of a line load moving with constant subsonic speed over the surface of a homogeneous elastic half-space by using integral transform method. A homogeneous 3D elastic half-space subjected to forces moving at constant speed was studied by Eason (1965) using the double Fourier transformation method. More recently, many researchers worked on the problems of traffic-load-induced dynamic responses by considering the road structures as different models such as thin plate, elastic medium or water saturated pore medium (Picoux 2005, Takemiya 2005 and Xu 2008, and among others).

The development of computational methods has opened the possibility to perform numerical simulations of the response of road structures under repeated load. By means of dynamic numerical analysis, Hyodo et al. (1996) investigated the deformation of the soft clay foundation of low embankment road under traffic loading. Kettil et al. (2007) presented modeling and simulation of inelastic deformation in road structures leading to rutting. In his paper, the modeling and calculating were performed in FE program Abaqus. Manzari and Prachathananukit et al. (2001) also calculated the cumulative plastic deformation of road structures subjected to cyclic mechanical loads by introducing different elastoplastic material models. Theoretically, explicit simulation is preferable. However, it would consist in implementing a conventional step-by-step elastoplastic procedure, requiring the discretization of each individual load cycle into sufficiently small number of cycles, will quite rapidly lead to considerable computational time and complete loss of numerical accuracy as the number of cycles increase.

It is noted that although many problems involving the dynamic behavior of road pavement and subgrade subjected to traffic loads have been investigated, the experiment or field measured data are still rather limited. In this paper, the somewhat complex phenomena involved in this kind of dynamic behavior is analyzed by using model experiments. Specific experimental devices are set up for testing representative sections of road structures subjected to repeated wheel loading. A three dimensional multi-layered computing model is also employed to compare the experimental results with theoretical ones.

2 EXPERIMENT DESCRIPTION

A laboratory experiment at full-scale which represents the typical road structures with multi-layers has been built in order to simulate the dynamic response of road subjected to vehicle traffic loads as shown in Fig. 1. The model includes, from bottom to top, a layer of subgrade (completely decomposed granite) of 0.9 m height compacted in three layers of 0.3 m each, a layer

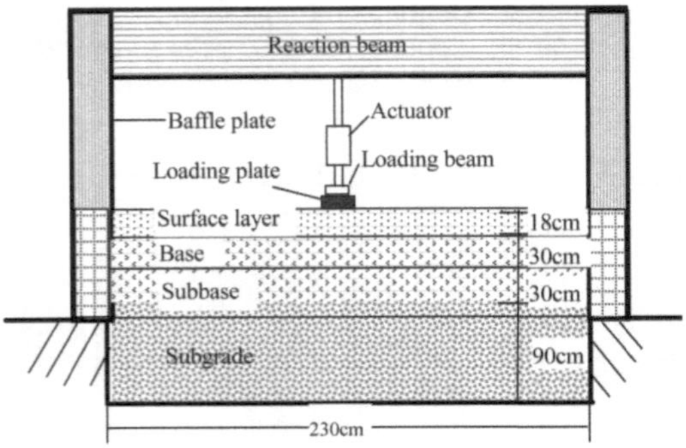

Figure 1. Representation of full-scale model.

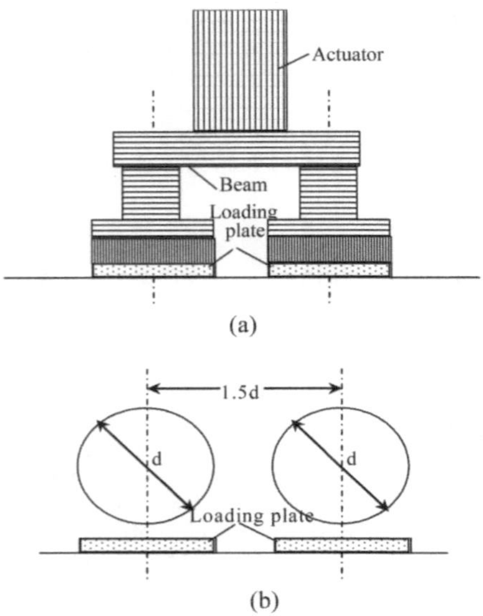

Figure 2. (a) Loading equipments; (b) distribution of the loads.

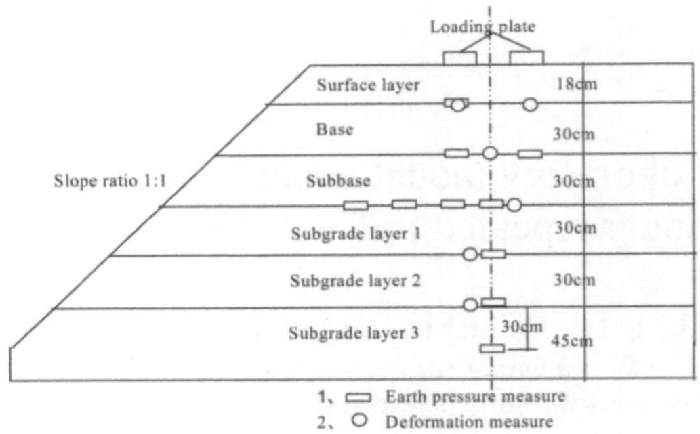

Figure 3. Elevation view drawing of measuring points.

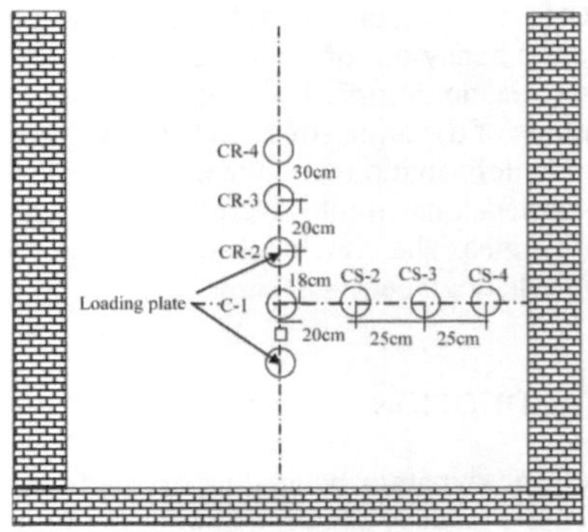

Figure 4. Plain view drawing of measuring points.

of subbase of 0.3 m height composed of cement stabilized gravels with 4% cement content, a layer of base of 0.3 m height composed of cement stabilized gravels with 6% cement content and an asphalt surface layer of 0.18 m height in order to simulate flexible pavement on which the loading equipments are placed with respect to the position of the loading plate. The structure layers are limited on two opposite sides by inclined planes (slopes) that represent free edges. The other sides are limited by vertical steel baffle plates representing the beginning and the end of the model. The loading equipments are fixed on a reaction beam which is bolted on the steel frame. All components of the experiment are located on a reinforced concrete foundation able to support the important weights of these components.

In this experiment, the effect of double wheels loading is taken into consideration to simulate the vehicle loads as shown in Fig. 2, and the distribution of the loads meets the requirements of Specifications for Design Highway Asphat Pavement (2006). The loading equipments are composed of, form bottom to top, rubber bearings, steel plates, loading beam and actuator. The shape of the tire-pavement contact area is assumed to be circular, and the load pressure within the contact area is assumed to be uniformly distributed. According to JTGD50 (2006), the diameter of contact circular d is 0.213 m, the force and pressure induced by single load is 25 KN and 0.7 MPa respectively, corresponding to the static conditions, the distance between two circulars is $1.5d$.

The detailed size of the model are as follows: the longitudinal length is 2.3 m, the horizontal width of the top is 2.0 m, the horizontal width of the bottom is 3.6 m, and the slope ratio is 1:1 as shown in Figs. 3–4. Different transducers are installed in the model of the road structures in order to measure stresses and displacements. All of the transducers are installed at the bottom of road structure layers as shown in Fig. 3. In the plain of the surface of subgrade, four stress transducers are placed along horizontal direction and longitudinal direction, respectively, as shown in Fig. 4 in order to derive the distribution of dynamic stresses. And the distances between each transducer are 0.20 m, 0.25 m and 0.25 m for horizontal direction, and 0.18 m, 0.20 m and 0.30 m for longitudinal direction. These accelerometers and their cables are protected by aluminum tubes to prevent damage from cyclic loadings.

Table 1. Load amplitude and frequency of multi-stage loading.

Load amplitude (kN)	10	15	20	30	40	45	50	60
Load frequency (Hz)	2	2	2	2	2	2	2	2
Load amplitude (kN)	70	75	75	75	75	75	75	
Load frequency (Hz)	2	2	3	4	5	6	7	

A typical experiment consists in generating signals of semi-sinusoidal wave in order to simulate the passage of traffic vehicles on the portion. If a truck is moving on a rough pavement surface, there will be a variation in the load amplitude imposed by the truck. According to Yao et al. (2009), the impact coefficient (the ratio of dynamic load to static load) of general highway pavement whose rough index is no more than IRI 2 mm is between 1.2~1.5. Therefore, we propose to choose the maximum dynamic load amplitude of the double wheels as 1.5 times than the static load. By multi-stage loading, the experiment contains several stages where every stage consists of 1000 cycles and corresponds to a frequency as shown in table 1. It is called "DT" whose purpose is to study the effect of traffic loads on the dynamic characteristics of road structure layers.

In order to obtain the cumulative deformation of road structures under long-term repeated vehicle loads, additional experiment called "CT" is performed by choosing the dynamic load amplitude as 75 KN and frequency as 7 Hz, and it consists of 10^6 cycles.

3 EXPERIMENT RESULTS

The dynamic load with signal of semi-sinusoidal wave is applied in this experiment. According to elasto-dynamical theory, the waveform of stress should be similar to that of load. Fig. 5 gives the measured stress waveforms of every layer's top of the model when the dynamic load amplitude is 75 KN and frequency is 7 Hz. It is shown that the stress waveforms are identical with semi-sinusoidal wave. The stress amplitudes increase gradually at initial several cycles since the hydraulic actuator needs adjusting pressure, and the waveforms keep steady after 20 cycles.

"DT" is conducted with respect to the effects of depth, load amplitude and load frequency on vertical stress and elastic deformation of road structure layers. Some test results are presented in Figs. 6–10. It can be seen from Fig. 6 that the vertical stress at the top of the base decrease with depth under different load. The curves decrease sharply in the initial range of depth of 0.8m, then gradually decrease to zero when depth is more than 0.8m. The ratios of the stresses of the base top, subbase top and subgrade top to the top of surface layer are 62.20%~81.45%, 20.12%~34.06% and 3.0%~4.63% respectively, which shows that the most part of stresses caused by vehicle loads are imposed or dispersed by road pavement.

The variations of vertical stresses at the subgrade top with load at different observation point (as shown in Fig. 4) are given in Fig. 7 and Fig. 8. It is seen that the vertical stresses of all observation points increase linearly with load increasing from 10 KN to 75 KN. In Fig. 9, the vertical stresses of different depth are

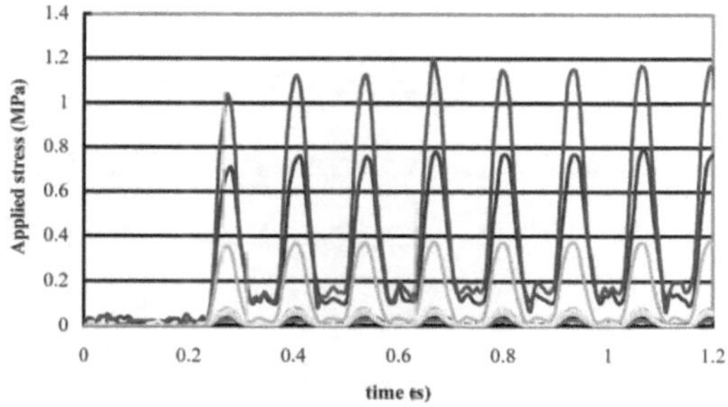

Figure 5. Applied stress at the top of every layer.

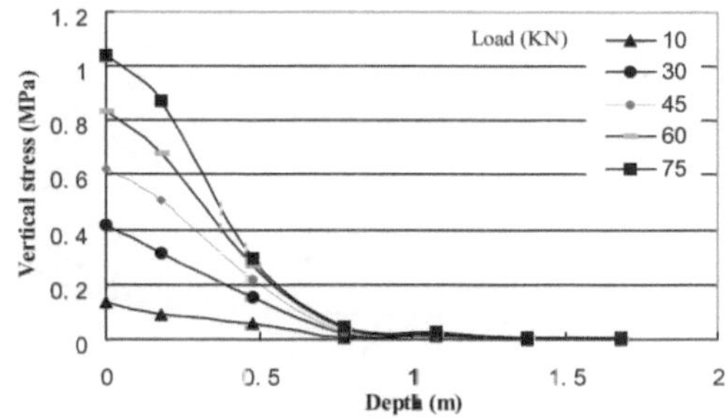

Figure 6. Variation of vertical stress with depth.

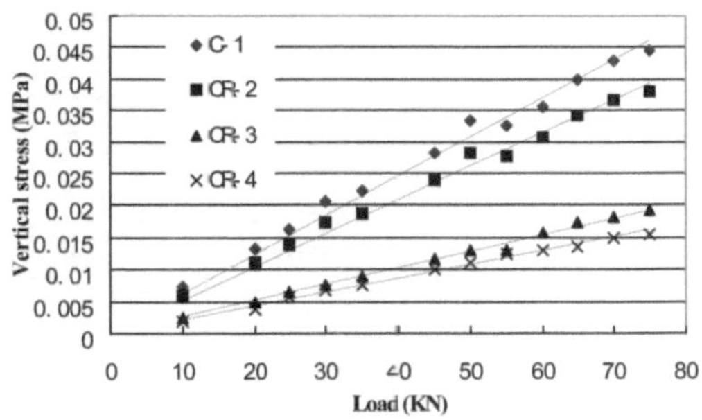

Figure 7. Variation of vertical stress with load.

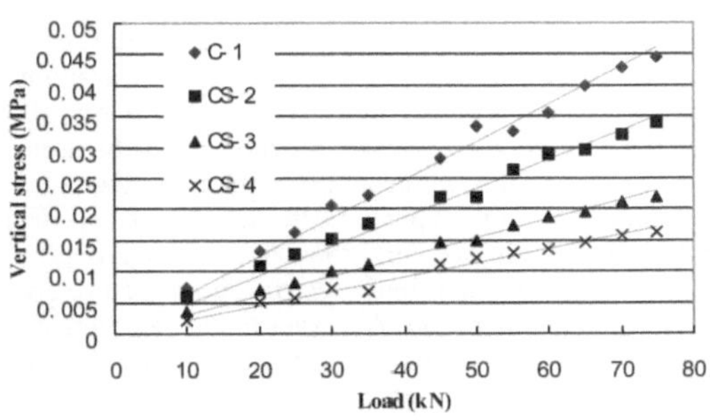

Figure 8. Variation of vertical stress with load.

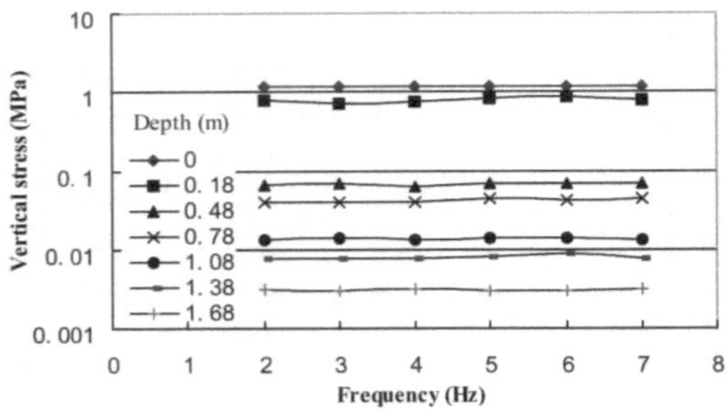

Figure 9. Variation of vertical stress with frequency.

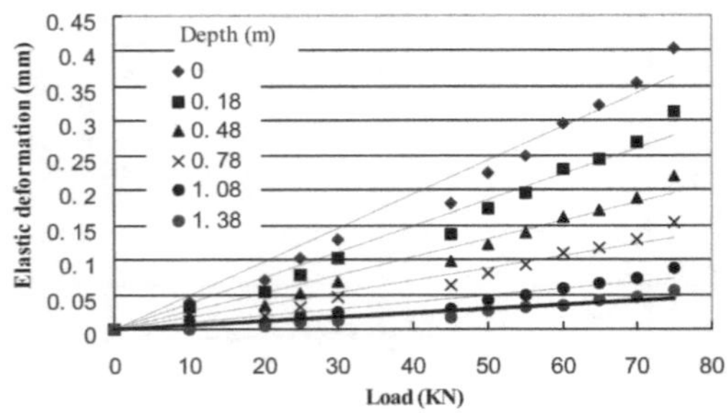

Figure 10. Variation of elastic deformation with load.

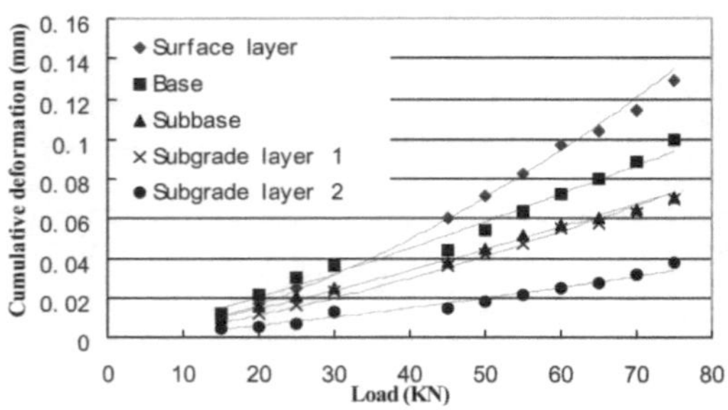

Figure 11. Variation of cumulative deformation with load.

plotted versus the loading frequency. As can be seen, the variation of vertical stress with frequency is slight when frequency varies from 2 Hz to 7 Hz. It reveals that the influence of loading frequency on dynamic characteristics of road structures can be neglected at low frequency. Fig. 10 presents the variation of elastic deformation with load amplitude. Six different depths corresponding to each interface of road structure are taken into consideration. An exact linear relationship between elastic deformation and load amplitude can be observed from Fig. 10.

The effects of load amplitude and the number of loading cycles on cumulative deformation (irreversible permanent deformation) of road structure layers are also investigated by using "CT", and are shown in Fig. 11 and Fig. 12. The cumulative deformation of each layer is considered as the difference between the cumulative deformation of top and bottom

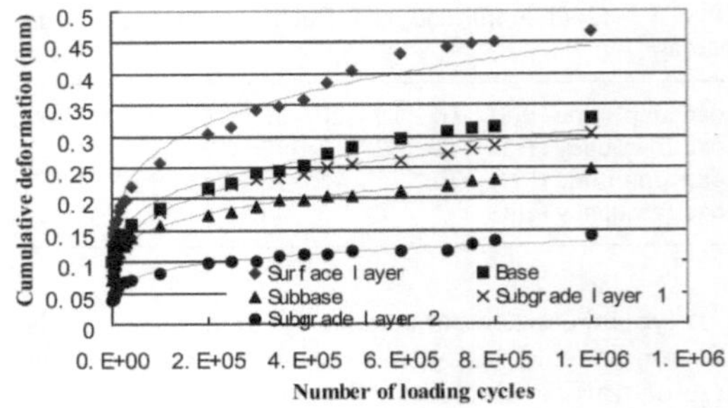

Figure 12. Variation of cumulative deformation with the number of cycles.

Table 2. Physical and mechanic characteristics of road structure layers.

Layers	Elastic modulus/ MPa	Poisson ratio	depth/ m	density/ (kg/m^3)
Surface layer	1 400.00	0.30	0.18	2 400
Base	734.63	0.30	0.30	2 201
Subbase	636.58	0.30	0.30	2 187
Subgrade layer 1	45.15	0.35	0.30	1 981
Subgrade layer 2	28.88	0.35	0.30	1 946
Subgrade layer 3	36.45	0.35	0.30	1 979

of the layer. We can see from Fig. 11 that the relationships between cumulative deformation and load amplitude of all layers are approximately linear, which follow on the same trends as those in Figs. 7–8 and Fig. 10. It is noted that the increasing law of cumulative deformation with the number of applied load cycles is very important in the study of the dynamic behavior of road structure. Therefore, the effect of the number of loading cycles on cumulative deformation is investigated as shown in Fig. 12. As can be seen, the cumulative deformation of each layer has the same tendency to increase with the number of cycles till one million cycles. The cumulative deformation is very fast for a short number of cycles and becomes much lower at other moments. These fast deformations come from the reorganization of grains of the road structure layers.

4 COMPARISONS OF MODEL EXPERIMENT RESULTS WITH THEORETICAL ONES

In order to compare theoretical results to experimental ones, a computing model of multi-layered road structure is also built by assuming that all of the road layers are considered as continuous and linear elastic medium. The physical and mechanical characteristics of road structure layers are obtained according laboratory tests, and are shown in table 2. In this part, the applied double-circular loads of experiment are simplified as equivalent double-rectangular loads by keeping the tire-pavement pressure consistent, and the moving load velocity is also obtained by using the

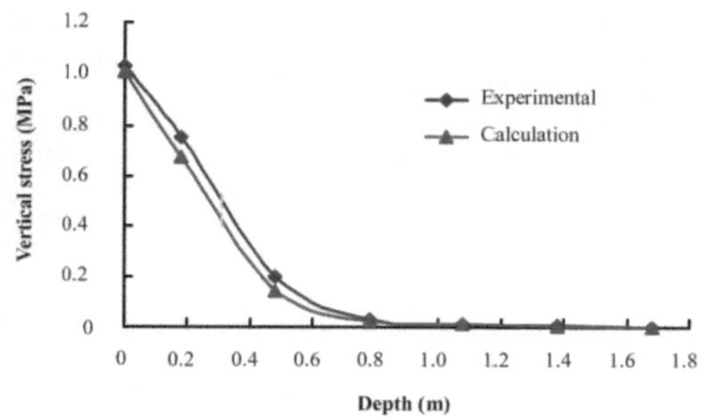

Figure 13. Comparison of calculated vertical stress with experimental one.

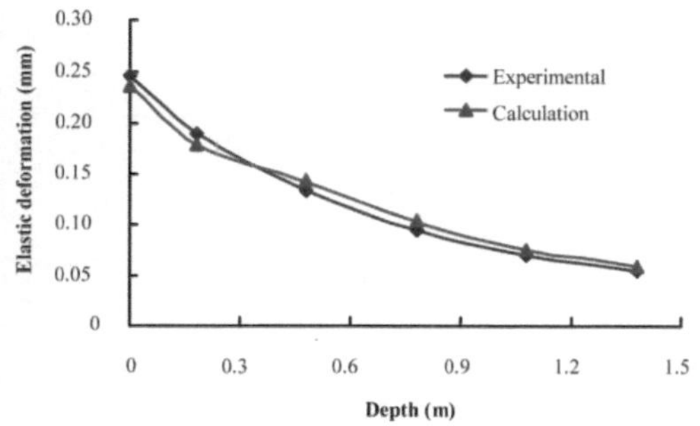

Figure 14. Comparison of calculated elastic deformation with experimental one.

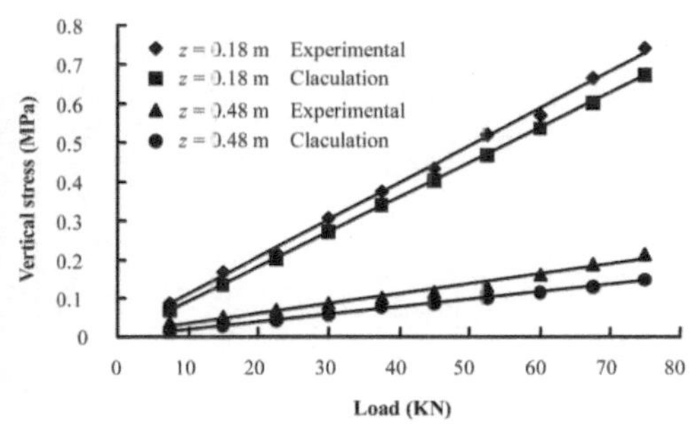

Figure 15. Comparison of calculated vertical stress with experimental one.

empirical relationship of load frequency and velocity given by Ling et al. (2002). Therefore, the computing model can be built in the rectangular Cartesian coordinate system based on the laboratory model, and the corresponding theoretical solutions have also been derived by employing the transfer matrix method illustrated by Lu et al. (2009).

The typical experiment "DT" is simulated and the numerical results are derived by solving the governing equations through the inverse fast Fourier transform method. Figs. 13–15 compare the vertical stress and elastic deformation obtained from experiment results with that calculated by theoretical solutions.

From Fig. 13 and Fig. 14, we can see that the vertical stress and elastic deformation of calculation are slightly less than those of experimental ones in superficial layer. This is because the applied double-circular loads of experiment are simplified as equivalent double-rectangular loads. The differences between experimental results and calculated results are obvious in the vicinity of loading area which can be explained by Saint-Venant principle. However, the total trends generated by theoretical calculation are in good agreement with those of experimental ones.

Comparison of the variation of vertical stress with increasing load amplitude of calculation and that of experimental results is presented in Fig. 15, where $z = 0.18$ m and $z = 0.48$ m are considered. As can be seen, the agreement between the two results, as far as the vertical stress is concerned, is excellent.

5 CONCLUSIONS

In this paper, a laboratory model experiment is presented in order to better understand the dynamic behaviour of road structures under repeated traffic loads. The following conclusions can be drawn from the studies:

1) The most part of dynamic stresses caused by vehicle loads are imposed by road pavement.
2) The relationship between dynamic responses of road structures under traffic loads and the load amplitude is exactly linear.
3) Comparisons of the calculated results with the laboratory model experiment ones indicate that the 3D elastic multilayered computing model can be used to study the dynamic responses of road structures under moving traffic loads.

REFERENCES

Eason G. 1965. The stresses produced in a semi-infinite solid by a moving surface force. *International Journal of Engineering Science*. 2:581–609.

Hyodo M., Yasuhara K., Murata. H. 1996. Deformation analysis of the soft clay foundation of low embankment road under traffic loading, *Pro 31th symposium of Japanese society of soil mechanics and foundation engineering*. 27–32.

JTGD50, 2006, Specifications for Design Highway Asphat Pavement, Beijing: *China Communications Press*. 45–59.

Kettil P. & Lenhof K. 2007. Simulation of inelastic deformation in road structures due to cyclic mechanical and thermal loads. *Computers and Structures*, 85, 59–70.

Ling J.M. & Wang W., 2002. On residual deformation of saturated clay subgrade under vehicle load. *Journal of Tongji University*, 30(11), 1315-1320.

Lu Z. & Yao H.L. 2009. 3D dynamic responses of layered ground under vehicle loads. *Rock and Soil Mechanics*, 30(10), 2965–2970.

Manzari M.T. & Prachathananukit R. 2001. On integration of a cyclic soil plasticity model. *Int J Numer Anal Methods Geomech*, 25, 525-549.

Picoux B., & Houedec D., 2005. Diagnosis and prediction of vibration from railway trains. *Soil Dynamics and Earthquake Engineering*, 25, 905–921.

Sneddon I.N. 1952. The stress produced by a pulse of pressure moving along the surface of a semi-infinite solid. *Rendiconti del Circolo Matematico di Palermo*. 2, 57–62.

Takemiya H. & Bian X.C. 2005. Substructure simulation of inhomogeneous track and layered ground dynamic interaction under train passage. *Journal of Engineering Mechanics, ASCE*, 131, 699–711.

Xu B & Lu J.F. 2008. Dynamic response of a layered water-saturated half-space to a moving load. *Computer and Geotechnics*, 35(1), 1–10.

Yao H.L., Lu Z., Luo H.N., et al. 2009. Dynamic response of rough pavement on Kelvin foundation subjected to traffic loads. *Rock and Soil Mechanics*, 30(4): 890–896.

Modelling and Computation in Engineering – Zhu (ed)
© 2011 Taylor & Francis Group, London, ISBN 978-0-415-61516-7

An agent-based distributed decision support system for fire rescue

Yongjun Wang, Wenwu Shao & Yongdong Wang
School of Administration, Shenyang Institute of Aeronautical Engineering, Shenyang, P.R.China

ABSTRACT: In real world fire rescue system is complex system involved in many people and equipments; it needs a decision support system to support interaction among different participants through IT. From the system modeling perspective, we present the architecture of fire rescue decision support system based on agents and describe the decision process based on interaction among agents. Then from the system implementation point view, we analysis the composition structure of agent and put forward the distributed computing platform based on RTI. Finally, the open feature of system is discussed.

1 INTRODUCTION

Fire rescue is one of the most important public safety activities. A typical scenario of the fire rescue process is depicted as follows. After the fire department gets a fire alarm call, it will send a fire fight team to the fire field. Normally, a fire fight team consists of one incident commander vehicle, two engine vehicles, one ladder vehicle, and the most important role, a set of firefighters, who are grouped as squads associating with one of the above vehicles. In the fire rescue spot, the incident commander is in charge of the whole fire rescue situation, including monitoring the fire field and making real-time schedule on firefighter assignment. The two engine vehicles carry water, which will be used in the case when water is lack near the fire field. And the ladder vehicle holds the utilities like ladders that are needed by the firefighters. The firefighters are organized into different squads based on their specialty and fight cooperatively to eliminate fire in the fire field; and near the rescue spot, there are medical rescue teams, police officers, person in charge of the fire area who are preparing to provide directed services in time; and in the fire rescue command center, command and decision-makers and experts in related fields are taking part in fire rescue in the same time who can put forward alternatives, coordinate distribution of relief resources, and control the progress of the fire rescue. So, fire rescue is complex system engineering involved in many people and equipments, and during the rescue, there are a large number of uncertain factors, fuzzy factors and even unknown factors needed to deal with. Emergency decisions in the process of fire rescue with the feature of real-time decision and non-programmed decision-making are different from the procedural decision making (Dai Ruwei. Metasynthetic 1991). When the fire rescue happening, a decision support system needs to share information resources form the internal and external fire spot among participants and to collaborate them to make a real-time decision. So, this article will describe this kind of fire rescue decision support system from system modeling, decision process and system implementation perspectives.

The rest of the paper is organized as follows. First we describe some related technical foundations to support decision support system like intelligent agent, HLA/RTI in section 2. Then we propose architecture of decision support system based on agent from system modeling point view and present the decision process based on interaction among agents in Section 3. In Section 4 we discuses the implementation of the above decision support system including the design of agent combination structure and the distributed computing platform to support a variety of agent running. Finally, we give a conclusion in section 5.

2 RELATED TECHNICAL FOUNDATIONS

2.1 *Intelligent agent (Michael Wooldridge 2003)*

The agent is an entity that reacts to external stimuli, meets the basic functions based on predefined scenarios, and generates an output based on knowledge held. There are several types of agents: agents based on reflexes, are represented by those entities that react after receiving stimuli from the environment; utility-based agents, are those agents who aim to achieve a specific action such as searching, scanning, or opening a door; agents that rely on an objective, are those agents that have a defined set of objectives and take all possible steps to achieve them; knowledge agents are materialized through entities who base their actions on a knowledge base; logical agents are that category of entities whose actions are made based on information obtained from the store of knowledge and logical reasoning implemented.

Given the complexity of data available, selecting information that is of interest is a laborious task requiring attention and thoroughness, the role of intelligent agents is to find the information needed identified using some well established criteria and present it to the user in a comprehensible and coherent way. Intelligent agents make the transition to knowledge economy easier by their ability to make decisions independently based on experience and knowledge bases available. Intelligent agents have environment that is made of all entities that surround them and interacting with the agent at some point; objective is the scope of the agent and what response is expected from the interaction with the environment; perceptions consist of the raw information that is processed by the agent; actions are based on information outputted by the agent, following the information obtained from the interaction with the environment.

Collaborative intelligent agents are implemented in simple collaborative applications to provide intelligent behavior. The main characteristics of this type of applications are intelligent behavior based on past actions and attained knowledge from the application's knowledge data base; optimized network communication infrastructure to allow fast data transfer between peers; high performance algorithms that boost processing speed and minimize the transaction completion time; provide a user customized work environment providing the user with necessary and desired tools to complete his tasks; implement a user friendly customizable interface that gives the system the ability to learn the patterns the users has when working such that it will be possible for the system to provide the user with the needed interfaces, applications or information he needs most. Distributed collaborative intelligent applications are to be the future of present collaborative applications by providing the user with better system capabilities, customized graphical user interfaces, helpful hints and most important with desired information by applying customize filters to present only the information the user would be interested in.

2.2 *High Level Architecture and RTI*

High Level Architecture (HLA) standard is a general architecture for simulation reuse and interoperability developed by the US Department of Defense. The conceptualization of this High Level Architecture led to the development of the Run-Time Infrastructure. This software implements an interface specification that represents one of the tangible products of the HLA. The aim of it is to develop, under the leadership of the Defense Modeling and Simulation Office (DMSO), reuse and interoperability across large numbers of different types of simulations developed and maintained. The HLA standard promotes reuse of simulations and their components. It attempts to specify the general structure of the interfaces between simulations without making specific demands on the implementation of each simulation. In the HLA baseline documents, The Interface Specification is a formal, functional description of the interface between the HLA application and the Run-Time Infrastructure. A set of technical rules is defined to which an HLA participant has to comply. The Object Model Templates are standardized formats to define the functionality of simulation models and the interaction between models. In this way the capabilities of all simulation models are defined before the actual simulation takes place.

The Run-Time Infrastructure (RTI) software implements the interface specification and represents one of the most tangible products of the HLA. It provides services in a manner that is analogous to the way a distributed operating system provides services to applications. The Run-Time Infrastructure supports HLA compliant simulations through a number of services. The main categories of services are: Federation Management: create/destroy execution (of a federation), join/resign execution (of a federate), pause/resume execution; Declaration Management: subscription/publication; Object Management: instantiate/delete (an object), update/reflect (an attribute), request/provide (an attribute), send/receive (an interaction); Ownership Management: transfer ownership (of an attribute) from one object to another; Time Management: handling of messages in different ways, depending on the requirements of their destinations, e.g., in receive order, in priority order, in causal order, or in timestamp order. Data Distribution Management: provides a flexible and extensive mechanism for further isolating publication and subscription interests – effectively extending the sophistication of the RTI's switching capability.

In this paper, we view many participants in fire rescue as different intelligent agents that have its own goal, responsibility and ability, from this point view we describe the system architecture. And based on the HLA/RTI, we can achieve the interaction among agents to share information resources and to coordinate between agents.

3 SYSTEM DESIGN

According to Fire Relief command structure, the Roles, responsibilities and authority of the subjects involved in fire relief are analyzed; these subjects are including command and decision-makers, experts in related fields, Fire commander, Medical Assistance Team, Police officers, and Person in charge of the fire area. During the fire rescue, different subjects are interconnected through information flow to implement intelligent decision support system with the participation of Relief experts, which can promote the real-time and scientific decision-making. Based on the Entity's objectives, tasks, responsibilities abstracted from the Realistic disaster relief command system, Fire rescue Decision Support System is forming.

3.1 *Composition structure of the system*

After analyzing several types of agents from technical perspective and different subjects involving in real fire rescue system from management point view,

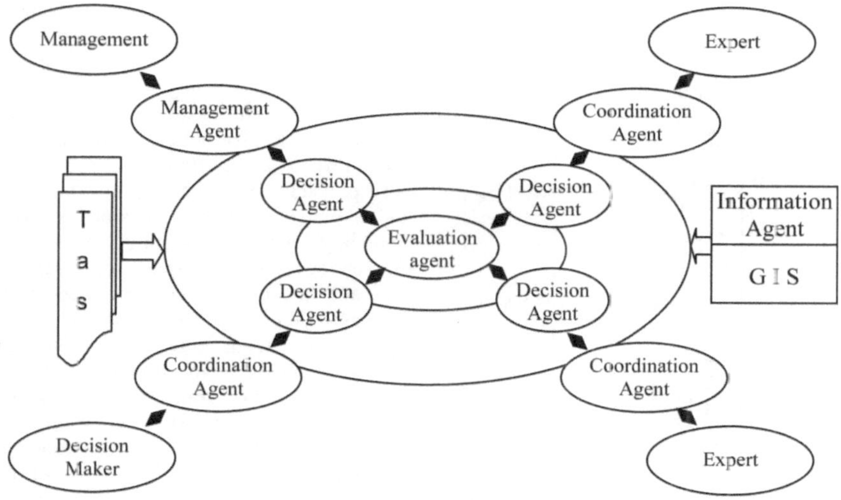

Figure 1. Architecture of fire rescue decision support system.

Fire rescue decision support system based on agent is composed of a variety of different types of software component Agent, which mainly including 1) management agent, responsible for the system management; 2) Interaction agent, providing a friendly interface between human and machine; 3) decision agent, solving the corresponding problem decision making; 4) evaluation agent, assessing decision-making programs; 5) combination agent, generating a new program by combined to existing programs; 6) information agent, providing the required information by collecting, storing and processing information (Hu Daiping, Wang Huanchen 2002). The entire system is composed of one management agent, more interactive agents, more decision agents and more information agents. And form the system user point view, system administrator, decision user and experts use their own interaction agent to connect with the rest agents. From the system architecture perspective, all of the agents including in system are deployed in network computers and can share a geographic information system. The composition diagram of fire rescue decision support system based agent is shown in fig. 1.

3.2 *Fire Rescue Decision-making process using the system*

Fire Rescue Decision-making process is a kind of human-computer interaction process to cooperate solving problems in this system based on agent (Hu Daiping, Wang Huanchen 2002), and specific decision making process is illustrated in fig. 2.

Decision-making users can input the decision problem into system by coordination agent; in the input process, users should provide qualitative and quantitative information as much as possible. The coordination agent provide a friendly user interface to prompt user the input information content and format, and show the results of decision-making to user by different forms.

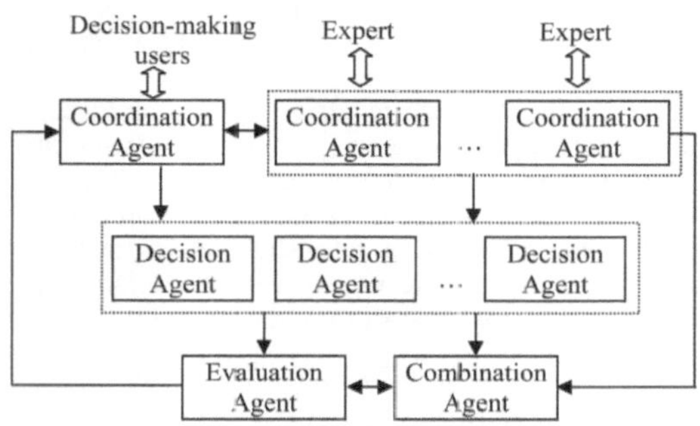

Figure 2. Decision process in fire rescue decision support system.

Based on the type of decision-making, the capacity of decision-making and the decision-making information, coordination agent can determine whether to accept the current decision-making tasks. When accepting the tasks, the decision agent can obtain the necessary data and information from the information agents, and make decision according to their method of decision-making, and then the decision result can be transferred to the combination agent. The combination agent can generate a new decision-making program by integrating programs coming from the decision agents with a certain method, and send the new program to evaluation agent. The evaluation agent accept decision-making programs coming from decision agents, combination agent and experts coordination agents, and assess the decision-making programs according to a certain rules, and finally Assessment results and the final program are transferred to user coordination agent and expert coordination agents. If the decision-making user gets satisfactory results, the decision-making process is completed, and if the decision-making user does not get satisfactory results, the system can start a new round of decision-making

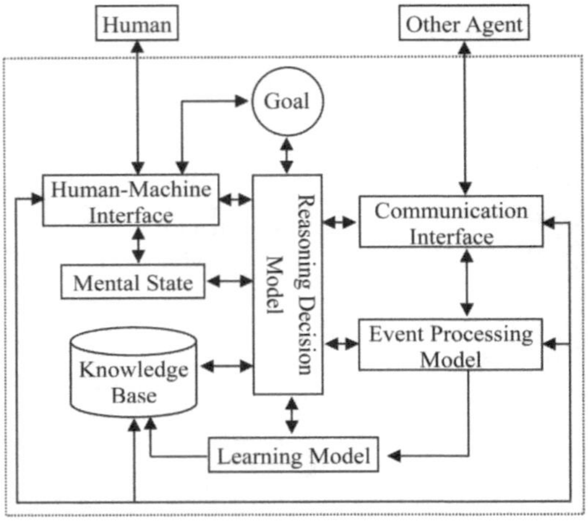

Figure 3. Agent composition structure based on goal achievement.

based on more information on related issues, up until a satisfactory program. (Jayendran Venkateswaran and Young Jun Son 2005).

During the decision-making process, authorized fire rescue specialists and experts in related fields can also directly interact with decision agent, combination agent, evaluation agent using their own coordination agent, guide their run, do heuristic decision making, and send the decision results to user interaction agent providing decision support for decision-making user.

4 SYSTEM IMPLEMENTATION

4.1 *Agent composition structure based on objectives achievement*

With regard to the research on the composition and structure design of agent, Shoham proposed design method based on mental state (Shu-Mei Guo, Chang-Shing Lee and Chin-Yuan Hsu 2005); Freeman put forward agent composition and structure based on knowledge (Shu-Mei Guo, Chang-Shing Lee and Chin-Yuan Hsu 2005); other researcher proposed the event handler-focused basic model (Shu-Mei Guo, Chang-Shing Lee and Chin-Yuan Hsu 2005). In this system, combining with the feature of fire rescue system, the method based on objectives achievement is adapted to design the agent composition structure.

As agent, targets can be seen as its pursuit, event handler can be considered as means to achieve goals for agent. The agent with objectives achievement is described by identity, goals, knowledge, mental status, etc. basic properties, and function models to implement communication, event processing, reasoning decision-making, learning and human-machine interaction. The agent composition structure is shown in fig. 3.

In fig. 3, agent identity including name and address is a kind of attribute that can separate a certain agent from other agents in a multi-agent system. Agent goal is the specific objectives to pursuit in continuous running, and agent's duties and obligations are identified by the agent objectives. Agent knowledge includes facts and rules and is stored in the agent knowledge base. Mental states include agent faith, commitment and ability to handle affairs. Communication interface is responsible for agent communication, acceptance and sending of a message, transferring task and results, achieving knowledge sharing, which includes communications executive, queue of accepting the message and queue of sending the message. Reasoning decision-making model can reason and make decision based on the agent target, knowledge, mental state, capability and latest information, and then effect on message processing, transaction processing, and mental status updates, etc. Transaction processing model is the main part to implement agent target and process transaction, and mainly composed of the transaction processing methods and some auxiliary methods. Learning model can summarize experiences, add new knowledge to the knowledge base, and improve the ability to adapt to environmental change. There are three ways to acquire knowledge in agent, that is, conviction transforming method, case-based learning method, and centralized learning method. Human-machine interface has a friendly graphical interface used by manager and decision-maker to maintain the agent property and to guide the agent's behavior.

4.2 *Method to implement interaction among agents in RTI*

Distributed fire rescue decision support system is a complex system involving many roles who interact and coordinate to achieve the goal. High Level Architecture (HLA) standard is a general architecture for simulation reuse and interoperability developed by the US Department of Defense. HLA provide a technology platform to solve interaction among agents' issues. This article selects RTI as interaction support platform, and implements the intercommunication and interoperability between multi-agents through the distributed computing platform shown in Fig. 4.

Fig. 4 illustrates the relationships between agents in decision support system. The agent can be implemented in different languages (e.g., JAVA, AgentSpeak, etc). The coordination agent can interact with other agents like decision agent, evaluation agent, information agent and other coordination agent through RTI. The direct interaction of the agent (both coordination agent software as well as other agent software) with the RTI (executable component of HLA) is quite complex and cumbersome. Therefore, we have developed a Distributed Simulation (DS) Adapter (Sebestyénová J.2005), based on COM/DCOM technology, to provide mechanisms for distributed application similar to those provided by the HLA RTI, but with a level of complexity that is manageable by the development resources available in the non-computer science communities. The

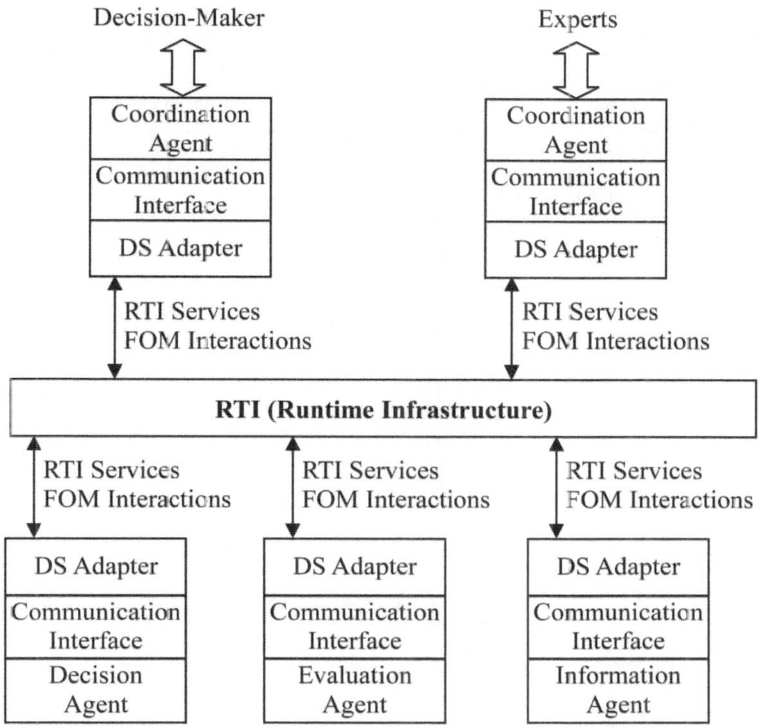

Figure 4. Distributed computing platform based on RTI and DS Adapter.

DS Adapter provides a simplified time management interface, automatic storage for local object instances, management of lists of remote object instances of interest, management and logging for interactions of interest, and simplified object and interaction filtering. And based on the DS adapter, from the agent perspective, we also design a Communication Interface (C. Muldoon, G.M.P. O'hare, R. Collier, M.J. O'grady 2006). Communication Interface is the interface for agent participating activity and perceiving external world, which includes four parts, that is, information receiver, message buffer queue, message processor and message transmitter. With communication interface, Sharing and exchange of knowledge among agents is achieved in RTI, other agent information such as state, service, and need to be dynamically understood in computer environment, and so, sharing and interoperability of system are achieved (Gao Zhinian, Xing Hancheng 2003).

5 CONCLUSION

In MAS-based distributed fire rescue decision support system, each agent is distributed in computer networks, and Shared services and interoperability among agents are achieved in RTI. At the same time, the system is an open structure which can absorb experts and new agents composed of new theory and method, and improve the decision support system's reusability, adaptability and scalability.

REFERENCES

C. Muldoon, G. M. P. O'hare, R. Collier, M. J. O'grady. (2006). Agent Factory Micro Edition: A Framework for Ambient Applications. In Proceedings of Intelligent Agents in Computing Systems Workshop (held in Conjunction with International Conference on Computational Science (ICCS)) Reading, UK. Lecture Notes in Computer Science (LNCS), Springer-Verlag Publishers, pp. 727–734

Dai Ruwei. Metasynthetic (1991). Technology from Qualitative to Quantitative [J], Pattern Recognition and Artificial Intelligence, 4(1): 5–10

Gao Zhinian, Xing Hancheng. (2003). A Multi-agent System Architecture Research Based on HLA [J], MINI-MICRO SYSTEMS, 24(3):337–339

Hu Daiping, Wang Huanchen. (2002). An Agent-based Macro-economy Decision Support System [J], SYSTEMS ENGINEERING—THEORY & PRACTICE, 22(1):33–37

Jayendran Venkateswaran and Young Jun Son. (2005). Hybrid System Dynamic – Discrete Event Simulation based Architecture for Hierarchical Production Planning, International Journal of Production Research., 43 (20), 4397–4429

Lean Yu, Shouyang Wang and Kin Keung Lai. (2009). An intelligent-agent-based fuzzy group decision making model for financial multi-criteria decision support: The

case of credit scoring, European Journal of Operational Research, Vol. 195, No. 3, June, pp. 942–959
M. H. Kuo. (2007). An intelligent agent-based collaborative information security framework, Expert Systems with Applications, Vol. 32, No. 2, February, pp. 585–598
Michael Wooldridge. (2003). An Introduction to Multi-Agent Systems [M]. BEIJING: Publishing House of Electronics Industry, 10
Sebestyénová J. (2005). Decision Support System for Modeling of Systems and Control Systems Design, in Proc. IEEE Int. Conf. on Computational Intelligence for Modeling Control and Automation CIMCA'2005, Ed. M. Mohammadian, Vol. 1, 28-30. Nov. Vienna, ISBN-13: 978-0-7695- 2504-4, ISBN-10: 0-7695-2504-0, pp. 70–75
Shu-Mei Guo, Chang-Shing Lee and Chin-Yuan Hsu. (2005). An intelligent image agent based on soft-computing techniques for color image processing, Expert Systems with Applications, Vol. 28, No. 3, April, pp. 483–494

Modelling and Computation in Engineering – Zhu (ed)
© 2011 Taylor & Francis Group, London, ISBN 978-0-415-61516-7

Optimum analysis of construction sequences in super large section and span tunnel

Dingheng Zhou
Technical University of Munch, Munich, Germany

Yongfeng Ma
China Petroleum East China Design Institute, Shandong, China

Liqiao Cao
China Railway Siyuan Survey and Design Group Co., Ltd, Hubei, China

Zhan Shi
Tongji University, Shanghai, China

ABSTRACT: Different from normal tunnel, the mechanical behaviors of super large section and span tunnel with two bores and eight lanes are more complex and the construction methods are also diversified. Based on the largest four-lane super large section and span highway tunnel in China, the dynamic tunnel construction are simulated and the excavation effects of three construction sequences are compared based on the displacements of surrounding rock mass, the preliminary lining and secondary lining. Some results have been drawn. The design method has advantages in coupling the surrounding rock and the preliminary lining into full play of supporting the tunnel, which will benefit the secondary lining with certain safety margin. The method with the construction of the right heading between the upper and down stages of the left heading has advantages in horizontal displacement control and vertical displacement control at the tunnel top. The experience and data presented in the paper can be refereed in the design, construction and research of similar tunnels.

1 INTRODUCTION

Recently, in order to suit for multiply-lane highway construction, many four-lane super large section tunnels have been built all over China. There are some developments about the stress, deformation and stability research on rockmass and lining in four-lane super large section tunnels. As to theory analysis aspect, Qu Haifeng proposed a new load mode for large cross-section and low flat-ratio tunnel. Some researchers such as Sun Xiang, Huang Lunhai, Wu Mengjun, Xu Chongbang and LI Zhigang have done some work on numerical analysis and model test. Those researches are mainly about model test and numerical analysis of dynamic construction process and the mechanical characteristics. Except that, Yuan Yong proposed a pre-built and prestressed innovative support method for four-lane super large section tunnel and made a validation about the pre-built and prestressed structure. Huang Chengzao proposed cross anchor to control the deformation of core rock in four-lane tunnel. As for in-situ test on four-lane tunnel, Chen Gengye analyzed the stress characteristics of four-lane tunnel based on stress monitoring. And Zhou Dingheng made a detailed analysis of rock deformation based on in-situ test of super large section and span tunnel. Based on one four-lane tunnel, Liu Heng finished some research about on the durability of tunnel structure. At present, the design and construction standard of four-lane super large section tunnels has not been formed and the related research is limited. Therefore, it is essential to do further research about four-lane tunnels. Except that, due to the characteristics of large excavation span and low flat-ratio, the construction sequences conversion is complicated and multiply blast has multiply disturbance to the surrounding rock. Especially, the core rock stress is very complicated. Thus, there tends to be many stability problems such as large section and tunnel collapse. Therefore, further research on super large section and span tunnel is an important content of rock tunnels.

Based on the largest four-lane super large section and span rock tunnel in China, the dynamic tunnel construction are simulated and the excavation effects of three different construction sequences are compared based on analysis of the displacements of surrounding rock mass, the preliminary lining and secondary lining.

Table 1. Physical-mechanical parameters.

Material	γ/ kN·m^{-1}	E/GPa	μ	φ/(°)	c/MPa
Gravelly clayey	16.6	0.14	0.38	24.5	40
Completely decomposed granite	19	0.2	0.35	100	21
Highly decomposed granite	21	0.65	0.3	30	130
Weakly decomposed granite	26.1	2.5	0.21	1055	47.6
Backfill	25	1.0	0.2	1500	50

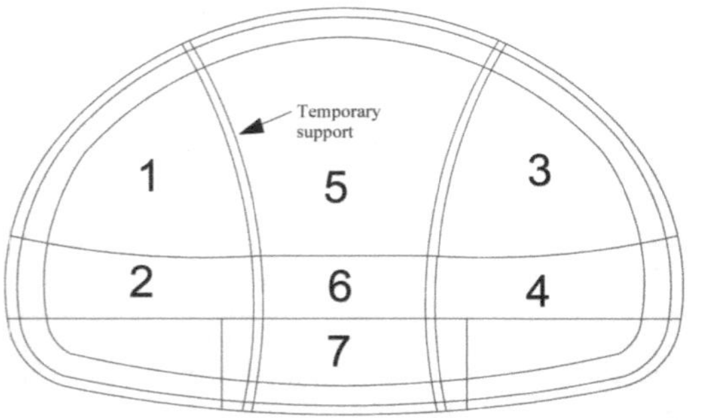

Figure 1. Design construction sequence.

2 BACKGROUNDS

The construction of this two bores and eight lanes tunnel began from January, 2005 to May, 2006, with the normal speed 100 km/h. The length of the left and right line is 1010 m and 1006 m with the distance of two tunnels 25 m. The rock includes second to fifth class rock with the maximum depth 98 m. The large excavation width and height are 20.7 m and 13.58 m. According to the engineering geological exploration, the rock layer is simple. The bedrock is granite intrusive mass with the rock type monzonitic granite. The rock parameters are listed in table 1.

The design construction sequence in soft rock is shown in figure 1. The firstly-excavated left heading and the secondly-excavated right heading can be divided into two parts. The excavation of the core rock has three stages. The structure is composite lining with preliminary supports such as anchor bolt and steel fiber reinforced concrete. Except that, other support measures such the steel support and pre-grouting with small ducts are also available.

3 NUMERICAL SIMULATION

3.1 *Numerical model*

The section k5 + 870 has been chosen as the simulation section. According to the incidence of tunnel excavation, the model size is described as following. The distance from the left or right boundary to the tunnel is 3 times of the span and the distance from the bottom boundary to the tunnel is also 3 times of tunnel span. The upper boundary is the ground surface. The physical-mechanical parameters are listed in table 1.

Because of the bad geotechnical condition and complicated rock stress field, the failure modes of rock have diversity, such as tension rupture, brittle shear failure or plastic failure. And those failure types are regarded as plastic failure. Therefore, the ideal elasto-plastic model and Drucker-Prager yield criterion are chosen in FEM analysis.

Boundary-restraint condition: 2-dimenssion plane strain model is used in the numerical simulation with the left and right boundary horizontal restraint and the bottom boundary vertical restraint. And the top is free with restraint. The mesh division is shown in figure 2.

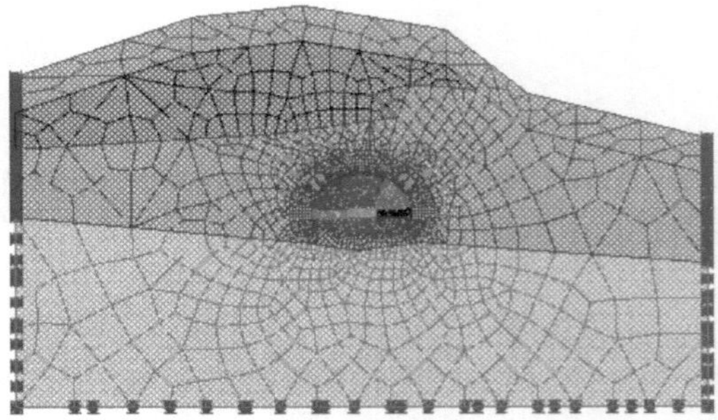

Figure 2. Mesh division of model.

Table 2. Supporting structures parameters.

Support structure	E/GPa	γ/(kN·m^{-1})	A/m^2	I/ m^4
System bolt	210	77	0.000491	/
Temporary bolt	210	77	0.000491	/
Preliminary lining	27.8	22	0.27	0.00164
Secondary lining	31	26	0.6	0.018

The construction process is achieved by the elements provide by some commands such as "Excavation", "Backfill" and "Copy of linear material". And the rock and backfill are simulated by 4-point solid elements. The anchor bolts, preliminary lining and spray concrete layer of core rock is simulated with 2-D bar element. As to the simulation of the secondary lining, steel arch and temporary steel arch and horizontal support of core rock, the 2-D beam element is adopted. The supporting structures parameters are listed in table 2.

3.2 *Dynamic simulation of construction scheme*

Three different construction sequences, the difference among which is the excavation sequence difference of

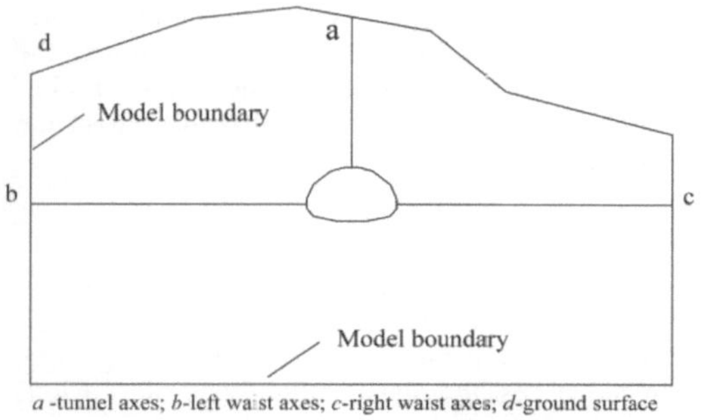

Figure 3. Schematic diagram of displacement acquisition.

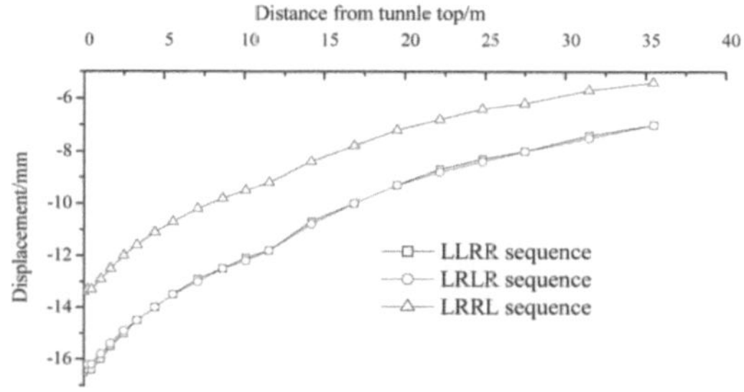

Figure 4. Comparison of rock displacement near arch crown.

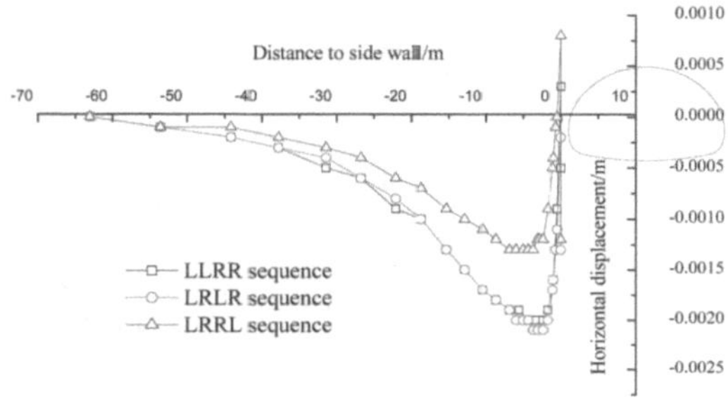

Figure 5. Comparison of horizontal displacement of left side.

upper-stage of the right heading, bottom-stage of right heading and bottom-stage of left heading, are chosen to make optimum analysis in simulation.

Sequence 1: upper-stage of left heading – upper-stage of right heading – bottom-stage of right heading – bottom-stage of left heading – core rock. Thus, this sequence is called LRRL sequence. The steps are described as following. (1) upper-stage of left heading; (2) preliminary lining and temporary support; (3) upper-stage of right heading; (4) preliminary lining and temporary support; (5) bottom-stage of right heading; (6) preliminary lining and temporary support; (7) secondary lining in invert arch; (8) backfill; (9) bottom-stage of left heading; (10) preliminary lining and temporary support; (11) secondary lining in invert arch; (12) backfill; (13) upper-part in core rock; (14) preliminary lining; (15) middle-part in core rock; (16) temporary support of core rock dismantling; (17) bottom-part in core rock; (18) preliminary lining; (19) secondary lining; (20) backfill; (21) secondary lining of other parts.

Sequence 2 (LRLR sequence): upper-stage of left heading – upper-stage of right heading – bottom-stage of left heading – bottom-stage of right heading – core rock. The detailed steps are similar to sequence 1.

Sequence 3 (LLRR sequence): upper-stage of left heading – bottom-stage of left heading – upper-stage of right heading – bottom-stage of right heading – core rock.

4 NUMERICAL RESULTS ANALYSIS

4.1 *Displacement analysis*

The effect of the core rock excavation and temporary support dismantling can be reflected with surrounding rock displacement. The displacement of those feature points is collected in numerical simulation. And the sign of surrounding rock is accorded with the model coordinate.

The rock displacement is closely related with construction methods and process. Construction methods can be judged by the surrounding rock displacement. And the sign of surrounding rock is accorded with that of the model coordinate.

4.1.1 *Rock displacement above the tunnel top*

The crown settlement is an important index for the stability of surrounding rock and safety evaluation of construction. There is no apparent difference between the vertical displacement of the axes direction above the tunnel top (figure 4) under LLRR and LRLR sequence. But the displacement of LRRL sequence is smaller than that of other two sequences. The vertical displacement of the tunnel top under LRRL is about 80% of that of other two sequences.

4.1.2 *Horizontal displacement*

The horizontal displacement of side walls (line b and c in figure 3) is shown in figure 5 and 6. Similar to the vertical displacement above the tunnel top, there is no apparent difference between the horizontal displacement of side walls under LLRR and LRLR sequence, which is larger than that of LRRL sequence. The maximum horizontal displacement of the left side is about 60% of that of other two sequences and the maximum horizontal displacement of the right side 65%. The horizontal displacement law under three sequences is: The rock beside the side walls expanses into the tunnel. With the distance to the side walls increasing, the horizontal displacement of rock is contrary to the tunnel direction. The displacement reaches the maximum at the place of 1/3 times of tunnel span. And outside that range the displacement decreases with distance.

Based on the results analysis of the vertical displacement above the tunnel top and horizontal displacement of side walls, the sequence with

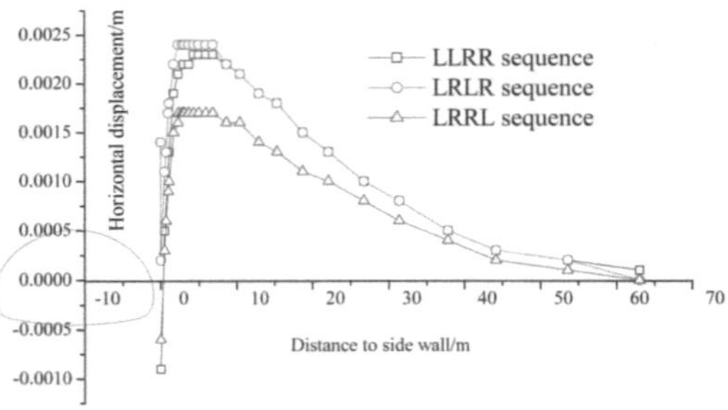

Figure 6. Comparison of horizontal displacement of right side.

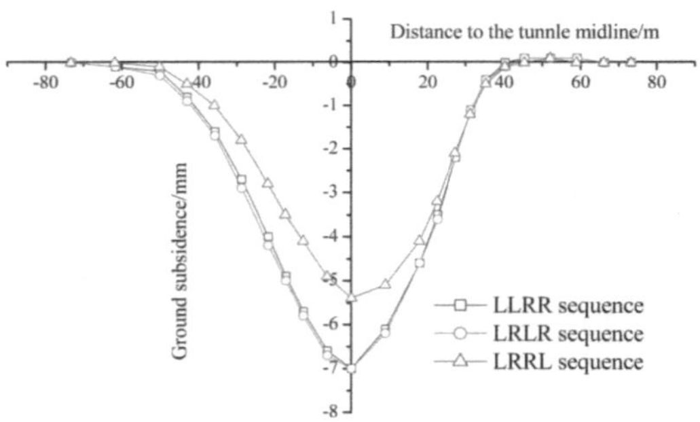

Figure 7. Comparison of surface settlement.

construction of the upper and bottom stage of after excavation of upper stage of left heading can controls the tunnel deformation better than other sequences. Except that, the deformation at the top is larger than that of side walls. Therefore, the control of tunnel top deformation should be strengthened.

4.1.3 *Ground subsidence*

The ground subsidence is shown in figure 7. The subsidence at the tunnel midline is the maximum. Based on the numerical results of three different sequences, LRRL sequence is benefit to controlling the tunnel subsidence. But the symmetry property of the subsidence curve is not good. Therefore, the sequence with construction of the upper and bottom stage of right heading after upper-stage of left heading controlling the biased characteristics is not better than other sequences.

With compassion of the vertical displacement above the tunnel top, horizontal displacement of side wall direction and ground subsidence, some conclusions has been drawn. LRRL sequence is beneficial to the rock deformation above the tunnel top and ground subsidence. And other sequences are beneficial to the horizontal deformation of side walls and up-deformation of the tunnel bottom.

4.2 *Stress analysis*

The effect of different construction sequences can be reflected by the rock stress distribution, the force of the preliminary lining and secondary lining. Therefore, the state of the rock stress distribution, the force of the preliminary lining and secondary lining under different sequences are compared in the following.

4.2.1 *Rock stress*

Tensile stress appears below the tunnel bottom under three sequences and the tensile stress zone has a certain distance from the tunnel bottom. Among those three sequences, the tensile stress zone of LLRR is the largest. Except that zone with tensile stress, the stress also exists in the place far from the tunnel. With the average tension stress of LLRR sequence as the standard, the average tension stress is 3 to 4 times and that of LLRR sequence 2 times. As to the distribution condition of compressive stress, the difference of the maximum compressive stress at the drift crown is small. The maximum stress of design sequence LLRR is the largest with that of LRLR sequence the smallest.

Based on the analysis of the distribution condition of rock compressive and tensile stress, the difference of rock stress distribution among three sequences is not apparent.

4.2.2 *Lining stress*

The support structure stress of the feature points is collected in numerical simulation, which is listed in table 3. With the design sequence LLRR as the standard, the axial force under LRLR sequence is the largest. At the place of the arch crown of left heading and right waist, the axial force under LRLR sequence increases by 40% and that of other places 10% to 20%. Compared with LLRR sequence, the axial force of LRRL sequence at left arch waist and spring reduces by about 30% and that of the arch crown of left heading and right waist increases by 20% to 45%. And the stress of other places is small. The effect on the axial force caused by three sequences is almost same.

As to the structure stress of preliminary lining, the stress at side walls is about 2 ~ 3 times of that at tunnel crown. From the aspect of lining stress controlling of super large section and span tunnel, the key is the control of side wall stress.

By comparing the rock and lining stress, the design sequence can strengthen the load self-bearing capacity of rock and avoid rock stress concentration. The load bearing capacity of the preliminary lining can play well. And the effect on the secondary lining caused by uneven distribution of rock stress reduces and the safety stock of secondary lining can be strengthened. In some extent, the sequence with the upper stage of left heading, the right heading and down stage of the left heading changes the stress condition of the former left heading but increases the stress of right heading. Comprehensive analysis, the effect on tunnel and rock stress under that sequence is the same to the design sequence.

Table 3. Comparison of axial force in preliminary lining.

Sequence		Axial force of preliminary lining/kN						
		Left spring	Left waist	Left heading top	Arch crown	Right heading top	Right waist	Right spring
LRRL	Value	651	1021	875	518	1027	1570	828
	Ratio/%	−31.3	−27.6	23.2	6.6	4.2	47.6	−1.7
LRLR	Value	1158	1679	1008	508	1105	1506	933
	Ratio/%	22.3	19.1	42.0	4.5	12.1	41.5	10.8
LLRR	Value	947	1410	710	486	986	1064	842

Ratio = (stress of sequence 1 or 2-stress of sequence 3)/stress of sequence 3.

Table 4. Comparison of negative moment of secondary lining.

Sequence		Negative moment of secondary lining/(kN.m)						
		Left spring	Left waist	Left heading top	Arch crown	Right heading top	Right waist	Right spring
LRRL	Value	53	/	38	20	13	/	71
	Ratio/%	−10.2	/	660	400	225	/	18.3
LRLR	Value	67	/	7	28	36	/	58
	Ratio/%	13.6	/	40	600	800	/	−3.3
LLRR	Value	59	/	5	4	4	/	60

5 CONCLUSIONS

Based on the numerical results comparison of the rock pressure, lining stress, tunnel deformation and ground subsidence under three different construction sequences, some conclusions can be drawn.

(1) Compared with other two sequences, the sequence with the left heading prior to the right heading can strengthen the load self-beading of surrounding rock, develop the load-bearing capacity of the preliminary lining and strengthen the safety stock of the secondary lining. It also reduces the construction disturbance to the core rock.
(2) During the construction of super large section and span tunnel, the sequence with upper-stage of left heading, right heading, bottom-stage of left heading and core rock is beneficial to the deformation control of tunnel and surrounding rock.
(3) With the existence of core rock, the displacement and stress of tunnel top is small. While the crown settlement of the left and right heading is large and the deformation of side walls is smaller. The preliminary lining stress of the side walls is larger than that of the heading and tunnel top. Thus, in order to control the deformation, the monitoring of the crown settlement of the left and right heading should be strengthened. At the same time, the side wall construction should be paid attention to for ensuring the safety of preliminary lining.

REFERENCES

Chengzao H, Zongxue Y & Xiaorong Z (2007). Cross anchor and its application in the fracture zone in highway tunnel. *Chinese Journal of Underground Space and Engineering* 3(5): 923–927.

Chongbang X, Caichu X & Hehua Z (2009). Optimum analysis of construction scheme of multi-arch tunnel with eight traffic lane. *Chinese Journal of Rock Mechanics and Engineering* 28(1): 66–73.

Dingheng Z, Haifeng Q & Yongchang C et al. (2009). In-situ test on surrounding rock deformation of super large section and span tunnel. *Chinese Journal of Rock Mechanics and Engineering* 8(9):1773–1782.

Gengye C, Bin L & Wan Mingfu et al. (2005). Analysis of stress monitoring of a large-span highway tunnel in Hanjiangling mountain. *Chinese Journal of Rock Mechanics and Engineering* 24(Supplement 2):5509–5515.

Lunhai H, Wei L & Mengjun W (2007). Model test on single four-lane highway tunnel excavation. *Highway Tunnel* (4): 10–15.

Mengjun W & Lunhai H (2006). Research on dynamic construction mechanics of four-lane highway tunnel. *Chinese Journal of Rock Mechanics and Engineering* 25 (Supplement 1): 3057–3062.

Mingfu W, Zhe H & Jianping L et al. (2007). Analysis of stability about excavation and support of super-span road tunnel. *Journal of Liaoning Technical University (Natural Science* 26(1):71–73.

Mingfu W, Hong H & Bin L (2007). Investigation on excavating a 4-lane road tunnel through indoor model test, *Journal of Northeastern University (Natural Science)* 28(2):266–269.

Yong Y & Shenghu W (2008). Numerical modeling of prebuilt and prestressed innovative support system for super cross-section highway tunnel with lower flat-ratio. *Rock and Soil Mechanics* 29(1): 240–244.

Zhigang L, Wenqi D & Zhongcun Y (2007). Simulation and analysis of core rock in flat and large span highway tunnel. *Chinese Journal of Underground Space and Engineering* 3(4): 627–632.

Modelling and Computation in Engineering – Zhu (ed)
© 2011 Taylor & Francis Group, London, ISBN 978-0-415-61516-7

Force mechanism and nonlinear finite element analysis on the behavior of CFDSST column-to-beam connections

Yi Li, Yi-Bin He, Jian Guo, Hai-Bing Zhou & Pin Huang
College of Civil Engineering, Hunan University, Changsha, Hunan, P.R. China

Bin Chen
School of Resources and Safety Engineering
Central South University, Changsha, Hunan, P.R. China

ABSTRACT: To study the mechanical properties of concrete-filled double skin steel tubular (CFDSST) column-to-H-beam connections assembled by T-stubs and high strength bolts, the research parameters are the thickness of T-stub flanges and webs and the diameter of bolts. In the paper, the forces mechanism analysis of composite joint under the antisymmetric loads at the cantilever end of H-beam, and the axial load at the top of CFDSST column, are presented. So the potential theoretical failure modes are found. To validate the theoretical results, the nonlinear finite element (FE) analysis on the behavior of CFDSST column-to-beam connections is conducted using the software ANSYS11.0. The analytical FE models take into account geometrical discontinuities, material nonlinearities, large displacements and contact problems. The results indicate that the FE method can effectively estimate the failure modes of composite joint, and the FE models results are good agreement with the theoretical ones. The thickness of T-stub flange and the diameter of bolt can markedly influence the carrying capacity of joint, but the thickness of T-stub web influences less. The FE model results also present that the theoretical carrying capacities of T-stub analogy calculated by Eurocode 3 are conservative, leading to limitation in predicting values for strength and stiffness of the connections. The tensile T-stub flange is in bidirectional flexural condition, to improve the whole mechanical performances of joint, the thicker T-stub flange should be adopted.

1 INTRODUCTION

The CFDSST column are provided with the performances of the high carrying capacity, good plasticity and tenacity, and convenient for construction, which can be found in ordinary concrete-filled steel tubes, in addition, the lighter deadweight, relatively expansion section, excellent flexible and fire-resistant characteristics are only equipped in CFDSST. The constructive reinforcing bar and moulding board can be replaced through the restriction actions that are brought by the double steel tubes to core concrete. At the same time, the inner steel tube can be used in anchor end of tensile cable and structural vertical passage, so the important applied values can be got in large scale bridge-tower and huge composite framework columns.

Over the past ten years, the mechanical properties of CFDSST column were studies in the United States, Australia, China Taiwan and Mainland (Wei 1995a,b, Elchalakani 2002, Cai 2001, Han 2006). And some practical engineering used CFDSST columns were found in Japan and China Mainland. In fact, for the structural members within the relatively slender, such as bridge pier, the bearing capacity will be controlled by the stiffness, the strength capacity of concrete-filled steel tube can not be used fully, but CFDSST column would have greater advantages.

Based on the systemic research on new type structural members, in the paper a new column-to-beam connection is put forward. The connections are made of CFDSST columns, H-beams, T-stubs and high strength frictional bolts. Based on the forces mechanism of joint, the possible failure modes can be got, and the nonlinear FE method is conducted by using software ANSYS version 11.0. The results which are got by theoretical method and FE method are compared and analyzed; some beneficial conclusions are drawn for the relevant engineering practices.

2 JOINT FORCE MECHANISM AND FAILURE MODES

2.1 *Transferring paths of force*

The connection is taken from the middle joint in frame structure under the horizontal loading action; the place is the reverse flexing point at column-to-beam joint, the constitution of composite joint is shown in Figure 1. In analysis, to assume that the column-to-beam connection is under the loading condition, which

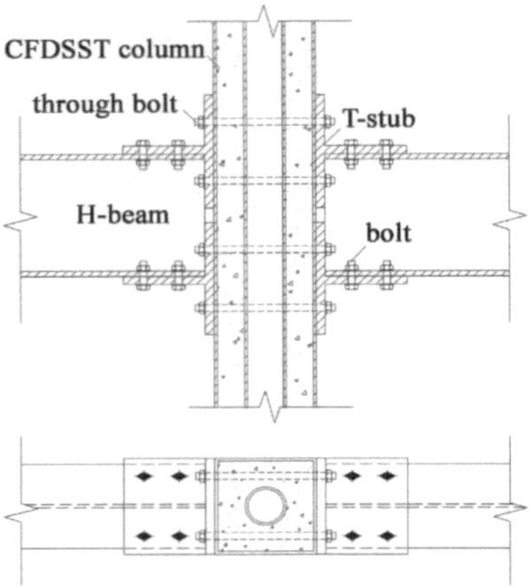

Figure 1. Constitution of composite joint.

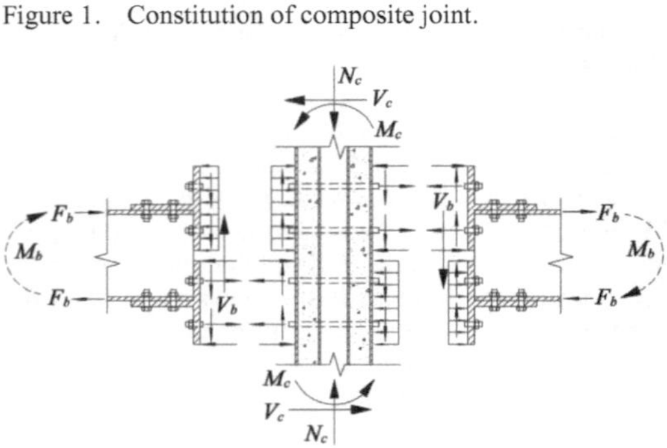

Figure 2. Inner forces in panel zone of composite joint.

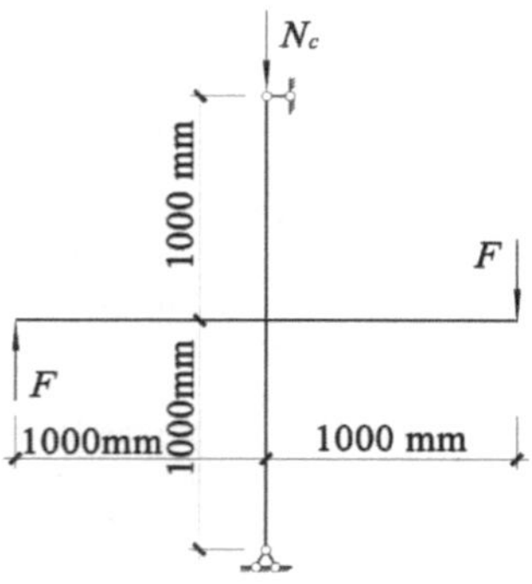

Figure 3. Geometrical size and simplified calculation figure.

the invariable axial force N_c is applied at the top of column, at one time, the antisymmetric loads F are applied at the cantilever end of H-beam, F can be transformed the shear force V_b and moment M_b at the section of H-beam. M_b can also be transformed a couple of forces F_b. The load forming of isolation partition in connection is shown in Figure 2, and the simplied calculation figure is shown in Figure 3.

The shear force and moment of H-beam end are transferred to column through the T-stubs, bolts and the contacting faces between the T-stub flange and column outer flange. With a view to the left-right symmetry of joint constitution, in analysis of the transferring forces process of H-beam, the right part of partition can be taken out. The tensile forces at the upper flange in H-beam are transferred to upper T-stub through the bolts, which is under shear condition and extrude the T-stub web, the contacting face friction between the H-beam upper flange and the T-stub web can also transfer the tensile force.

Then the loads go shares by the two sides through bolts and are transferred to column. The compressive forces at the lower flange in H-beam are transferred to lower T-stub through the bolts and contacting face friction, then the compressive forces are transferred to column through the contacting face between the lower T-stub flange and column outer flange. However, the shear force at H-beam is transferred to H-beam web, to the H-beam flange, and to the bolts thirdly, and then to the T-stub web, at last, the shear force is transferred to column through the contacting face friction between the T-stub flange and column outer flange. The contacting extrusion action and friction forces are mainly concentrated at the place around the through bolts, next actions are the place around the compressive H-beam flange and the edge of extending board at tensile T-stub flange, and the forces at the other parts are less.

According to the Figure 2, the CFDSST column at the panel zone, on the one hand, is applied to the axial force N_c, the horizontal shear force V_c and the moment M_c, which are caused by reaction forces. On the other hand, the CFDSST column at the panel zone, is applied to the vertical shear force V_b and a couple of forces F_b. All the forces are transferred through the adhesion slippage action among the inner steel tube, core concrete and outer steel tube.

2.2 *Different failure conditions*

The joint will lose its bearing capacities and take place local or whole destructions, when one or more possible failure modes may occur simultaneously.

(1) A tensile or shear failure will be taken place at the through bolts or bolts connected with H-beam flanges; (2) A bend yield occurs at T-stub flange; (3) A extrusion failure appears at T-stub web; (4) A compressive yield or buckling will arise at H-beam flange or web; (5) The outer or inner steel tube will appear the compressive yield or buckling and shear yield; (6) The core concrete will take place the local compressive or shear failures.

3 FE MODEL

In FE analysis, the software ANSYS11.0 is introduced. The 3D entitative element models are established, taking into account the geometrical discontinuities, material nonlinearities, large displacements and contact problems.

Table 1. Joint model mainly variational parameters.

No.	T-stub mm	Bolt mm
DS1	180×230×12×14	M16
DS2	180×230×12×18	M20
DS3	180×230×16×18	M20
DS4	180×230×18×20	M22

Some assumptions in FE model are made: (1) ignoring the surface contact among the bolt head, nut and T-stub; (2) ignoring the interaction between bolt shank and hole wall; (3) ignoring the adhesion slippage action between steel tubes and core concrete.

3.1 *Model geometrical parameters*

The basal geometrical parameters in joint FE model are shown in Figure 3. All the H-beam sections are 250 mm × 175 mm × 7 mm × 11 mm; the circular steel tube sections are 80 mm × 5 mm; the square steel tube sections are 200 mm × 200 mm × 6 mm; the T-stub sections and the bolt diameters are shown in Table 1. Using the asymmetry of structure, taking the H-beam web mid-plane as centro-symmetric axis, and establishing a half model for structural analysis.

3.2 *Material physical parameters*

The materials of steel tube, H-beam and T-stub are adopted by Q235B, and the constitutive relationship of steel is ideally elastic-plastic characteristic. The yield strength $f_y = 330$ MPa, the elastic modulus $E_s = 194707$ MPa. The constitutive relationship for high strength bolt is trilinear model, and the yield strength $f_{by} = 940$ MPa, and the ultimate tensile strength $f_{bu} = 1140$ MPa. The grade level of all frictional type bolts is 10.9 s, and the pretension forces of M16, M20 and M22 bolt are respectively 100 kN, 155 kN and 190 kN. All metallic materials are isotropic, their Poisson's ratio $\nu_s = 0.3$, Von Mises yield criterion with kinematic hardening rule is adopted to model steel material.

In FE model, the axial compression ratio of column is 0.3. The core concrete is under triaxial compression, and its constitutive relationship should take into account the restriction effect (Han 2000), its compressive strength $f_c = 62.7$ MPa, the elastic modulus $E_c = 27977$ MPa, the Poisson's ratio $\nu_c = 0.167$. The William-Warnker concrete model is adopted as the failure criterion for core concrete.

3.3 *Element types and meshing*

SOLID45 element is adopted in steel tube, H-beam, T-stub and bolt. Moreover, SOLID65 element is adopted to model the core concrete. The pretension forces in bolts are achieved by PRETS179 element, which is applied by PSMESH command.

The contacts between the T-stub flange and column flange and the T-stub web and H-beam flange are simulated by creating the 3D contacting pairs. The T-stub flange or web surfaces are defined as contacting faces and simulated by CONTA174 element. And the column flange or H-beam flange surface are defined as target faces and simulated by TARGE170 element. The friction coefficient at contacting face is 0.45.

In meshing, the key or irregular shape parts, where the mesh size should be divided less, leading to transit naturally at irregular irregular shape parts. For the concrete element around the bolt hole, the mesh size should be taken larger, to prevent the local stress of concrete to exceed its compressive strength, which will not lead to the FE results converge.

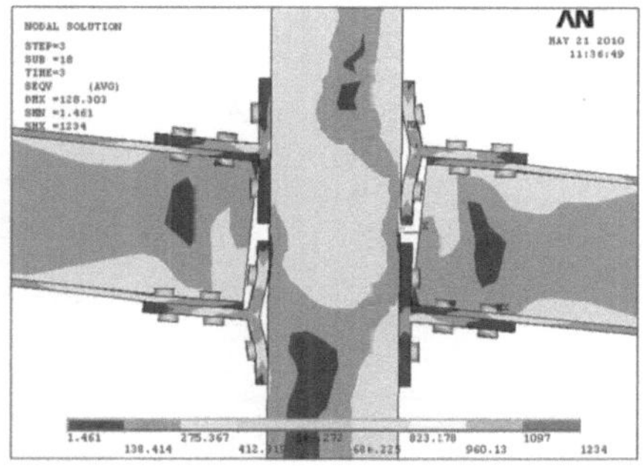

Figure 4. Von Mises stress and deformation in DS3 panel zone.

3.4 *Boundary conditions and solve process*

The symmetrical constrainer is applied at H-beam web mid-plane. The displacements in the x, y, and z directions at the bottom of the column are constrained. Also the displacements in the x and z directions at the top of the column are constrained. The pretensions of bolts are forced by SLOAD command and take small static analysis at first. And then to lock the displacement caused by bolt pretension. At last, the antisymmetric monotonic loads are applied at the H-beam end, where the distances to column flange are 1000 mm. The Newton-Raphson equilibrium iteration method is used to solve these nonlinear problems. The conjugate grads solver with pre-set condition is chosen. The automatic time step control and linear search are opened. The double convergence criterion is adopted.

4 NUMERICAL RESULTS

4.1 *Numerical failure modes*

To take the joint DS3 as example, the stress distribution and deformation near the joint panel zone at the ultimate state are shown in Figure 4. The joint failure modes are not the beam hinge failure modes, but the joint shear failure modes or the local large deformation failure modes.

4.2 *F–Δ curves at H-beam end*

The relationship between the H-beam load F and its displacement Δ are shown in Figure 5. The bearing

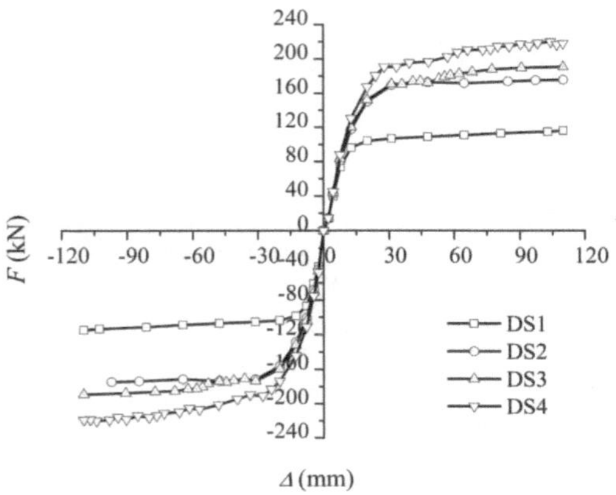

Figure 5. F–Δ curves at H-beam end.

Table 2. FE results of H-beam end under monotonic loading.

No.	Load direction	F_{by} kN	F_y kN	F_{bu} kN	F_u kN
DS1	vertical down	96.0	110.9		
	vertical up	98.4	97.2	108.7	109.8
DS2	vertical down	150.0	171.7		
	vertical up	156.2	153.1	171.5	171.6
DS3	vertical down	152.1	182.6		
	vertical up	159.8	156.0	182.7	182.7
DS4	vertical down	180.3	207.5		
	vertical up	183.0	181.7	205.4	206.5

capacities are calculated in Table 2. The yield strength F_{by} is obtained by a graphical method (Tang 1989). The corresponding load when the story rotation reached 0.06rad is taken as the ultimate strength F_{bu}. For convenient analysis, the means are taken as yield strength F_y and ultimate strength F_u.

Comparing DS1 and DS2, increase in T-stub flange thickness and bolt diameter, the obvious increase in the yield strength and ultimate strength at H-beam end, the values are respectively 57.5% and 56.3%. Comparing DS2 and DS3, only increase in T-stub web thickness, there is no obvious change in the yield strength, but little raise in the ultimate strength, the value is 6.5%. Comparing DS3 and DS4, increases in thickness of T-stub flange, web and bolt diameter, the yield strength rises 16.5% and the ultimate strength rises 13.0%.The bearing capacity is distinctly influenced by T-stub flange thickness and bolt diameter, but the T-stub web thickness.

4.3 *Computational formulas of T-stub bearing capacity*

To aim at the T-stub connected with the tensile H-beam flange, according to Eurocode 3 (EC3 2005), there are three typical failure modes to calculate the bearing

Table 3. Comparison of computational methods of T-stub.

No.	F_{Tu1} kN	F_{Tu2} kN	F_{Tu3} kN	F_{Tu} kN	F_{Tu}/F_{Tu1}	F_{Tu}/F_{Tu2}	F_{Tu}/F_{Tu1}
DS1	294	243	320	385	1.310	1.584	1.203
DS2	486	384	496	587	1.208	1.529	1.183
DS3	509	392	496	600	1.179	1.531	1.210
DS4	644	486	608	651	1.011	1.340	1.071

capacity of T-stub. Mode 1: complete yielding of the T-stub flange or column flange near the bolts; Mode 2: bolt failure with yielding of the T-stub or column flange; Mode 3: bolt failure. The three computational formulas used to T-stub are following.

$$F_{Tu1}=\frac{4M_u}{m} \tag{1}$$

$$F_{Tu2}=\frac{4M_u+n\sum B_u}{m+n} \tag{2}$$

$$F_{Tu3}=\sum B_u \tag{3}$$

where F_{Tu1}, F_{Tu2} and F_{Tu3} are the bearing capacities of T-stub; M_u = plastic moment of T-stub flange, $M_u = bt_f^2f_y/4$; b = width of T-stub flange; t_f = thickness of T-stub flange; f_y = yield strength of T-stub; m = distance between the edge of welding line and the center of bolt; n = distance from the edge of T-stub flange to the center of bolt; B_u = ultimate tensile strength of single bolt, $B_u = 0.8P$; P = pretension of single bolt.

The computational results according to EC3 are shown in Table 3. In FE model, to assume that the load at H-beam end will be turned into the tensile force at H-beam flange, this load will be transferred to T-stub completely (Song 2008), so the results F_{Tu}will be got from FE model.

Table 3 shows that the EC3 results are conservative compared with FE models, especially when the thickness of T-stub flange and diameter of bolt are thinner. With the increases in the thickness of T-stub flange and diameter of bolt, the results from FE models are close to the EC3 results. The thickness of T-stub web has little influence to the computation. In fact, on one hand, the assumption neglects the weakening action in forces transfer for the deformation of bolts, on the other hand, for the reason of the EC3 results don't take into account the eccentricity between the location of bolt hole and T-stub flange or web, so that the results got from EC3 are less.

4.4 *T-stub deformation and contact analysis*

In order to show the stress or displacement, whose variational states are along a certain scheduled path in entitative mode, the POST1 in ANSYS11.0 can achieve this objective (Maggi 2005, He 2009). We

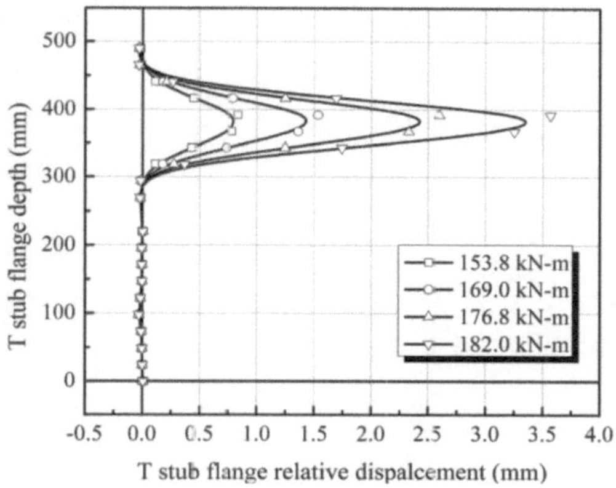

Figure 6. T-stub flange relative displacements-AA' in DS3.

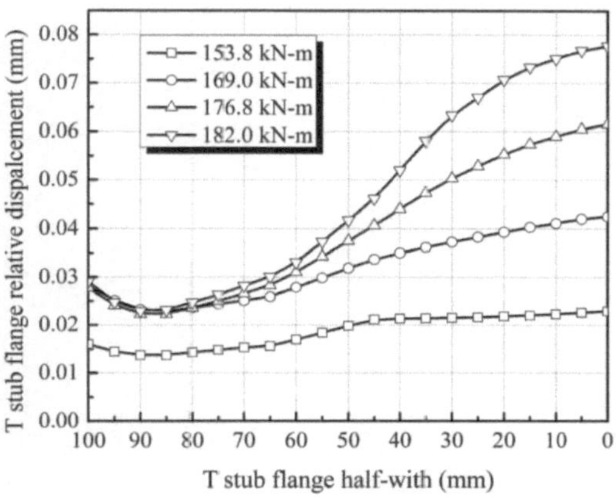

Figure 7. T-stub flange relative displacements-BB' in DS3.

define the path along T-stub flange vertical outer edge, which is named path AA'; the other path long T-stub flange vertical mid-plane edge, which is named path BB'.

With the variational load, the relative displacements of T-stub to column flange are shown in Figure 6 and Figure 7. Figure 6 shows that the relative displacements of tensile T-stub flange change acutely with the load, but the others have little changes. Figure 7 shows that the relative displacements of T-stub flange end are small, the prying actions with column flange always exist, and the deformation of T-stub flange near the H-beam mid-plane is more obvious with the increase of load. The tensile T-stub flange is under double bending condition, and the compressive T-stub flange always keeps in touch with column flange.

5 CONCLUSIONS

A new type of column-to-beam connection is introduced. In analysis, based on the force mechanism, some possible failure modes are got and nonlinear FE analysis is conducted by using ANSYS11.0. And to compare the results of the theory and FE method, the main conclusions can be drawn as follows:

(1) The failure modes of the connections are not the beam hinge failure modes, but the joint shear failure modes or the local large deformation failure modes. So the results got from theoretical analysis are close to the FE method, which shows the FE method is effective to forecast the failures modes;
(2) The carrying capacities of joint are influenced obviously by the thickness of T-stub flanges and the diameter of bolts, but less the thickness of T-stub webs, so the thinner T-stub flange and bolts should be used restrainedly ;
(3) The FE model results also present that the theoretical carrying capacities of T-stub analogy calculated by EC 3 are conservative, leading to limitation in predicting values for strength and stiffness of the connections;
(4) The tensile T-stub flange is in bidirectional flexural condition, to improve the whole mechanical performances of joint, the thicker T-stub flange should be adopted.

REFERENCES

ANSYS. 2008. *Structural Analysis Guide 11.0*. Pittsburgh: Swanson Analysis Systems Inc.

Cai, K. Q. 2001. Hollow steel column to the foundation of concrete actions with the experiment. *The Second session of the Taiwan straits and the Hongkong steel technology to exchange; Proc. intern. symp., Taibei, 29–30 November 2001*. Taiwan: The Chinese Institute of Civil and Hydraulic Engineering.

Elchalakani, M. 2002. Tests on concrete filled double-skin composite short columns under axial compression. *Thin-Walled Structures*. 40(5): 415–441.

Eurocode 3 (BS EN 1993-1-8). 2005. *Design of steel structures, Part 1–8: Design of joints*. London: British Standards Institution.

Han, L. H. 2000. *Concrete-filled steel tubular strutures*. Beijing: Science Press.

Han, L. H. 2006. Concrete-filled double skin steel tubular (CFDST) beam-columns subjected to cyclic bending. *Engineering Structures*. 28 (12): 1698–1714.

He, Y. B. 2009. Finite element analysis for extended endplate connections. *J. Hunan. Univer.* 36(5):1–6.

Maggi, Y.I. 2005. Parametric analysis of steel bolted end plate connections using finite element modeling. *J. Construct. Steel Res.* 61(5): 689–708.

Song, X. G. 2008. Hysteretic behavior study and design suggestions on steel beam-to-column T-stud connections. *Steel Construction*. 4(23): 18–23.

Tang, J. R. 1989. *Seismic resistance of joints in reinforced concrete frames*. Nanjing: Southeast University Press.

Wei, S. 1995a. Performance of new sandwich tube under axial loading: experiment. *J. Struct. Engrg.* 121(12): 1806–1814.

Wei, S. 1995b. Performance of new sandwich tube under axial loading: analysis. *J. Struct. Engrg.* 121(12): 1815–1821.

Modelling and Computation in Engineering – Zhu (ed)
© 2011 Taylor & Francis Group, London, ISBN 978-0-415-61516-7

Seismic isolation of the railway bridge with tall piers using controlled rocking approach

Xiushen Xia, Xingchong Chen & Xiaoyang Li
School of Civil Engineering, Lanzhou Jiaotong University, Lanzhou, China

ABSTRACT: A self-centering seismic isolation mechanism for bridge with tall piers is proposed. With a discontinuous interface between the pier and the pile foundation, a rocking system is designed, which can rock intermittently as the seismic overturning moment exceeds the restoring moment provided by the gravity. The simplified analysis model and some formulations of the bridge with tall piers using rocking mechanism are presented. For a railway bridge with tall piers, the effects due to controlled rocking mechanism under strong earthquake is discussed, which are compared with the bridge with lead rubber bearing on the top of the piers. The results show that the gravity can counteract the seismic action in bridge with tall piers using controlled rocking mechanism, and the seismic isolation effects of controlled rocking for the bridge with tall piers is more efficient and more stable than those with the lead rubber bearing on the top of piers.

1 INTRODUCTION

As development of rail transport to the western region, a large number of tall piers need to build in high seismicity regions of China. Under the influence of higher order vibration modes, two or more plastic hinges may be formed in a tall pier, and the location of plastic hinges are distributed and uncertain (Xia 2008). The current ductile seismic design method is only applied to the bridge with specific potential plastic hinge. Therefore, the ductile design for tall pier is very difficult, and seismic performance of tall piers with multi-plastic hinges may be unreliable. In order to improve seismic performance of railway bridge with tall piers, seismic isolation technology has become essential aseismatic measure.

There are many research literatures on bearing seismic isolation technology of bridge with short pier, but less literatures about bearing seismic isolation for the railway bridge with tall piers, and none about the tall piers using controlled rocking approach in China now (Chen et al. 2008). Therefore, seismic isolation bearing for railway bridge with tall piers in China is on trial phase. South Rangitikei tall pier railway bridge (Chen et al. 2006) of New Zealand completed in 1981 had used controlled rocking mechanism for seismic isolation in the lateral direction. With the progress of seismic technology and the requirements for improvement of seismic performance, to ensure the integrity of bridge structures post the strong earthquake is the world trend. Therefore, in order to avoid bridge damage during the earthquake, the researchers (Alessandro 2005 & Dimitra 2006) have focused on isolation technology of base in United States, Japan, Europe and Taiwan.

Research on seismic isolation of the tall piers is not enough, the seismic design for the railway bridge with tall piers has not yet been solved. In this contribution, seismic reductions of controlled rocking and bearing seismic isolation are discussed based on a railway bridge with tall piers in southwest China.

2 CONTROLLED ROCKING MECHANISMS FOR TALL PIER

Spread foundation as a common foundation form, is widely used for bridge structure with well soil condation. Ensuring stability and anti-overturning particularly stability against overturning under the interaction of earthquake act and gravity, are the essential seismic behaviors of the spread foundation. Existing literatures (Chen 2005 & Zheng 2009) show that the lift up of spread foundation can decrease seismic moment at the pier bottom. The investigation of post earthquake (Panagiotis 2005) also shows that the lift-up of spread foundation has prominent seismic isolation behavior, so the bridges with spread foundation can avoid serious damage.

The seismic design concept for railway bridge with tall piers using controlled rocking approach is proposed in this contribution. Controlled rocking mechanism for tall pier is performed by improving the current pile foundation design. With a discontinuous interface between the pier and the plie foundation,a rocking system is designed to rock intermittently as the seismic overturning moment exceeds the restoring moment provided by gravity.

The design principle for controlled rocking mechanism between tall piers and the pile foundation is that the normal train working, wind load and low-level earthquake can not make relative displacement on the contact surface, but the pier can rotate under strong earthquakes. Controlled rocking mechanism for tall pier is shown in Figure 1. The stability of anti-overturning of the pier under normal use and low-level earthquake can be designed as spread foundation. Controlled rocking seismic isolation has the following advantages:

(i) Pier and foundation will be better integrity and little damage when using controlled rocking mechanism. And the pier can be restored under the interaction of pier self-weight, load of superstructure and earthquake action.
(ii) The difference between the design of controlled rocking mechanism and the existing conventional design of tall pier is very little, and it is easy to implement for the design of controlled rocking mechanism, futhermore, it needs not more investment due to the seismic design.
(iii) Displacement and rotation angle of pier top will not be big enough to make the pier overturn after using steel cable or reinforced bar in the controlled rocking mechanism.
(iv) Controlled rocking mechanism is applicable to both along and across the bridge.

3 LAYOUT OF ANALYTICAL MODEL OF TALL PIER WITH CONTROLLED ROCKING MECHANISM

Analytical model for tall pier with controlled rocking mechanism is a cantilever with rotational spring at the pier bottom and concentrated mass at the top of pier shown in Figure 2. Rotational spring is used to simulate the nonlinear rotation between the pier and foundation. The mass of superstructure is equivalent to concentrated mass at the top of pier. When the pier is lift up and rotated, the elastic deformation is smaller compared with the rotational rigid displacement. the earthquake-response of the tall pier can be divided into stick stage and rocking stage as the earthquake-response of rigid body. Because the bottom moment

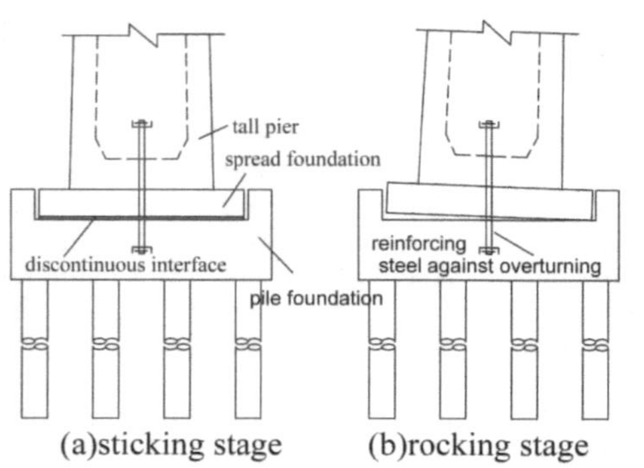

Figure 1. Tall pier using controlled rocking approach.

no longer increases after the pier is lift up and rotated, so the controlled rocking system of tall pier performances as nonlinear elastic properties. The hysteretic of rotational spring is shown in Figure 3. The moment M_y (Figure 3) is related to the axial force N of superstructure dead load, weight G of pier and the action point. The distance between the rotation center point O and the action point of the vertical force N and G is B/2 (Figure 4). The elastic deformation happened before the pier is lift-up and rotated is far shorter than B/2. If ignoring the effect of the elastic deformation on the action point of the vertical force N and G, M_y can be calculated by balance of force as follows:

$$M_y = F \cdot H = (N + G) \cdot B/2 \tag{1}$$

In the figure and formula above, N for the weight of superstructure, G for the dead weight of pier, B for

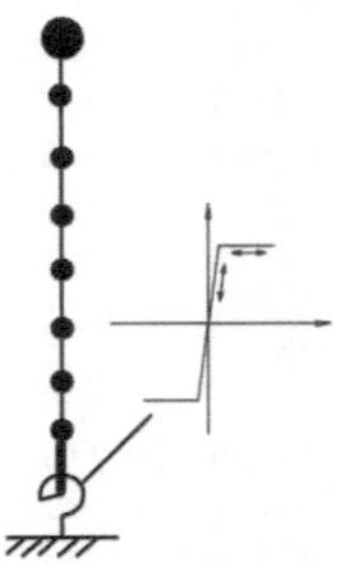

Figure 2. Analytical model of tall pier with controlled rocking mechanism.

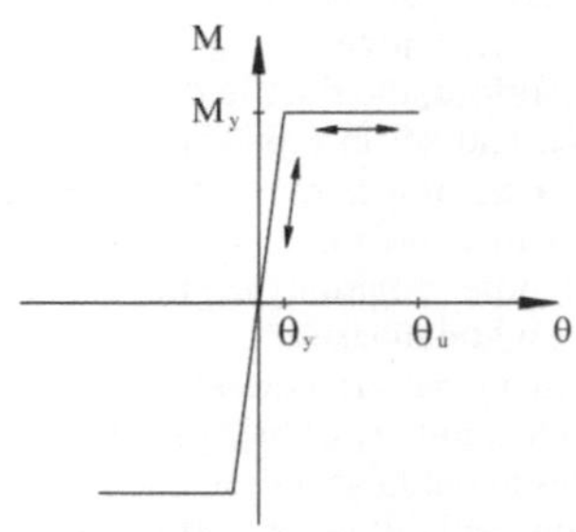

Figure 3. Hysteretic of rotational spring.

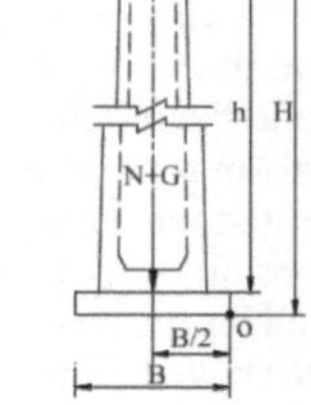

Figure 4. Calculated overturning moment.

width of the spread foundation along calculating direction. For the security and other factors, in the actual engineering, the allowed rotation angle $[\upsilon_u]$ of pier is much less than the ultimate rotation angle υ_u, so ultimate rotation angle υ_u only makes sense in theory. Currently, the experimental study of ultimate rotation angle υ_u is not enough in home and abroad. So it is difficult to accurately analyze ultimate rotation angle υ_u for tall pier. In this paper, the approximate estimation formula of ultimate rotation angle υ_u is given as follows:

$$\theta_u = B/(2H) \tag{2}$$

4 TIME HISTORY ANALYSIS OF TALL PIERS WITH CONTROLLED ROCKING MECHANISM

The example is a single line railway bridge. The superstructure are simply supported box girders with span length of 32 meters and the substructure are round-shaped hollow piers with pile foundation, which general arrangement is shown in Figure 6. The pier No.15 and No.18 are selected in the research of reduction for controlled rocking at pier base and bearing isolation at pier top. The first period of pier No.18 is 0.95 second in axial direction before seismic isolation. In order to observe the discreteness of seismic reduction, three earthquake records (El-centro, Taft and Northridge) are applied and their amplitude is unified to 0.57 g.

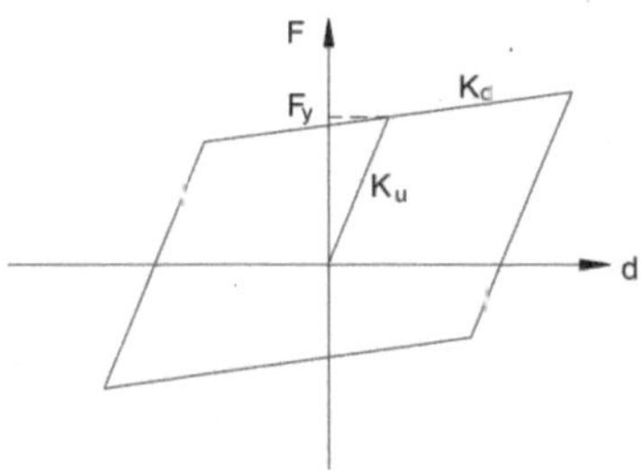

Figure 5. Hysteretic of lead rubber bearing.

4.1 *Analysis of the tall piers with controlled rocking mechanism*

The seismic analysis results of controlled rocking are shown below. From table 1, table 2 and Figure 7, Figure 8, it is found that the variation of displacement is complex. The maximum displacements of pier top under the action of the three earthquake records are different greatly. It slightly decreases when using El-centro wave, and increases more than doubled when using Northridge wave. This shows that the displacement reduction of pier top is influenced greatly by the spectrum properties.

Table 1 and table 2 show that the reduction of moment of pier bottom and pile cap bottom is significant, the reduction is best when using El-centro wave and it is close when using the other waves, and that the taller the pier is the better the reduction of pier bottom moment will be. From figure 9, it is found that reduction of controlled rocking for tall pier is intermittent rather than continuous, the pier bottom begin to rotate and lift up when seismic moment exceeds the restoring moment provided by the gravity, then the seismic moment of the pier bottom no longer increases. After the earthquake, the pier will be self-centring as a result of the balance of the vertical load of superstructure and the pier weight.

4.2 *Analysis of the tall piers with LRB on pier top*

The seismic analysis results of pier top lead rubber bearing (LRB) seismic isolation are shown in table 3, table 4, Figure 10, Figure 11. From table 3, table 4 and figure 10, figure 11, it is found that LRB can reduce

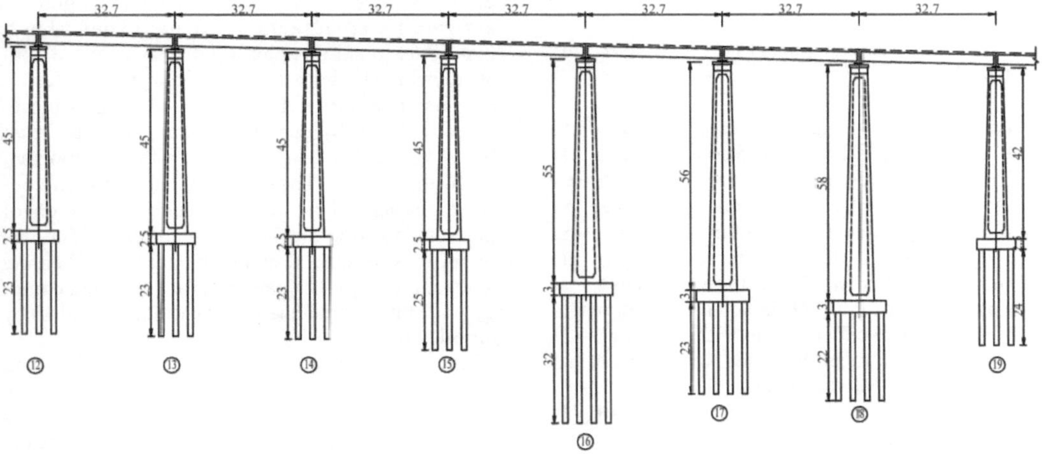

Figure 6. Front view of a typical railway bridge with tall piers (unit:m).

Table 1. Effectiveness assessment of controlled rocking (pier No.15).

	Pier top displacement			Moment of pier bottom			Moment of pile cap		
Earthquake record	Conventional cm	Rocking cm	Reduction	Conventional kN · m	Rocking kN · m	Reduction	Conventional kN · m	Rocking kN · m	Reduction
El-centro	23.9	20.8	13%	373473	131539	65%	397411	140185	65%
Taft	14.9	16.9	−13%	224395	133744	40%	233710	136658	42%
Northridge	15.6	33.7	−116%	266367	148508	44%	288865	157284	46%

Table 2. Effectiveness assessment of Controlled rocking (pier No.18).

	Pier top displacement			Moment of pier bottom			Moment of pile cap bottom		
Earthquake record	Conventional cm	Rocking cm	Reduction	Conventional kN · m	Rocking kN · m	Reduction	Conventional kN · m	Rocking kN · m	Reduction
El-centro	26.3	21.4	19%	671118	182319	73%	708502	192924	73%
Taft	16.4	22.1	−35%	413299	192169	54%	440781	192317	56%
Northridge	16.6	42.1	−154%	484037	210569	56%	522129	219431	58%

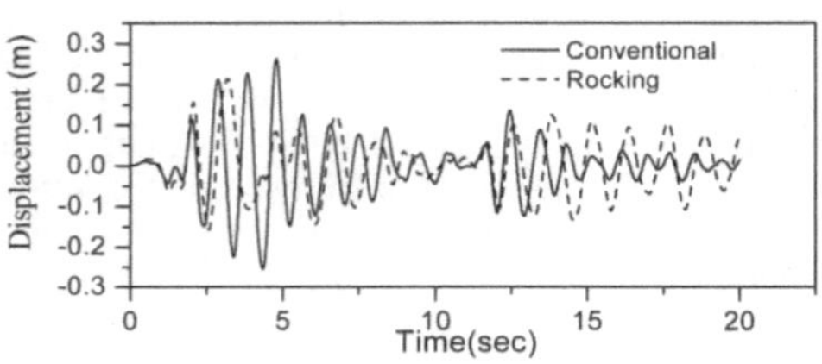

Figure 7. Displacement time-history of pier top (El-centro).

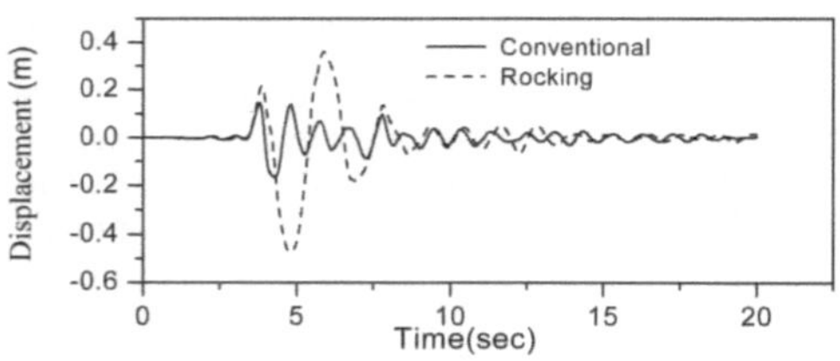

Figure 8. Displacement time-history of pier top (Northridge).

pier top displacement caused by earthquake action to a extent, and that the maximum displacements under the action of the three earthquake records are different greatly. The displacement reduction of pier top is best for using El-centro wave but worst for using Northridge wave, thus it is influenced greatly by the spectrum characteristics. Table 3, table 4 show that the displacement of the LRB is about 10 centimeters during the earthquake. Because though the initial stiffness, yield strength meet the requirement for normal use, they are still relatively small.

Table 3 and table 4 show that the moments of pier bottom and pile cap bottom are also reduced to a extent. The moment reduction of three earthquake records is different and has the trend that the taller the pier is the

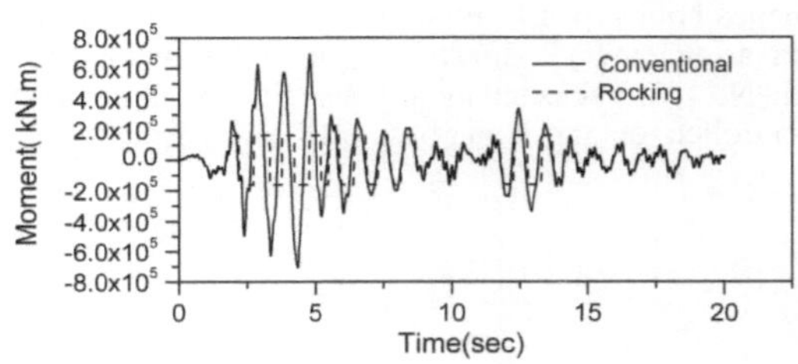

Figure 9. Moment time-history of pier base (El-centro).

more different it will be. The reduction of moment has the same variation as displacement of pier top.

4.3 *Comparison of results*

The moment time-history of pier base and displacement time-history of pier top are shown in figure 12 and figure 13. From figure 12, figure 13 and table 1, table 4, it is found that pier top displacement of controlled rocking is obviously greater than that of pier top LRB seismic isolation. Because the pier top displacement is reduced by damping energy of LRB. From table 1 to table 4 and figure 13 show that the moment time-history of pier base are different between controlled rocking and pier top LRB seismic isolation, of which the isolation mechanism is different. The former reduces seismic response by means of bearing sliding and energy dissipation through bearing damping, while the latter filters larger earthquake pulses by controlled rocking.

From table 1 to table 4 and figure 7, it is found that the displacement of pier top for controlled rocking seismic isolation is greater but can be restricted by displacement limit mechanism without influencing the seismic reduction, compared with the pier top seismic isolation.

Table 3. Effectiveness assessment of LRB (pier No.15).

	Pier top displacement			Moment of pier bottom			Moment of pile cap bottom		
Earthquake record	Conventional cm	LRB cm	Reduction	Conventional kN · m	LRB kN · m	Reduction	Conventional kN · m	LRB kN · m	Reduction
El-centro	20.8	8.3	60%	366679	166352	55%	388354	182863	53%
Taft	13.6	6.1	55%	221376	119759	46%	229856	137438	40%
Northridge	14.0	10.0	29%	261686	221658	15%	282565	245277	13%

Table 4. Effectiveness assessment of LRB (pier No.18).

	Pier top displacement			Moment of pier bottom			Moment of pile cap bottom		
Earthquake record	Conventional cm	LRB cm	Reduction	Conventional kN · m	LRB kN · m	Reduction	Conventional kN · m	LRB kN · m	Reduction
El-centro	25.3	9.8	61%	633519	270922	57%	677982	300891	56%
Taft	16.7	8.9	47%	404978	268651	34%	436273	302569	31%
Northridge	15.3	13.5	12%	450728	388190	14%	509532	426622	16%

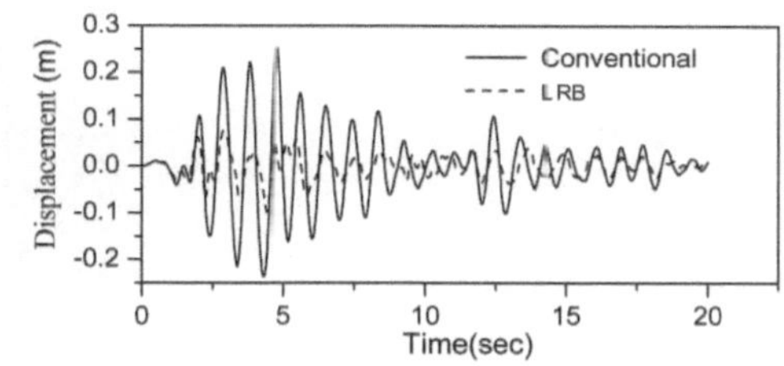

Figure 10. Displacement time-history of pier top (El-centro).

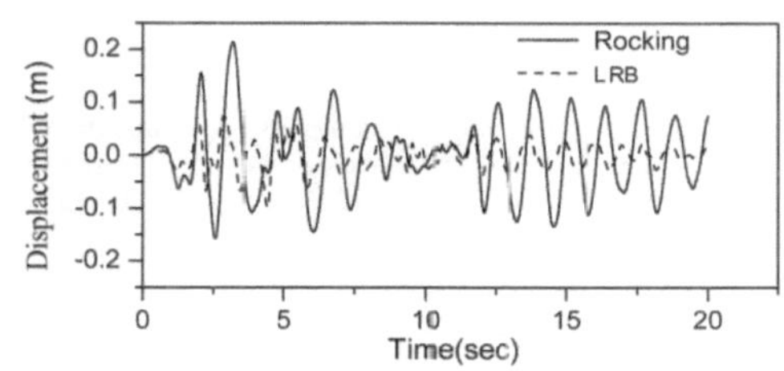

Figure 12. Comparison of displacement of pier top.

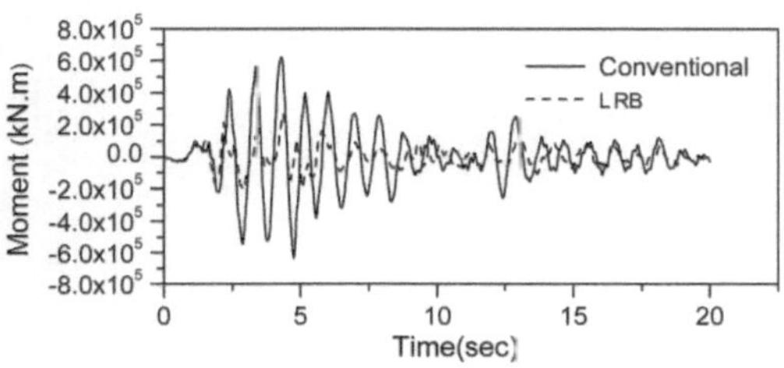

Figure 11. Moment time-history of pier base (El-centro).

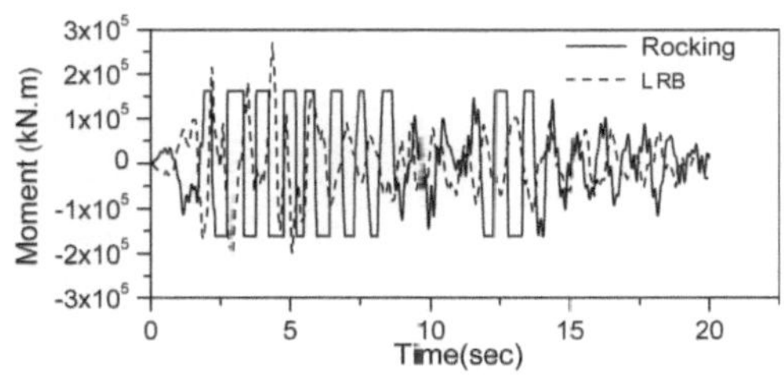

Figure 13. Comparison of moment at pier base.

For the force of pier base, the seismic efficiency of controlled rocking seismic isolation is apparently better than pier top LRB seismic isolation and the discreteness of controlled rocking seismic isolation is less. In addition, the seismic force is greatly reduced when transferring from the pier base to the foundation because the pier and the foundation will be separated in earthquake. So it can protect not only the pier but also the foundation which can not be compared with pier top seismic isolation.

The analysis above shows that the controlled rocking mechanism proposed in this paper can reduce the input seismic energy basically by separating the pier and the foundation and the maximum seismic moment is independent of spectral characteristics of the input ground motions. For controlled rocking mechanism, the seismic efficiency is significant. It can protect not only the pier but also the foundation. Therefore, there has the prospect of wide application for controlled rocking mechanism.

5 CONCLUSIONS

Both controlled rocking and LRB can reduce the seismic action of pier base. Comparatively, the reduction of controlled rocking is better and more stable. The isolation mechanism of controlled rocking mechanism is different from lead rubber bearing on the top of piers, which is a intermittent nonlinear rotation. In terms of seismic ductile design, it is easier to implement. The pier and the foundation will be separated in earthquake by using controlled rocking seismic isolation, so both the pier and foundation can be protected at post earthquake. The reduction of controlled rocking mechanism is more significant and stable than the lead rubber bearing on the top of piers. Further more, the maximum seismic moment of pier base for controlled rocking mechanism is independent of spectral characteristics of the input ground motions.

ACKNOWLEDGEMENTS

This research is sponsored by the Ministry of Railways, Housing and Urban-Rural Development of the Republic of China under contract No. 2008G028-C, 2010-K5-28.

REFERENCES

Alessandro Palermo & Stefano Pampanin & Gian Michele Calvi. 2005. Concept and development of hybrid solutions for seismic resistant bridge systems. *Journal of Earthquake Engineering*; 9(6):899–921.

Chen Xingchon, Shang Yaozhao, Zhang Yongliang et al. 2008. Analysis of seismic reduction performance on the high pier and long span railway simply-supported steel truss girders. *World Earthquake Engineering*;24(1):6–11.

Chen Xingchong & Zheng Yue. 2005. Seismic response of double-column bridge piers on elasto-plastic winkler foundation. *Engineering Mechanics*;22(3):112–125.

Dimitra Sakellaraki & Kazuhiko Kawashima. 2006. Effectiveness of seismic rocking isolation of bridges based on shake table test. *Proc. of First European Conference on Earthquake Engineering and Seismology*, Geneva, Switzerland.

Panagiotis elia mergos & Kazuhiko kawashima. 2005. Rocking isolation of a typical bridge pier on spread foundation. *Journal of Earthquake Engineering*; 9(2):395–414.

Xia Xiushen, Chen Xingchong & Wang Changfeng. 2008. Nonlinear seismic response of the tall RC piers of railway bridges. *World Earthquake Engineering*; 24(2):117–121.

ChenYi-Hsuan, LiaoWei-Hsin, Chien-LiangLee et al. 2006. Seismic isolation of viaduct piers by means of a rocking mechanism. *Earthquake Engng Struct. Dyn*; 35:713–736.

Zheng Yue & Chen Xingchong. 2009. Seismic response of double-column bridge piers considering uplift and yield of foundation. *Journal of Earthquake Engineering and Engineering Vibration*; 29(4):125–131.

Modelling and Computation in Engineering – Zhu (ed)
© 2011 Taylor & Francis Group, London, ISBN 978-0-415-61516-7

Application of Miner's Law to the remaining service life prediction of airport cement concrete pavement

Ji Feng Liang & Xie Dong Zhang
School of Transportation, Wuhan University of Technology, Wuhan, Hubei, China
Department of Airport Engineering, Xuzhou Air Force College, Xuzhou, Jiangsu, China

Zhen Qiang Zhu
The Second General Team of Chinese Airfield Construction, Xuzhou, Jiangsu, China

ABSTRACT: Referring to the *Specification of Concrete Pavement Design for Civil Airport*, this article applies Miner's Law to the remaining service life prediction of cement concrete pavement and suggests that the prediction should be based on the fatigue loss of the pavement. The calculation procedure is simple and the result is easy to read. Moreover, when the type of aircrafts taking off and landing on the airport varies, the calculation can be conveniently adjusted by introducing new data of slab edge stresses and equivalent flight frequencies.

1 INTRODUCTION

Remaining service life prediction is one of the important criteria for pavement service behavior evaluation (Ling & Zheng 2001). At present, the remaining service life of the pavement is indicated by the equivalent flight frequency, which is inconvenient for practical use because it is not a direct visual representation of pavement remaining service life. In essence, the prediction of pavement remaining service life is an inverse process of concrete pavement design. That is, we can find out the total permitted loading frequency remained of the existing pavement according to the mechanical parameters and the traffic history. Based on Miner' Law, the article establishes a pavement remaining service life calculation model with service time (year) as the index by referring to the *Specification of Concrete Pavement Design for Civil Airport* (MHJ 5004-95 1995).

2 FUNDAMENTALS OF MINER'S LAW

Usually, a specific airport will serve a variety of aircrafts and the loading effects of the aircrafts on the pavement vary from type to type. Therefore, it's not proper to add simply the loading frequency of different types of aircrafts. We can apply Palmgren-Miner's Law to the determination of the accumulated effects of all types of aircrafts on the pavement. That is, the total fatigue losses of the pavement under repeated loading is linearly accumulated and the unexhausted fatigue resistance resulted from a previous load can be made use of by the next load (Yao 1999). Suppose the fatigue service life (the maximum loading frequency) of the pavement for each type of aircrafts is as follows: N_{max1}, N_{max2}……, and N_{maxm}, then, when flight frequency of each type is N_1, N_2, ……, and N_m, the fatigue life consumed would be N_1/N_{max1}, N_2/N_{max2},……, N_m/N_{maxm}, and the accumulated fatigue life consumption is (Wang 2004):

$$X = \sum_{i=1}^{m} = x_i = \sum_{i=1}^{m} \frac{N_i}{N_{\max i}} \tag{1}$$

3 PREDICTING METHOD

3.1 *Annual loading frequency calculation of design aircraft type*

We can calculate the annual loading frequency of a certain type of craft by the following equation (Len & Weng 1995):

$$N_e = \frac{N_j \eta N W_t}{100T} \tag{2}$$

In it, N_j is the annual flight frequency of the aircraft, η is pavement occupation ratio, which is usually set at 0.85, N is the number of tyres on a main landing chassis, N_t is the width of the tyre, T is the passage width of the pavement, which is usually set at 5 m for the runway, and 1.5 m for the taxiway.

3.2 *Calculation of slab edge stress applied by the aircraft tyres*

The slab edge stress applied by the aircraft tyres can be calculated by the finite element computation program considering joint load-transfer. As for the slab edge stress applied by common type aircrafts, we can refer to the stress calculation chart appended to the *Specification of Concrete Pavement Design for Civil Airport.* The parameters we need include the thickness of the slab, the ratio of anti-fracture elastic modulus to base course top surface resilience modulus E_c / E_s, and load transfer coefficient T_w.

3.3 *Calculation of pavement remaining service life*

Based on the cement concrete pavement fatigue test data under repeated flight loads during 1943 to 1973 by the USA COE, Salsilli et al put forward a fatigue equation for concrete slab under different failure probabilities. From this field testing pavement fatigue equation (Yao 1999), we can derive the computing equation for maximum loading frequency of a certain aircraft type under constant slab edge stress:

$$\lg N_f = \left[\frac{-S^{-5.367} \lg(1-P_r)}{0.0032} \right]^{1/4.394} \tag{3}$$

Where N_f is the accumulated loading frequency when the pavement fractures out of fatigue; $S = \frac{\sigma_{max}}{f_r}$ is the ratio between slab edge stress and the anti-fractures strength; and P_r is the reliability. Related literatures show that the reliability of cement concrete pavement usually ranges from 0.5808 to 0.7130 (Jiang 2001). In this paper, we set that at 0.65. Based on equation (3), we get

$$\lg N_f = 3.0914\left(\frac{f_r}{\sigma_{max}}\right)^{1.2214} \tag{4}$$

According to Portland Cement Association (PCA), when stress ratio $S \leq 0.5$, loading frequency can be limitless (Len & Weng 1995), while in equation (4), $S > 0.5$. When $S \leq 0.5$, N_f is infinite, that is, we can ignore the fatigue loss.

The fatigue loss of a certain type of aircraft within a certain period of time can be calculated by the following equation:

$$X_i = \frac{N_i}{N_f} \tag{5}$$

And, the annual fatigue loss of a certain type of aircraft:

$$\overline{X_i} = \frac{\overline{N_i}}{N_f} \tag{6}$$

The total fatigue loss of all types of aircrafts within a certain period of time can be computed by the following equation:

$$X = \sum X_i = \sum \frac{N_i}{N_f} \tag{7}$$

And the annual total fatigue losses:

$$\overline{X} = \sum \overline{X_i} = \sum \frac{\overline{N_i}}{N_f} \tag{8}$$

The pavement remaining fatigue strength:

$$Y = 1 - X \tag{9}$$

The remaining service life of the pavement:

$$n = \frac{Y}{\overline{X}} \tag{10}$$

4 CALCULATION PROCEDURES

4.1 *The calculation of past total fatigue loss of the pavement*

(i) Finding out the annual flight frequency and total flight frequency of each type of aircraft in the past.
(ii) Referring to the *Specification* and converting the annual flight frequency of each type to the equivalent flight frequency of the design aircraft type.
(iii) Determine the following parameters: thickness of the pavement slab, load transfer coefficient, modulus ratio and the main wheel load of each type of aircraft.
(iv) Look up the slab edge stress applied by each type of aircraft in the stress calculation chart appended to the *Specification.*
(v) Calculating the maximum permitted loading frequency of each of the yielded slab edge stress by equation (4).
(vi) Calculating the fatigue loss of each type of aircraft by equation (5).
(vii) Adding the yielded fatigue losses with equation (7).
(viii) Calculating the remaining fatigue strength of the pavement by equation (9).

4.2 *Calculation of the remaining service life of the pavement according to the future service types and their annual flight frequency*

(i) Predicting the future annual flight frequency of each type and their total flight frequency.
(ii) Converting the flight frequency of each type of aircraft to the equivalent flight frequency of the designing aircraft.
(iii) Determine the following parameters: thickness of the pavement slab, load transfer coefficient, modulus ratio and the main wheel load of each type of aircraft.
(iv) Look up the slab edge stress applied by each type of aircraft in the stress calculation chart appended to the *Specification.*

Table 1. Servicing aircraft types and the flight frequency in the past and in the future.

Servicing types in the past	A	B	C	
Annual flight frequency	16,500	400	900	
Total flight frequency	214,500	5200	11,700	
Servicing types in the future	A	B	D	E
Annual flight frequency	16,500	400	2000	400

Table 2. Accumulated fatigue loss of the pavement in the past.

Type	A	B	C
Total flight frequency (13y)	214,500	52,000	11,700
Equivalent flight frequency	214,500	36,348	18,941
Main wheel load (kN)	103.14	112.44	138.25
Slab edge stress (MPa)	2.20	2.24	3.0
Maximum loading frequency	∞	∞	587,854
Single fatigue loss	0	0	1.7×10^{-6}
Accumulated fatigue loss	0	0	0. 03222
Total fatigue loss	0. 03222		
Remaining fatigue loss	0.96778		

Table 3. Pavement fatigue loss caused by the future serving aircraft types.

Aircraft type	A	B	D	E
Future annual flight frequency	16,500	4000	2000	300
Equivalent annual frequency	16,500	2796	563	5508
Main wheel load (kN)	103.14	112.44	82.31	235.24
Slab edge stress (MPa)	2.20	2.24	2.35	3.55
Maximum permitted flight frequency	∞	∞	∞	49,783
Fatigue loss per flight	0	0	0	2.0×10^{-5}
Annual fatigue losses	0	0	0	0.110645
Annual fatigue losses	0.110645			

(v) Calculating the maximum permitted loading frequency of each of the yielded slab edge stress by equation (4).

(vi) Adding the annual fatigue loss of each type of aircraft.

(vii) Calculating the remaining service life of the pavement by equation (10) (with Y equaling to the yielded result).

5 THE APPLICATION AND ANALYSIS

As is recorded, in the past 13 years, the types of aircrafts taking off and landing on a certain airport and their flight frequencies are presented in Table 1. During these years, the pavement has undergone no major repairs. Field reconnaissance and testing show that the thickness of the pavement slab is 26 cm, its flexural-tensile elastic modulus is 36,000 MPa, its anti-fracture strength is 5 MPa, its base course top surface resilience modulus E_s is 300 MPa, and $E_c/E_s = 120$. The predicted service types and annual flight frequency in the future years are also listed in Table 1. And, we are going to find out the remaining service life of the pavement under such traffic volume.

Following the procedures in 4.1, we get the fatigue losses accumulated in the past, as is shown in Table 2. The calculation results show that the fatigue loss is 0.03222, while the remaining fatigue strength is 0.96778.

Following the procedures in 4.2, we find out the annual service fatigue loss of the future servicing aircraft types, as is shown in Table 3. The annual fatigue losses of flight type A, B, and D are all 0, while fatigue loss per flight of E is 2.0×10^{-5}. According to Table 1, the annual flight frequency of E is 300, thus the annual fatigue losses is 0.110645. Therefore, the total annual fatigue losses of all types of aircraft are 0.110645. By equation (8), we find out that the remaining service life of the pavement is 8.7467 years.

Through the foregoing analysis, we find that First, for A, B and D, $S \leq 0.5$, and their fatigue loss are negligible.

Therefore, the load applied by E determines the remaining service life of the pavement. The main wheel load of E is the largest among all the servicing types, which is set at 235.24 kN, and S for it is 0.71 > 0.5. Under the loads equivalent to that of E, the remaining service life of the runway is estimated at around 9 years. Hence the aircraft like E which can cause high slab edge stress will greatly reduce the remaining service life of the pavement, which should arouse the attention of airport administrations.

What's more, for pavement not yet withstood any major repairs, we can figure out its remaining service life relying on the field collected data and the traffic history. As a matter to be noted, the accuracy of the traffic history data is crucial for the accuracy of remaining service life prediction.

6 CONCLUSION

- By reversing the pavement designing procedure, the paper sets up a pavement remaining service life prediction model based on the fatigue damage property of the cement concrete pavement. To predict the remaining service life of the pavement is of great significance to the operation of the airport and will help the management to make optimized decision.
- The result of this calculation model is direct and easy to read. We can tell easily the effects of different types of aircrafts on the service life of the pavement. Moreover, when the flight frequency of each servicing type varies, the calculation can be easily adjusted.
- Aircraft which yield high slab edge stress will greatly shorten the remaining service life of the pavement, and the airport administrations should pay special attention when such aircrafts are landing and taking off on the airport.
- It is crucial for the accuracy of the calculation to obtain accurate traffic history data. Errors are sure to arise if the data are not accurate.

REFERENCES

Jiang, H.Q. 2001. Compute the reliability of cement concrete pavement. *Hunan Communication Science and Technology* 27(2): 12–13.

Len, P.Y. & Weng, X.Z. 1995. *Airport Pavement Design*. Beijing: China Communication Press.

Ling, J.M. & Zheng, Y.F. 2001. On airport pavement evaluation system. *Journal of Traffic and Transportation Engineering* 1(1): 29–33.

MHJ 5004-95. 1995. *Specification of Concrete Pavement Design for Civil Airport*. Beijing: China Civil Aviation Administration.

Wang, W. 2004. A method for computing remaining service life of civil airport rigid pavement based on fatigue loss. *Highway* 1: 10–14.

Yao, Z.K. 1999. *Cement Concrete Pavement Design*. Hefei: Anhui Science and Technology Press.

Modelling and Computation in Engineering – Zhu (ed)
 ISBN 978-0-415-61516-7

Structural robust design based on Info-Gap model

R. Xu & H. Tang
Research Institute of Structural Engineering and Disaster Reduction, Tongji University, Shanghai, China

S. Xue
Department of Architecture, Tohoku Institute of Technology, Sendai, Japan

ABSTRACT: A novel Info-Gap robust design concept to structural robust optimization under severe uncertainties is presented in this paper. This Info-Gap model is a non-probabilistic method for the problem considering severe uncertainties using the Info-Gap Decision Theory (IGDT). IGDT models the clustering of uncertain events in families of nested sets instead of assuming a probability structure, which only require the nominal estimate of uncertain parameters to be known before analysis or use in optimization. The heuristic algorithm is applied to the nested optimizations in IGDT and simulation results show that the proposed approach can solve complex problems effectively.

1 INTRODUCTION

Uncertainty is prevalent in constructional engineering. So when these uncertainties could not be ignored, conventional design based on deterministic assumptions is not a rational method. Recent prevailing methodologies for uncertainty are possibility theory and fuzzy set theory. All of the preceding methods rely on either multiple data samples or subjectively defined distributions or membership functions based on expert knowledge. However, enough data samples can not be gathered in engineering design usually. Several studies have proved that both probability and fuzzy-logic models are quite information-intensive, and even a little error in statistical information would make distribution or membership function fluctuate acutely, especially in the tails of the distribution (Ben-Haim & Elishakoff 1990, Ben-Haim 1994a, Moens & Vandepitte 2005). The sensitivity of the tails of the input probability density models to their parameters undoubtedly affects the value of the failure probability.

There are circumstances, however, where there is not enough available information of the uncertainties to define probability distributions other than both upper and lower bounds on either side of these uncertainties. In 1990s, Ben-Haim and Elishakoff, established the foundation for the convex model theory and application (Ben-Haim & Elishakoff 1990, Elishakoff et al. 1994). Convex models are especially useful for problems where data on the uncertain parameters is scarce, as in the case of severe uncertainties.

Recently, the Info-Gap Decision Theory (IGDT) proposed as a non-probabilistic decision theory (Ben-Haim 1994b) under uncertainties has steadily evolved over 15 years from a body of work on convex set-based models of uncertainty (Ben-Haim 2006), and has been applied to wide fields including neural networks (Pierce et al. 2006), biological conservation (Duncan et al. 2008), financial economics (Ben-Haim 2005), etc. Some researchers have tried to apply IGDT to structural design (Takewaki & Ben-Haim 2005, Ben-Haim 2004), however, these attempts only limited to the problems of structural design which the robustness function of structures could be computed analytically. Thus, to sum up the above arguments, this paper is first to suggest a new approach which combines IGDT and heuristic algorithm, and apply this approach to complex problems of structural design effectively.

2 INFO-GAP ROBUST DESIGN

2.1 *Severe uncertainty in structural design*

Because civil structures are not mass-produced, the probabilistic representation of these uncertainties seems to be difficult in most cases (Takewaki & Ben-Haim 2005). Thus there are circumstances in structural design, where there is nothing available to describe an uncertain variable other than a nominal estimate (e.g., an approximation, a comparable baseline, expert estimate, etc.). Common responses to such severely deficient information include postponing decision making entirely, haphazardly collecting more data without a clear understanding of its value, or even relying on unwarranted assumptions that fill in missing information. Due to awareness of the limitations of probability in condition of such severely deficient information, Ben-Haim has defined the concept of 'severe uncertainty', and proposed IGDT as an alternative approach. In real engineering, 'severe

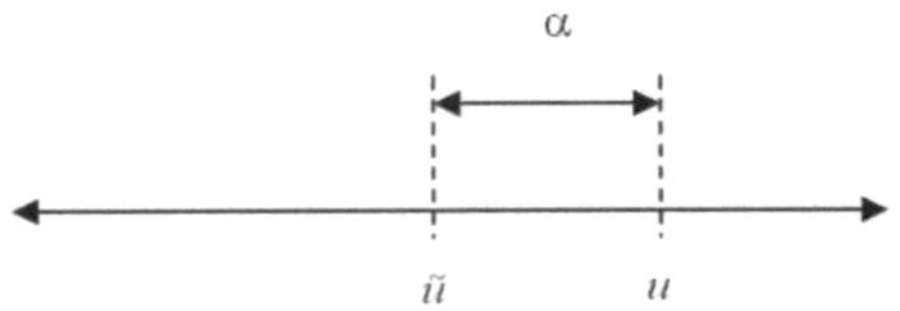

Figure 1. Info-Gap model of severe uncertainty.

uncertainty' is the state of deficient information where one knows that there is uncertainty associated with a particular quantity, and knows a nominal estimate for that quantity, but does not know the size of the uncertainty for that quantity.

As Figure 1 depicted, the simplest form of severe uncertainty models, is the discrepancy between an uncertain quantity's available (but suspect) nominal value and the quantity's true value, which could be known but is not.

In fig. 1, u is the uncertain quantity's available (but suspect) nominal value, but the uncertain quantity's true value, u, is unknown. Moreover the discrepancy between u and u also can not be seized. Because the size of this gap is unknown, it is represented mathematically using an uncertainty parameter, α, sometimes referred to as the horizon of uncertainty. In IGDT, the decision maker does not wish to assume more information than is available, so α remains unspecified. Mathematically, this sparse information can be formulated into an info-gap model of uncertainty, defined as:

$$U(\alpha,\tilde{u}) = \{u : |u-\tilde{u}| \le \alpha\} \ , \qquad \alpha \ge 0 \tag{1}$$

So far, IGDT is the only uncertainty formalism allow any of their parameters (e.g., mean, standard deviation, interval bounds, etc.) to remain unspecified before analysis or use in optimization (Duncan et al. 2008).

2.2 *IGDT components and equation*

Using IGDT, a decision maker confronts uncertainty by maximizing robustness to it, seeking reasonably satisfactory performance over the widest range of unknown uncertainty, i.e., the largest horizon of uncertainty, α. Performance is not optimized but satisfied. Satisficing generally means accepting 'good enough' performance in order to afford the achievement of other objectives, especially when only idealized models or limited information is available (Takewaki & Ben-Haim 2005). Using IGDT, the decision maker chooses a satisficing level of critical performance, r_c, to be satisfied by all design alternatives. The info-gap robustness, $\hat{\alpha}(q, r_c)$, of a design alternative q is the largest horizon of uncertainty, α, that the design can withstand while still guaranteeing better than critical performance r_c. In other words, info-gap robustness is the greatest amount of error in the nominal that a design can endure and still perform well (but not necessarily optimally). The robust-satisficing decision rule used in IGDT prefers the design with the largest info-gap robustness at a given r_c, i.e., $\hat{\alpha}(q, r_c)$ is maximized over the set of design options. Settling for a less demanding r_c often affords a design more info-gap robustness, and in some cases, changes which design is most preferred under the robust-satisficing decision rule. Thus, the decision maker may need to develop a preference for a robustness-performance trade-off to be able to select between design alternatives.

The mathematics of info-gap uncertainty models, in their simplest form, as well as the info-gap robustness function are presented subsequently. The three components needed for an info-gap analysis are (Duncan et al. 2008):

1) $R(q, u)$, a performance model (referred to interchangeably as a 'reward' model) of system response that is a function of an uncertain variable, u, and (continuous or discrete) design option(s), q; and whose output is a performance attribute of interest.
2) u, the uncertain variable, representable as an info-gap and relating to (1) above. Actually, u could also be a model or function, though that will not be considered in this paper.
3) r_c, a critical satisficing level of performance that must be guaranteed exceeded; alternatively considered a failure criterion.

From the three IGDT components, an info-gap robustness function can be defined that maximizes the size that the uncertainty parameter α can take and still satisfy a critical constraint. When an increase in $R(q, u)$ is desirable, the critical constraint equation is:

$$R(q,u) \le r_c \tag{2}$$

This constraint is embedded into the info-gap robustness function, defined mathematically as an optimization problem:

$$\hat{\alpha} = \max\{\alpha : (\max R(q,u) \le r_c)\} \tag{3}$$

If the info-gap robustness for a given critical reward could be known, the corresponding decision (or decisions) is the optimal design alternative:

$$\hat{q}(r_c) = \arg\max_{q \in Q}(\hat{\alpha}(q, r_c)) \tag{4}$$

2.3 *Nested optimization using heuristic algorithm*

In IGDT, the robustness function plays a key role as a measure of robustness. In the optimization procedure of equation (3), at each step of the first optimization process to maximize the robustness, the second optimization process to find the worst condition produced on the constraints expressed by r_c, is solved. Therefore, the procedure of two optimizations is nested. Takewaki and Ben-Haim computed the robustness function of structures in a particular case where the worst case can be obtained analytically (Takewaki & Ben-Haim 2005). Unfortunately it is difficult to compute exactly the robustness function of structures. Thus, many complex problems of structural design could not be solved using IGDT. To overcome the difficulty, we utilize the heuristic algorithm, which is an alternative to solve the model above with simplicity and immune to the local optima.

Among the heuristic algorithms, Particle Swarm Optimisation (PSO) was originally introduced by Kennedy and Eberhart (Kennedy & Eberhart 1995). Based on swarm intelligence, PSO is the intelligent collective behaviour of organisms in a swarm (e.g. a flock of birds migrating), while the behaviour of a single organism in the swarm may seem totally inefficient. And on this basis, the probability of premature convergence of the PSO is remedied by the extension to the comprehensive learning particle swarm optimization (CLPSO), which is utilized to estimate robustness function and performance in equation (3) in this paper. The success and simplicity of CLPSO in civil engineering was experimentally verified (Tang et al. 2008). The theory and manipulation of CLPSO can be obtained from correlative study of Liang (Liang et al. 2006).

3 ILLUSTRATIVE EXAMPLE

3.1 *Design problem scenario*

The ten-bar truss problem has effectively become a benchmark problem in the field of structural optimization. Cross-sectional areas may vary from $64.516\times10^{-6}\,\text{m}^2$ to $2.5806\times10^{-2}\,\text{m}^2$. The material has a modulus of elasticity of 68947.57 MPa and a mass density of 2767.99 kg/m^3. The design constraints are: The maximum allowable stress in any member of the truss is ±172.252 MPa; and the maximum deflection of any node in both the vertical and horizontal direction is ±50.8 mm. Fig. 2 shows the configuration of 10-bar plane truss including its node and element numbering schemes, and in Fig. 2 external load P is 444.52 kN and structural height L is 9.144 m.

Low-weight designs have been developed for this ten-bar truss problem by the different algorithms, including GA (Camp et al. 1998) with a weight of 2302.576 kg, PSO (Schutte & Groenwold 2003) with a weight of 2295.110 kg and BBBC (Charles & Camp 2007) with a weight of 2278.236 kg.

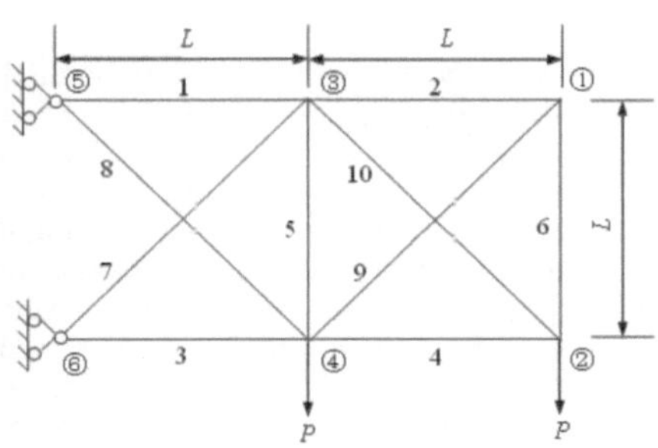

Figure 2. 10-bar plane truss.

3.2 *Info-Gap problem formulation and solution*

In this section, we continue to use above example in Section 3.1, and assume the true value of the external load in node 2 is uncertain. Like equation (1), from this sparse information, the info-gap model for this truss design can be defined as:

$$\mathcal{P}(\alpha,\widetilde{P}) = \{P : \left|\frac{P-\widetilde{P}}{\widetilde{P}}\right| \le \alpha\}\ , \qquad \alpha \ge 0 \tag{5}$$

For this example, the total mass of this truss is chosen as the reward function, expressed in the form of equation (2) as:

$$R(q,u) = Mass_{totcl} = \sum_{i=1:10} m_i = \sum_{i=1:10} \rho \times q_i \times l_i \tag{6}$$

where ρ is the mass density, q_i is cross-sectional area of each bar, and l_i is length of each bar.

Next, the critical performance constraint is:

$$Mass_{total} \le r_c \tag{7}$$

4 RESULTS AND REMARKS

4.1 *Analysis of the structural designs in IGDT*

Main interest in IGDT is what amount of robustness, $\hat{\alpha}(q, r_c)$, can be achieved for each design alternative. To reiterate again, this robustness is the largest size in uncertainty parameter α that can be sustained and still guarantee no worse than the chosen critical performance level r_c.

Table 1 lists the final solutions developed by IGDT. In table 1, the designs for different critical performance are defined as different 'Case', and the 'Robustness' is the percentage increases relative to the nominal estimate value of the external load, as equation (5) shown.

According to the table 1, it is noted that the robustness improves as the critical performance requirement becomes less stringent. Moreover, by analysis of the structural designs in table 1, we can find which member is the key one to improve the structural robustness. Thus according to table 1, a plot of cross-sectional areas vs. robustness is shown in Figure 3.

As Fig. 3 shown, under different case, the cross-sectional areas of members signed No. 1, No. 3, No. 4, No. 7 and No. 10 are increasing observably. Therefore, in real engineering, it is advisable to reinforce these sensitive members to increase the robustness of this 10-bar plane truss by IGDT.

4.2 *Trade-off between performance and robustness to uncertainty*

Preferences for the trade-off between robustness and acceptable performance can be elicited by examining a trade-off plot. In most info-gap analyses in the

Table 1. Info-Gap optimal results of 10-bar plane truss.

	Cross-sectional areas (mm^2)										
Members	Case 1	Case 2	Case 3	Case 4	Case 5	Case 6	Case 7	Case 8	Case 9	Case 10	Case 11
1	19565.4	20018.4	20189.3	20640.5	21227.6	21168.4	22765.1	23037.5	25682.3	25805.3	25714.9
2	66.5	67.6	65.8	68.2	66.6	65.0	66.0	82.4	83.8	85.0	84.89
3	15343.3	15529.3	15148.1	15309.2	15655.6	15938.9	16993.9	16887.9	17133.0	17650.6	18325.5
4	9847.7	9890.8	10283.7	9741.9	10393.2	11196.0	10801.5	12314.5	12836.8	12658.1	14351.8
5	64.7	64.8	65.0	64.6	64.7	64.5	64.5	64.9	64.7	65.8	65.6
6	332.6	332.5	377.1	392.2	310.2	444.5	322.3	348.2	329.2	301.0	328.4
7	13740.1	13740.1	13790.8	14585.5	14701.3	15323.4	15690.3	16633.4	16592.5	17696.5	18389.3
8	4816.6	4816.6	4852.5	4895.1	4854.6	4839.7	4835.1	4824.6	4915.9	5116.3	4881.9
9	64.8	64.5	64.5	64.6	64.6	64.6	64.5	72.9	73.7	70.9	73.3
10	13631.8	13728.3	14057.0	14842.3	15084.5	15042.7	15978.5	16612.5	16792.6	18148.9	18853.0
Mass (kg)	2299	2320	2340	2400	2450	2500	2600	2700	2799	2900	3000
Robustness (%)	0.13	1.20	2.29	5.40	8.06	10.02	15.91	21.05	26.05	31.49	36.68

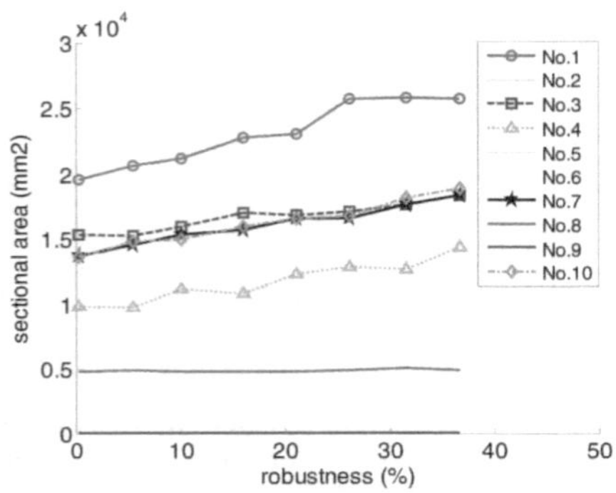

Figure 3. Illustration of the optimum sectional area in each case.

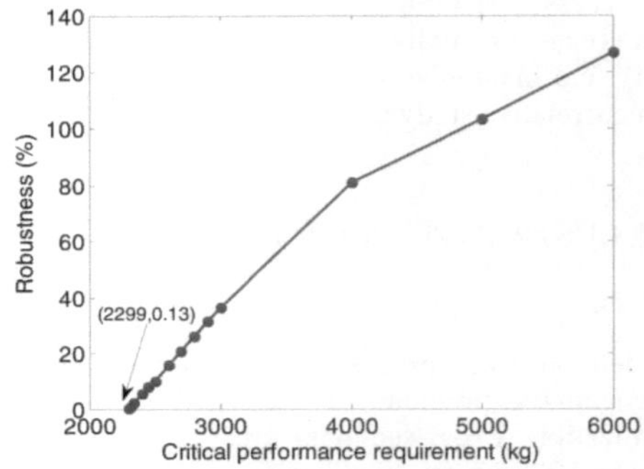

Figure 4. Illustration of the trade-off between robustness and performance.

literature, this has meant plotting the robustness function, $\hat{\alpha}(q, r_c)$, with r_c as the dependent variable to generate a robustness-reward plot. Whichever the orientation, the key idea is that info-gap robustness and critical performance – regardless of what axis they are on – are related through a function that can be evaluated graphically to help elicit the trade-off preferences necessary to a decision. According to table 1 and sequential designs for more critical performances, a plot of performance vs. robustness is shown in Figure 4.

The first thing to note about Fig. 4 is that the robustness improves ($\hat{\alpha}(q, r_c)$ increases) as the performance requirement becomes less stringent (r_c gets larger). This expresses the inevitable trade-off between robustness-to-uncertainty and functional performance which was noted in Fig. 4: either one can be improved only by degrading the other.

In real engineering, the info-gap decision analysis approach can eliminate the need for further data collection by facilitating decision making under extreme uncertainty (i.e., when estimation error cannot be quantified). For instance, if switching one's design choice (e.g., some designs in Fig. 4) requires a small sacrifice in guaranteed performance yet affords a reasonably large amount of extra robustness to error in an estimate, one could decide to switch without needing to know more, and benefit from percentage increases of the structural robustness. However, the increase rate of robustness is slowing down gradually as the performance requirement becomes less stringent, which is worth noting for structural designers to estimate gain and loss.

4.3 *Difference between structural robust-satisficing optimal design and structural performance optimal design*

As seen in Fig. 4, the info-gap robust-satisficing optimal design defined in equation (3) is very different, conceptually and practically, from performance-optimization of design, as we have seen in the example in Section 3.1. We will now more thoroughly explain the difference, and why the robust-satisficing approach is preferable if there is severe uncertainty.

As shown in Section 3.1, the usual approach to design optimization is based on a best-estimate of the model. In optimizing the performance, one seeks the design which minimizes the performance function, $R(q, u)$, based on this best model. In fact, structural performance optimal design is only a special case of the robust-satisficing optimal design. And when the robustness function of the robust-satisficing optimal design is zero, it is equivalence between the two kinds of above optimal design concepts. Therefore, the result of the lowest point in the trade-off curve of Fig. 4 approximately equals the design in Section 3.1. However, the robustness-to-uncertainty is zero at the level of performance predicted by the best-estimated model. What this means is that no design can be depended upon to perform at the level which is predicted by the best-estimate of the model.

5 CONCLUSIONS

This paper has developed a new methodology for design of complex structures subject to severe uncertainties. The main ideas of the paper are followed.

1) Conventional uncertain optimization methods represented by probability method are difficult to deal with the severe uncertainty, which are usually encountered in practical engineering. In this paper, IGDT is applied efficiently to solve the problem of optimization of structures with severe uncertainty.
2) Up to now, most structural designs followed the methodology as performance-optimal design by directly optimizing structural performance. However, the approach in this paper is seeking reasonably satisfactory performance over the widest range of unknown uncertainty and maximizing the largest horizon of uncertainty, and this robust-satisficing design can observably increase structural robustness.
3) As Fig. 4 shown, the unavoidable trade-off between the functional performance of a structure and its robustness to uncertainties allows structural designers to choose preferred design by weighing the need of performance and robustness, and this design using IGDT induces the designer to satisfy rather than optimize structural performance.
4) As we have seen, the difficulty in structural robustness function calculation can be overcome by introducing the heuristic algorithm in IGDT. Thus, many complex problems of structural design in real engineering which had never be solved, can be solved by this new method of combining IGDT and heuristic algorithm.

REFERENCES

Ben-Haim, Y. & Elishakoff, I. 1990. Convex Models of Uncertainty in Applied Mechanics. *Elsevier Science Publishers*, Cambridge, MA.

Ben-Haim, Y. 1994. Convex models of uncertainty: Applications and Implications, Erkenntnis: *An International Journal of Analytic Philosophy*, 41:139–156.

Ben-Haim, Y. 1994. A non-probabilistic concept of reliability. *Structural Safety*, 14 (4): 227-245.

Ben-Haim, Y. 2004. Uncertainty, probability, and information gaps, *Reliability Engineering and System Safety*, Vol. 85, pp.249–266.

Ben-Haim, Y. 2005. Value at risk with Info-gap uncertainty. *Journal of Risk Finance*, 6:388–403.

Ben-Haim, Y. 2006. Info-Gap Theory: Decisions Under Severe Uncertainty, 2nd edition, Academic Press, London.

Camp C., Pezeshk S. & Cao G. 1998. Optimized design of two-dimensional structures using a genetic algorithm. J. *Struct. Eng.*, 124(5): 551–559.

Charles, V. & Camp. 2007. Design of space trusses using big bang-big crunch optimization. *Journal of Structural Engineering*, 999–1008

Duncan S.J., Bras B. & Paredis C.J.J. 2008. An approach to robust decision making under severe uncertainty in life cycle design. *IJSDes* 1(1):45–59

Elishakoff, I., Elisseeff, P & Glegg, S.A.L. 1994. Non-probabilistic, convex-theoretic modeling of scatter in material properties. *AIAA Journal*, 4(4).

Kennedy, J & Eberhart, R.C. 1995. Particle swarm optimization. *Proceedings of the IEEE International Conference on Neural Networks*, IV, pp 1942–1948.

Liang J.J., Qin A.K, et al. 2006. Comprehensive learning particle swarm optimizer for global optimization of multimodal functions. *IEEE Transactions on Evolutionary Computation*, 10(3): 281–295.

Moens, D. & Vandepitte, D. 2005. A survey of non-probabilistic uncertainty treatment in finite element analysis, *Computer methods in applied mechanics and engineering*. 194:1527–1555.

Pierce, S.G., Worden, K. & Manson, G. 2006. A novel information-gap technique to assess reliability of neural network-based damage detection. *Journal of Sound and Vibration*, 293:96–111.

Schutte J.J. & Groenwold A.A. 2003. Sizing design of truss structures using particle swarms. *Struct. Multidiscip. Optim*, 25: 261–269

Tang, H., Zhang, W., Fan, C. & Xue, S. 2008. Parameter estimation using a CLPSO strategy. *IEEE World Congress on Computational Intelligence*, Hong Kong: IEEE press, 70–74.

Takewaki, I. & Ben-Haim, Y. 2005. Info-gap robust design with load and model uncertainties, *Journal of Sound and Vibration*, Vol. 288, No. 3, pp.551–570.

Modelling and Computation in Engineering – Zhu (ed)
© 2011 Taylor & Francis Group, London, ISBN 978-0-415-61516-7

Mathematical modeling method of trawler equipped with CPP

Xiao-Feng Sun, Yong Yin, He-Long Shen & Xin-Yu Zhang
Dalian Maritime University, Dalian, China

ABSTRACT: To study the maneuverability of trawler under cruising and trawling conditions, the paper proposes a mathematical modeling method of trawler equipped with Controllable Pitch Propeller (CPP) according to separate modeling theory. Based on Netherlands Ship Model Basin's open-water tests, CPP's thrust and torque coefficients in four-quadrant under different pitch ratio of screw are obtained, hence CPP's thrusts and torques are gained. To reflect influence of fishing gear acting on trawler under trawling condition, the paper establishes the hydrodynamic model of fishing gear by take a single-boat mid-water trawl system for example. The model is divided into models of warp and trawl, which connected by boundary conditions, and can also be used to reflect the influence of fishing gear acting on trawler during the process of drawing in and giving out the trawl. The model's rationality is verified by comparisons of simulated results and that of sea-trial tests.

1 INTRODUCTION

Trawler is the power provider of the trawl system, and its maneuverability has a great significance of the security of trawl fishing. Trawlers have two basic modes of operation: cruising and trawling.

The mathematical model of trawler under cruising condition can be established through the modeling method of CPP (Controllable Pitch Propeller) ship, and the maneuverability of trawler can be analyzed. Under trawling condition, fishing gear will act external forces on trawler, and affect the maneuvering motion of trawl. To analyze the maneuverability of trawler under trawling condition, the paper establishes the mathematical model of fishing gear taking the single-boat mid-water trawl system for example, and the influence of fishing gear acting on trawler is reflected through external force and moment.

Fishing gear includes trawl, cables and accessories etc. To reflect the influence of fishing gear acting on trawler in the process of drawing in and giving out the trawl, the hydrodynamic model of fishing gear is divided into hydrodynamic models of warp and trawl, which are connected by boundary conditions reflecting the influence of each other.

2 MATHEMATICAL MODEL OF TRAWLER UNDER CRUISING CONDITION

The mathematical model of trawler under cruising condition had been described in previous papers (SunXiaofeng, YinYong & ZhangXiufeng 2005, SUN Xiaofeng, YIN Yong & ZHANG Xiufeng 2006), so this paper only gives the outline.

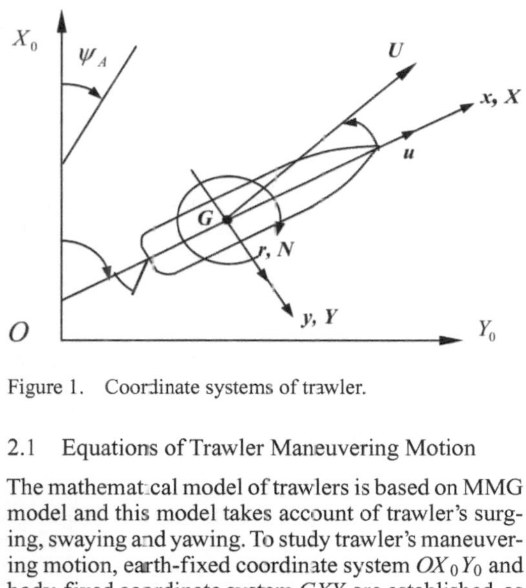

Figure 1. Coordinate systems of trawler.

2.1 Equations of Trawler Maneuvering Motion

The mathematical model of trawlers is based on MMG model and this model takes account of trawler's surging, swaying and yawing. To study trawler's maneuvering motion, earth-fixed coordinate system OX_0Y_0 and body-fixed coordinate system GXY are established, as shown in Figure 1.

Trawler's maneuvering equations can be described as follows in body-fixed coordinate system:

$$\begin{cases} (m+m_x)\dot{u}-(m+m_y)vr = X_H - X_P + X_R \\ (m+m_y)\dot{v}-(m+m_x)ur = Y_H + Y_P + Y_R \\ (I_{ZZ}+J_{ZZ})\dot{r} = N_H + N_P + N_R - Y_H x_C \end{cases} \quad (1)$$

Where m, m_x, m_y is mass and added mass of ship; u, v, r is longitude and lateral velocities and yaw rate of ship; X, Y, N is longitude and lateral forces and moment acting on trawler; I_{zz}, J_{zz} are moment of inertia and added moment of inertia of ship; Subscripts "$_H$", "$_P$", "$_R$", represent the forces or moments that are created by hull, propeller, rudder.

In general, the modeling of propeller is based on propeller's characteristic curve and propeller's thrust and torque can be expressed as follows:

$$\begin{cases} X_P = \frac{\rho}{2}[V_A^2 + (0.7\pi nD)^2]\frac{\pi}{4}D^2 \cdot C_T^* \\ Q_P = \frac{\rho}{2}[V_A^2 + (0.7\pi nD)^2]\frac{\pi}{4}D^2 \cdot D \cdot C_Q^* \end{cases} \tag{2}$$

Where X_P is thrust; Q_P is torque; ρ is water density; D is propeller diameter; V_A is undisturbed stream velocity; n is propeller revolution; C_T^*, C_Q^* are thrust coefficient and torque coefficient.

In the modeling of CPP, each variation of propeller's pitch can be considered that the propeller's characteristic has changed, and the CPP model must be achieved according to the changed propeller's characteristic. The paper gains thrust and torque of CPP under different conditions by using Netherlands Ship Model Basin's open water tests (Assemberg 1984), described as follows:

$$\begin{cases} C_T^* = \frac{1}{100}\sum_{k=0}^{30}[A(k)\cos k\beta + B(k)\sin k\beta] \\ C_Q^* = \frac{-1}{1000}\sum_{k=0}^{30}[C(k)\cos k\beta + D(k)\sin k\beta] \end{cases} \tag{3}$$

Where *A(k), B(k), C(k) and D(k)* are regression coefficients, relating to number of blades Z, blade area ratio λ, and pitch ratio P/D; β is hydrodynamic pitch angle at 0.7 tip radius of blade, and:

$$\beta = \arctan(V_A / 0.7\pi nD) \tag{4}$$

The computation method of above variables can be seen in our present papers (SunXiaofeng, YinYong & ZhangXiufeng 2005, SUN Xiaofeng, YIN Yong & ZHANG Xiufeng 2006).

3 HYDRODYNAMIC MODEL OF FISHING GEAR

3.1 *Hydrodynamic model of warp*

To analyze the hydrodynamic behavior of warp, inertial coordinate system O_T-ijk and local coordinate system P-tnb are adopted in this paper, as shown in Figure 2. In system O_T-ijk, O_T is set to trawl point; axes O_T-i and O_T-j are parallel to heading and starboard direction separately, and in the opposite direction; axes O_T-ijk and is vertical to horizontal plane. In system P-tnb, P is a point in warp; axes Pt points to tangent direction; axis Pn and Pb point to the two normal directions.

The two coordinate systems can be related by rotations through the Euler angles, as shown in Figure 3.

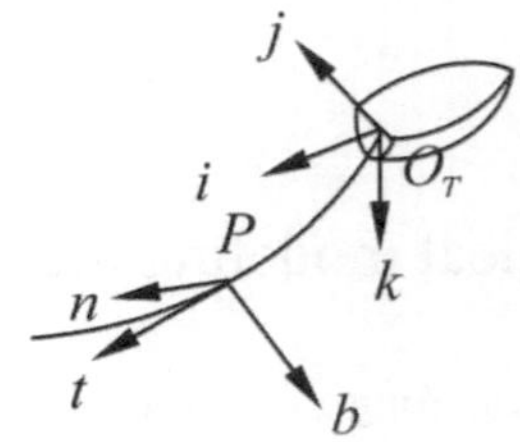

Figure 2. Coordinate systems of warp.

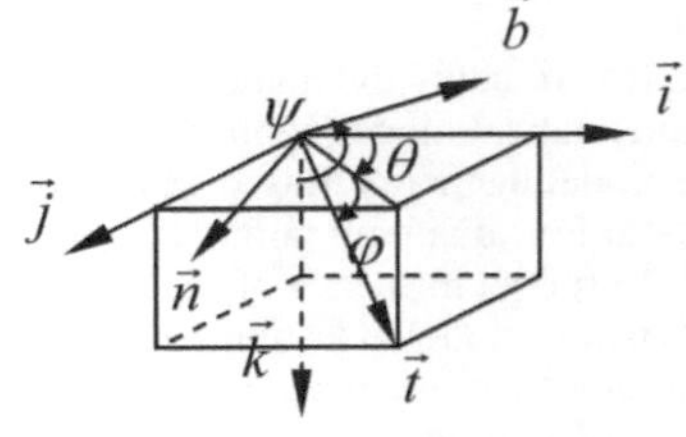

Figure 3. Definition of Euler angles.

In this paper, we assumes that ψ is a constant, and $\psi = \pi/2$.

The balance equation at infinitesimal section of warp can be described as follows (Abolw, C. W & Schechter, S 1983):

$$\frac{\partial \vec{T}}{\partial S} + \vec{W} + \vec{F} + \vec{B} = 0 \tag{5}$$

Where $\overline{T}$ is the tension; $\overline{S}$ is distance along the stretched warp; $\overline{W}$ is the weight less buoyancy per unit length; $\overline{F}$ is the force exerted on warp by the fluid per unit length; $\overline{B}$ is d'Alembert force of warp motion per unit length. The computation method of above variables can be seen in our present paper (Sun Xiaofeng, Yin Yong & Zhang Xiufeng 2007).

3.2 *Hydrodynamic model of trawl*

To describe the curve shapes of bars, the paper adopts the method of Suzuki, K., Takagi, T., Shimizu, T., Hiraishi, T., Yamamoto, K. & Nashimoto, K. (2003) to establish the hydrodynamic model of the mid-water trawl. Based on lumped mass model, mid-water trawl is divided into finite mass points which are connected with springs without mass.

To analyze the forces impacted on knots and bars, spatial coordinate *oxyz* and bar coordinate $km_1\xi\eta\varsigma$ are established, as in Figure 4. The original point of spatial coordinate *oxyz* can be any point in space. Axes *ox* and *oy* are in horizontal plane, and axis *oz* is perpendicular to horizontal plane pointing to the center of the earth. The original point of bar coordinate $km_1\xi\eta\varsigma$ is in the position of knot mass point km_1 connected with bar mass point b_m. Axis $km_1\xi$ is in the direction of km_1 to another mass point km_2 connected with b_m. Axis $km_1\eta$ is in horizontal plane and is perpendicular to axis $km_1\xi$. Axis $km_1\varsigma$ is perpendicular to axis $km_1\xi$ and axis $km_1\eta$.

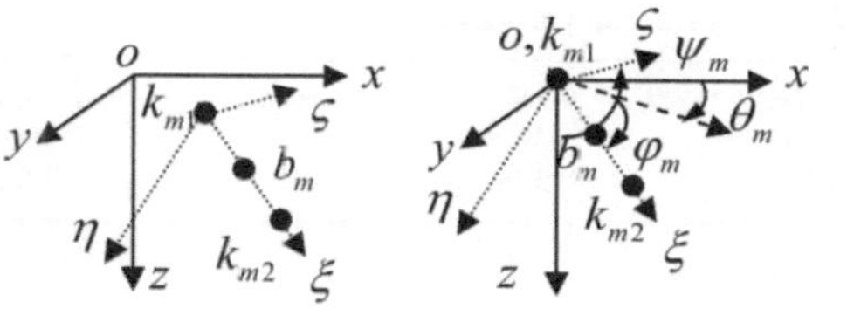

Figure 4. Coordinate systems and definition of Euler angles.

3.2.1 *Equations of knots*

Equations of knots can be described as follows according to Newton's Second Law:

$$\left.\begin{aligned}(m_i+\Delta m_i)\ddot{x}_i &= T_x + F_x\\ (m_i+\Delta m_i)\ddot{y}_i &= T_y + F_y\\ (m_i+\Delta m_i)\ddot{z}_i &= T_z + F_z + W\end{aligned}\right\} \tag{6}$$

where $m_i, \Delta m_i$ are mass and added mass of knot i; x_i, y_i, z_i are coordinates of knot i in spatial coordinate; T, F are elastic force and drag force impact on knot i. Subscripts "$_x$", "$_y$", "$_z$", represent components of each force along axis ox, oy and oz respectively; W is weight in fluid of knot i. The computation method of above variables can be seen in our present paper (Sun Xiaofeng, Yin Yong & Zhang Xiufeng 2009).

3.2.2 *Equations of bars*

Ignoring the influence of bar's curved shape to the hydrodynamic force, equations of bars can be describes as follows according to Newton's Second Law:

$$\left.\begin{aligned}(m_i+\Delta m_{i\xi})\ddot{\xi}_i &= T_\xi + F_\xi + W_\xi\\ (m_i+\Delta m_{i\eta})\ddot{\eta}_i &= T_\eta + F_\eta + W_\eta\\ (m_i+\Delta m_{i\varsigma})\ddot{\varsigma}_i &= T_\varsigma + F_\varsigma + W_\varsigma\end{aligned}\right\} \tag{7}$$

Where m_i is mass of bar i; ξ_i, η_i, ς_i are coordinates of bar i in the bar coordinate system; $\Delta m_{i\xi}$, $\Delta m_{i\eta}$, $\Delta m_{i\varsigma}$ are added mass of bar i in ξ, η, ς direction respectively; T, F are the elastic force and drag force impacted on knot i; W is weight in fluid of bar i Subscripts "$_\xi$", "$_\eta$", "$_\varsigma$" represent components of each force in axis $k_{m1\xi}$, $k_{m1\eta}$ and $k_{m1\varsigma}$ respectively.

3.3 *Boundary conditions*

Under trawling condition, the otter door connects warp and bridles of trawl, as shown in Figure 5.

According to D'Alembert Principal, dynamic equation of otter door can be written as follows:

$$\vec{W}+\vec{T}+\vec{F}+\vec{B}+\vec{F}_{bridle1}+\vec{F}_{bridle2}=0 \tag{8}$$

Where $\vec{W}$ is the weight less buoyancy of otter door; $\vec{T}$ is the tension of warp; $\vec{B}$ is inertia force of the otter door; $\vec{F}$ is the force exerted on otter door by the fluid;

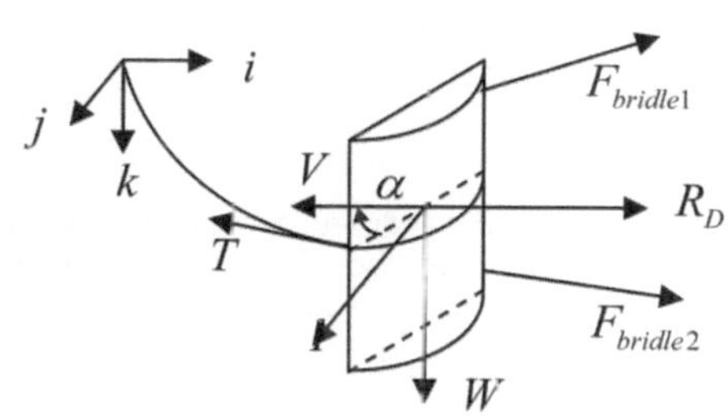

Figure 5. Analysis of forces acting on otter door.

$\vec{F}_{bridle1}, \vec{F}_{bridle2}$ are the forces created by the bridles of trawl.

Taking the otter door as the end node of warp, forces acting on otter door can be described as follows:

3.3.1 *Otter door weight*

$$\begin{aligned}\vec{W} &= (m_d-\rho V_d)g\vec{k}\\ &= (m_d-\rho V_d)g\cdot(\sin\phi_N\vec{t}+\cos\phi_N\vec{n})\end{aligned} \tag{9}$$

Where m_d is the mass of otter door; V_d is the volume of otter door.

3.3.2 *Tension of warp*

$$\vec{T} = -T_N\vec{t} \tag{10}$$

Where T_N is the tension of warp's end point.

3.3.3 *Fluid force*

The force exerted on otter door by the fluid can be divided into lift R_L and resistance R_D, as follows:

$$\begin{cases}R_L = 1/2\cdot\rho SC_L V^2\\ R_D = 1/2\cdot\rho SC_D V^2\end{cases} \tag{11}$$

Where ρ is the fluid density; S is side area of otter door; V is current speed; C_L and C_D are lift coefficient and resistance coefficient.

In this paper, the calculation of fluid force of otter door is simplified, as follows:

$$\begin{cases}\vec{R}_D = -\dfrac{1}{2}\rho SC_D[u_t|u_t|\vec{t}+u_n|u_n|\vec{n}+u_b|u_b|\vec{b}]\\ R_L = \dfrac{1}{2}\rho SC_L\sqrt{u_i^2+u_j^2}(u_j\vec{i}-u_i\vec{j})\end{cases} \tag{12}$$

3.3.4 *Inertia force*

$$\vec{B} = -(m+\Delta m)(\dot{v}_t\vec{t}+\dot{v}_n\vec{n}+\dot{v}_b\vec{b}) \tag{13}$$

Where Δm is added mass of otter door.

3.3.5 *Bridle tension*

The forces created by the bridles of trawl $\vec{F}_{bridle1}, \vec{F}_{bridle2}$ can be calculated by the hydrodynamic model of trawl.

The force R_L and $\vec{F}_{bridle1}, \vec{F}_{bridle2}$ can be described in the warp local coordinate system, as follows:

$$\vec{F}_{bridle1}+\vec{F}_{bridle2}+\vec{R}_L = Q_t\vec{t}+Q_n\vec{n}+Q_b\vec{b} \tag{14}$$

4 MATHEMATICAL MODEL OF TRAWLER UNDER TRAWLING CONDITION

4.1 *Equations of trawler maneuvering motion*

While trawling, fishing gear will act external force on trawler. Under this condition, trawler's maneuvering equations can be described as follows:

$$\left.\begin{aligned}(m+m_x)\dot{u}-(m+m_y)vr &= X_H+X_P+X_R+X_N\\(m+m_y)\dot{v}+(m+m_x)ur &= Y_H+Y_P+Y_R+Y_N\\(I_{ZZ}+J_{ZZ})\dot{r} &= N_H+N_P+N_R-Y_Hx_C+N_N\end{aligned}\right\} \quad (15)$$

Where X_N, Y_N and N_N are the forces and moment created by fishing gear.

In case that trawler connects with trawl directly, the forces and moments created by fishing gear can be described as follows:

$$\left.\begin{aligned}X_N &= \sum_{n=1}^{2} F_{xn}\\Y_N &= \sum_{n=1}^{2} F_{yn}\\N_N &= \sum_{n=1}^{2}(-F_{xn}\cdot y_w+F_{yn}\cdot x_w)\end{aligned}\right\} \quad (16)$$

And

$$\begin{cases}F_{xn}=F_{Xn}\cdot\cos\varphi+F_{Yn}\cdot\sin\varphi\\F_{yn}=-F_{Xn}\cdot\sin\varphi+F_{Yn}\cdot\cos\varphi\end{cases} \quad (17)$$

Where F_{xn}, F_{yn} can be acquired through the hydrodynamic model of trawl.

In the case that trawler connects with trawl through warp, the forces acting on trawler can be described as follows in warp inertial coordinate system:

$$\begin{cases}F_{Xn}=F_{in}\cdot\cos\vartheta-F_{jn}\cdot\sin\vartheta\\F_{Yn}=F_{in}\cdot\sin\vartheta+F_{jn}\cdot\cos\vartheta\end{cases} \quad (18)$$

And:

$$\begin{cases}F_{in}=T\cos\phi\cos\theta\\F_{jn}=T\cos\phi\sin\theta\\F_{kn}=T\sin\phi\end{cases} \quad (19)$$

5 SIMULATION RESULTS

The paper does some researches on an ocean-going trawler, "LiaoYu18". The principal particulars of trawler are shown in Table 1. This trawler equips with a mid-water trawl ZT8909(52◇×26 m), and parameters of warp and otter door are shown in Table 2.

Table 1. Principal particulars of LiaoYu18.

Loa (*m*)	93.5	B (*m*)	15.6
dm (*m*)	4.42	D (*t*)	3511
Cb	0.598	Cp	0.628

Table 2. Parameters of Warp and Otter Door.

Warp	Weight in fluid per length unit (*N*)	130. 5
	Diameter (*cm*)	5.0
	Elastic modulus (*GPa*)	150
Otter Door	Weight in fluid (*N*)	45.685
	Side area (*m2*)	2.3

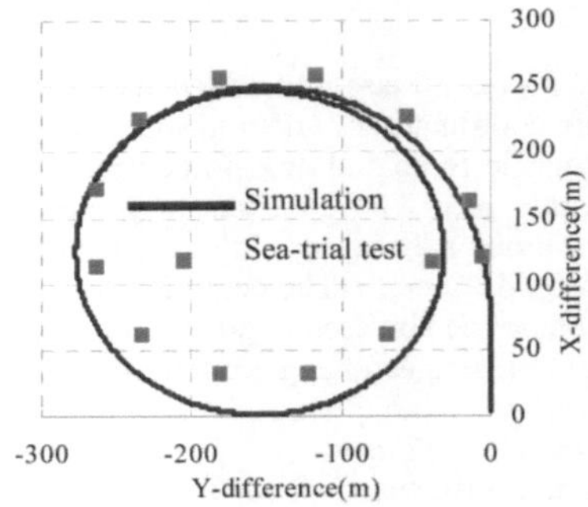

Figure 6. Comparisons of port turning circle tracks.

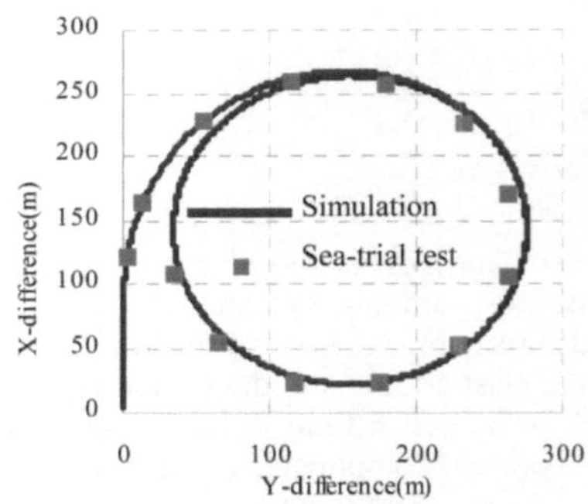

Figure 7. Comparisons of starboard turning circle tracks.

5.1 *Simulation under cruising condition*

Figures 6–7 shows port & starboard turning trajectories comparisons of the trawler between simulation results and sea-trial tests at full rudder angle (35 degree).

Table 3 shows the comparisons of speed of the trawler between the simulation results and sea-trial tests at different combinations of propeller's revolution and pitch angle.

Table 3. Comparisons of trawler speeds.

Pitch angle (*deg*)	Propeller revolution (*RPM*)	Simulation (*Knot*)	Sea-trial (*Knot*)	Error
17.5	201	15.17	15.685	−3.3%
17.5	192.9	14.65	15.219	−3.7%
17.5	160.7	12.57	13.18	−4.6%
17.5	137.3	11.07	11.66	−5.1%
17.5	75	5.52	5.94	−7.1%
12	200	11.78	12	−1.8%
7	200	8.57	8	7.1%
3	200	4.18	4	4.5%

Table 4. Different equipments of the trawl.

No	1	2	3	4
head line length (*m*)	101.4	101.4	101.4	101.4
over sweep line length (*m*)	160.0	160.0	160.0	220.0
under sweep line length (*m*)	165.0	165.0	165.0	225.0
Cross rop length (*m*)	30.0	30.0	30.0	30.0
Float force (*Kg*)	800.0	1200.0	1200.0	1200.0
Sink force (*Kg*)	3320.0	3320	4320.0	4320.0

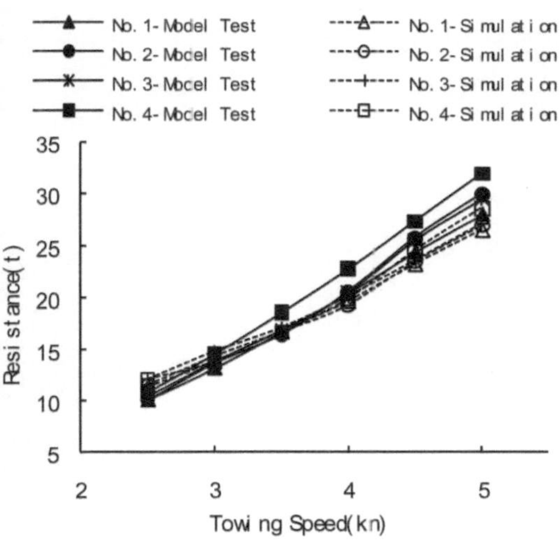

Figure 8. Curves of net resistance with towing speed

5.2 *Simulation under trawling condition*

Some flume experiments of mid-water trawl ZT8909 (52◇×26 m) had been done by East China Sea Fisheries Research Institute (East China Fishery Research Institute, 1989), which are shown in Table 4. This paper does some simulation researches on this mid-water trawl and compares the simulated results with that of the flume experiments, as shown in Figure 8.

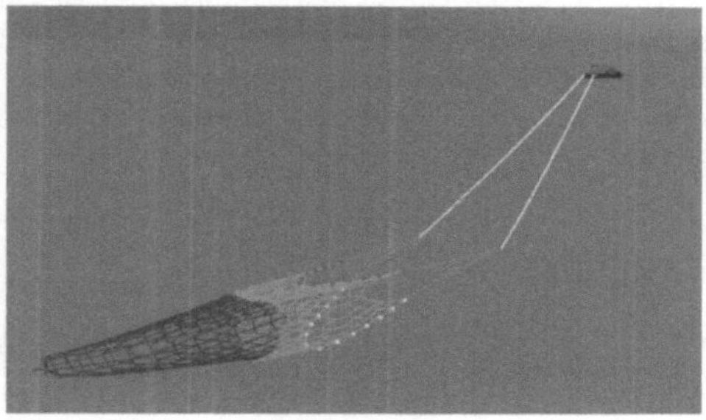

Figure 9. Visualization of single boat mid-water trawl system.

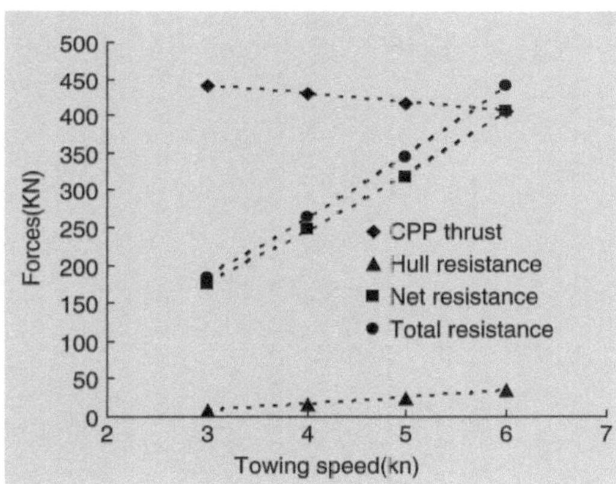

Figure 10. Curves of thrust and resistance of a trawler with towing speed.

In the condition of rated revolution, length of warp is 500 *m*, float force is 800 *Kg* and sink force is 3200 *Kg*, the visual result of the system is shown in Figure 9, and the curves of CPP thrust, net resistance and hull resistance with towing speed are shown in Figure 10.

6 CONCLUSIONS

To study trawler's maneuverability under cruising and trawling conditions, this paper proposes a mathematical modeling method of trawler equipped with

Controllable Pitch Propeller. The influence of fishing gear acting on trawler is reflected through external force and moment. To reflect influence of fishing gear acting on trawler under trawling condition, the hydrodynamic model of fishing gear is divided into mathematical models of warp and trawl, which are connected by boundary conditions reflecting the influence of each other. Comparisons of simulated results and that of sea-trial tests show that the proposed model is rational.

ACKNOWLEDGEMENT

The paper is supported by the National Program on Key Basic Research Project, 973 Program, (Grant No. 2009CB320805) and project of Dalian Maritime University (Grant No.2009QN012), and we appreciate their help.

REFERENCES

Abolw, C. W & Schechter, S. 1983. Numerical Simulation of Undersea Cable Systems. *Ocean Engng* 10(6): 443–457.

Assemberg. 1984. Virerkwadrant Schroefkaracteristieken Voor B-Berir Schroeven Fourier-reeks Onlwikkling en Oppraional Genbraik. *MARIN Rapport* No.60482-1-MS: 1–30.

East China Fishery Research Insititute. 1989. Report on model experiment of fishing gear. *East China Fishery Research Institute* DT8916.

Suzuki, K., Takagi, T., Shimizu, T., Hiraishi, T, Yamamoto, K. & Nashimoto, K. 2003. Validity and visualization of a numerical model used to determine dynamic configurations of fishing nets. *Fisheries Science* 69(4): 695–705.

SunXiaofeng, YinYong & ZhangXiufeng. 2005. Rsearch on key technologies of fishing vessels simulator. The Sixth International Conference on System Simulation and Scientific Computing. Beijing: International Academic Publishers World Publishing Corporation: 36–40.

SUN Xiaofeng, YIN Yong & ZHANG Xiufeng. 2006. Study on maneuvering mathematical model for trawl. Proceeding of the International Conference MARSIM' 2006. Terschelling: Maritime Institute Willem Barentsz: M-30-1~M-30-5.

Sun Xiaofeng, Yin Yong & Zhang Xiufeng. 2007. Research on Dynamic Model of Trawl Warp in Fishing Simulator. *Journal of simulation system* 19(Suppl.2): 14–16.

Sun Xiaofeng, Yin Yong & Zhang Xiufeng, 2009. Research on dynamic simulation and visualization of fishing net based on physical model. *Journal of Dalian Fisheries University* 24(6): 563–567.

Modelling and Computation in Engineering – Zhu (ed)
© 2011 Taylor & Francis Group, London, ISBN 978-0-415-61516-7

Modeling preferential zone-regulated freshwater-saltwater mixing zone

Y.Q. Xia
Key Laboratory of Biogeology and Environmental Geology of the Ministry of Education, School of Environmental Studies, China University of Geosciences, Wuhan, PR China
Center for Natural Resources Development and Protection (NRDP), Department of Civil and Environmental Engineering, Temple University, Philadelphia, PA, USA

M.C. Boufadel
Center for Natural Resources Development and Protection (NRDP), Department of Civil and Environmental Engineering, Temple University, Philadelphia, PA, USA

ABSTRACT: The freshwater-saltwater mixing zone is of great importance in the study of the land-ocean interaction processes and coastal groundwater environments. This paper considers a coastal multilayered system which consists of an unconfined aquifer and a confined aquifer with an aquitard between them. The aquitard is assumed to have a breached zone (preferential zone). 2D density-dependent numerical simulations with the MARUN code were conduced to explore effects of the preferential zone on the movements of the freshwater-saltwater mixing zone. The results show that an inland preferential zone caused the freshwater-saltwater mixing zone in the confined aquifer to move landward, and induced seawater upcoming around the zone. An intertidal preferential zone caused the freshwater-saltwater mixing zone in the confined aquifer to move close to the preferential zone, while an offshore preferential zone would enhance the water exchange in the system. The findings provide insights into the management of coastal environment.

1 INTRODUCTION

The coastal aquifer or "subterranean estuary" (Moore 1999) is defined as the coastal regions where ground water derived from land drainage measurably dilutes sea water that has invaded the aquifer through a free connection to the sea. The groundwater level (hydraulic head or water table) in coastal aquifer fluctuates with time in response to the water level fluctuations of the tidal water body (sea or river/stream). Since 1950s, the theoretical research and field investigation of groundwater flow and salinity distribution in the coastal aquifer systems have been conducted by many hydrogeologists (e.g., Nielsen 1990, Kim et al. 2006, Xia et al. 2007, Brovelli et al. 2007, Li & Boufadel 2010).

Both the field monitoring and numerical simulations have shown that groundwater dynamics, saltwater intrusion and groundwater-seawater interaction in a coastal aquifer are influenced by tidal fluctuations. Ataie-Ashtiani et al. (1999) used a variable-density groundwater model to analyze the effects of tidal fluctuations on seawater intrusion in an unconfined aquifer. To investigate the tidal effect on groundwater-seawater flow in the coastal aquifer, Kim et al. (2006) conducted various field studies at the seawater intrusion monitoring wells located in the eastern part of Jeju Island, Korea. Based on Xia et al. (2007)'s analytical solution, Xia & Li (2009) estimated aquifer parameter values using tidal effect in a coastal aquifer in Beihai Peninsula where seawater intrusion occurred. Actually, saltwater intrusion, i.e., the displacement of fresh groundwater by groundwater with a higher salinity, is one of the freshwater-saltwater interactions in coastal aquifers, which has become a hot topic in the field of coastal hydrogeology (Post & Abarca 2010). The freshwater-saltwater mixing zone is of great importance in the study of freshwater-saltwater interactions.

Based on measurements of hydraulic gradients and offshore fluxes taken at Waquoit Bay, Massachusetts, together with a modelling study of a generalized coastal groundwater system, Michael et al. (2005) found a shift in the freshwater-saltwater interface controlled by seasonal changes in water table elevation, which can explain large saline discharges that lag inland recharge cycles. Guo & Jiao (2007) developed analytical solutions to investigate alteration of groundwater table and freshwater-saltwater interface in response to land reclamation. Robinson et al. (2007) investigated the freshwater-saltwater dynamics over a spring-neap tidal cycle based on field measurements and numerical simulations of pore water salinities and groundwater flow in the intertidal zone of an unconfined coastal aquifer. Falgas et al. (2009) detected dynamic changes in the seawater-freshwater interface in the deltaic zone of the Tordera River used the audiomagnetotelluric method.

Zhou & Wang (2009) briefly reviewed several methods for estimation of the location of

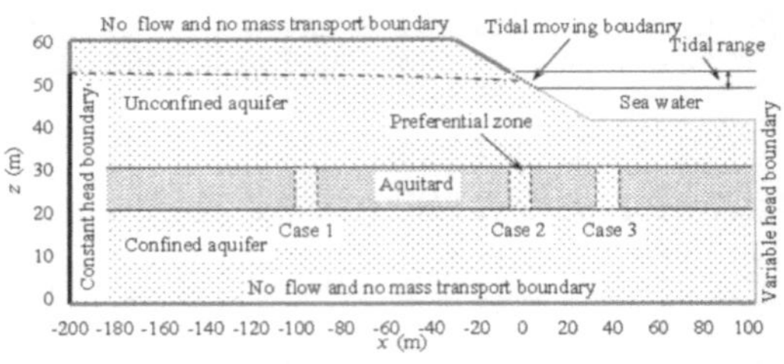

Figure 1. Schematic of a multilayered aquifer system cross-section vertical to the shoreline. Four boundary conditions are shown in this figure using different lines with the names of boundary conditions.

freshwater-saltwater interface in coastal aquifers. If the freshwater and saltwater zones are assumed to do not mix (immiscible), they have a sharp interface. If the freshwater and saltwater are considered being density-dependent fluids, they have a mixing zone.

This paper conducts four density-dependent two-dimensional numerical simulations in a coastal multilayered system to investigate the alteration of the freshwater-saltwater mixing zone regulated by a preferential zone.

2 MODEL SETUP

2.1 *Domain selection*

A schematic representation of the coastal multilayered aquifer system is shown in Figure 1. This system consists of an unconfined aquifer and a confined aquifer with an aquitard between them. The aquitard is assumed to have a breached zone, referred as to be "preferential zone". The coastal aquifer system is 300 m long and 60 m deep with a 1:3 beach slope for the beach zone (ranges from −30 m to 30 m). The thicknesses of the confined aquifer and aquitard are 20 m and 10 m, respectively. The system is assumed to be piecewise homogeneous within each of the three layers and the preferential zone. The aquifers (unconfined aquifer and confined aquifer) are assumed to have the same permeability. The preferential zone is assumed to be 10 m long, completely penetrates through the aquitard and has the same permeability as the aquifers.

The horizontal dimensions of the mesh are 1 m from −200 m to −30 m and 0.5 m in seaward area of −30 m. The vertical dimensions of the mesh are 1 m from −200 m to −30 m and vary from 1.0 m near −30 m to 0.7 m in seaward parts of the beach zone. The numbers of nodes and elements are 26291 and 51600, respectively. The seaward boundary is subjected to tide, which is described by a mean sea level of 50 m, on which is superimposed a sinusoidal diurnal tide with amplitude of 1 m and frequency of 0.243 rad/h.

2.2 *Numerical model*

The MARine UNsaturated (MARUN) model was employed to conduct two-dimensional simulations, which can simulate two components (one is salinity and another could be nutrient or tracer concentration) in variably saturated media, taking into account the effects of salt concentration on water density and water viscosity (Boufadel et al. 1999, Boufadel 2000). The validity of the MARUN model has been verified by many case studies or experiments (e.g., Li et al. 2008, Li & Boufadel 2010).

The equation for the conservation of the water and salt can be written as

$$\beta\phi\frac{\partial S}{\partial t}+\beta S_0 S\frac{\partial \psi}{\partial t}+\phi S\frac{\partial \beta}{\partial t} = \frac{\partial\left(\beta\delta K_x\frac{\partial\psi}{\partial x}\right)}{\partial x}+\frac{\partial\left(\beta\delta K_z\frac{\partial\psi}{\partial z}\right)}{\partial z}+\frac{\partial(\beta^2\delta K_z)}{\partial z} \quad (1)$$

where β is the density ratio [−] and δ is the dynamic viscosity ratio [−]. ϕ is the porosity of the porous medium [−], S is the soil moisture ratio [−], S_0 is the specific storage [L^{-1}], ψ is the pressure head [L], and K_x and K_z are the horizontal and vertical freshwater hydraulic conductivities [LT^{-1}].

The soil moisture ratio and the freshwater hydraulic conductivity are correlated by the van Genuchten (1980) model:

$$\text{for } \psi \geq 0,\ S = 1.0,\ K_x = K_{x0},\ K_z = K_{z0} \quad (2)$$

$$\text{for } \psi < 0,\ S = (1-S_r)S_e + S_r \quad (3)$$

where K_{x0} and K_{z0} are the saturated horizontal and vertical hydraulic conductivity for freshwater, respectively. S_r is the residual saturation ratio, α represents the characteristic pore size of the beach soil [L^{-1}]. n represents the uniformity of the pores and $m = 1 - 1/n$. S_e is the effective saturation ratio, defined as

$$S_e = \left[\frac{1}{1+(\alpha\,|\psi|)^n}\right]^m \quad (4)$$

The relative hydraulic conductivity k_r is given by

$$k_r = K_i / K_{i0} = \sqrt{S_e}\,[1-(1-S_e^{1/m})^m]^2,\ i=(x,z) \quad (5)$$

Following Boufadel et al. (1999) and Boufadel (2000), the solute transport equation (convection-dispersion equation) can be written as

$$\phi S\frac{\partial c}{\partial t} = \beta\nabla\cdot(\phi SD\cdot\nabla c) - \mathbf{q}\cdot\nabla c \quad (6)$$

where c is the salt concentration expressed in grams of salt per liter of solution. $\mathbf{q} = (q_x, q_z)$ is the Darcy flux vector given by

$$\mathbf{q} = (q_x, q_z) = -K_i\delta\left(\frac{\partial\psi}{\partial x}, \frac{\partial\psi}{\partial z}+\beta\right),\ i=(x,z) \quad (7)$$

The term D represents the physical dispersion tensor written as

$$D=\frac{1}{\|\mathbf{q}\|}\begin{pmatrix} \alpha_L q_x^2-\alpha_T q_z^2 & (\alpha_L-\alpha_T)q_x q_z \\ (\alpha_L-\alpha_T)q_x q_z & \alpha_T q_x^2+\alpha_L q_z^2 \end{pmatrix} \quad (8)$$

where $||\mathbf{q}|| = \sqrt{q_x^2 + q_z^2}$, α_L and α_T are the longitudinal and transverse dispersivities [L], respectively.

2.3 *Boundary and initial conditions*

Figure 1 depicts the boundary conditions used in this paper. A specified constant head of 51 m (1 m above the mean sea level, i.e. high tide line) is assigned at the saturated part of landward boundary of the domain. No-flow (a zero flux for water and salt) boundary condition is assigned on the domain bottom and top, and the boundaries of the unsaturated zone, which include the unsaturated (upper) part of the landward boundary and the beach surface above sea level (see Fig. 1). The water pressure on the submerged beach surface was determined by the tidal seawater column above the beach surface. Therefore, the sea boundary condition was moving with time, which was named "tidal moving boundary". At each time step, the portion of the beach surface submerged by tide was updated by comparing the tidal level and beach surface elevation. The salt concentrations depended on the flow direction. At locations where seawater was entering the beach, the salt concentration was set equal to that of the sea, whereas a Neumann boundary condition with zero dispersive flux was used at locations where the water leaves the beach. This is known as the outflowing boundary condition (Galeati et al. 1992, Boufadel 2000). To simple the simulation, the seepage face simulation module within MARUN was inactivated.

The initial condition was a hydrostatic state at low tide with a sharp freshwater-seawater interface as given by the Ghyben-Herzberg approximation, and 1800 tidal cycles (~five years) were run to obtain the quasi-steady state numerical solution, such that the "signature" of the initial condition disappeared.

2.4 *Numerical implementation*

The convergence criterion of the Picard iteration for solving the nonlinear groundwater flow equation was 10^{-5} m. In order to eliminate numerical oscillations, a variable time step was used. The upper limit of the time step was set such that the grid Courant number ($C_r = v\Delta t/\Delta l$, where v is the Darcy velocity, Δl is the maximum of Δx and Δz of the mesh) remained less than 0.95. Dispersivity values were chosen to guarantee a small safe value for the grid Péclet number (Zheng & Bennett 2002, Brovelli et al. 2007). The ratio of transverse to longitudinal dispersivity was 0.1. Table 1 summarized the model parameter values used in the simulations. Four cases considering different locations of the preferential zone were considered as below.

Base case: The case without preferential zone was taken as the base case for comparison with three

Table 1. Parameters values used in the numerical simulation.

Symbol	Definition	Value
K_0	Saturated freshwater hydraulic conductivity*	10^{-3} or 10^{-5} m/s
α	Parameter of van Genuchten (1980)	$20\ m^{-1}$
n	Parameter of van Genuchten (1980)	4
α_L	Longitudinal dispersivity	0.5 m
α_T	Transverse dispersivity	0.05 m
ε	Density-concentration relationship	6.46×10^{-4} L/g
ξ	Viscosity-concentration relationship	1.56×10^{-3} L/g
S_0	Specific storability	$0\ m^{-1}$
S_r	Residual soil saturation	0.01
θ	Porosity	0.3
τD_m	Product of tortuosity and diffusion coefficient	$10^{-9}\ m^2/s$
COVNP	The convergence criterion of pressure head in the Picard iterative scheme of MARUN code	10^{-5} m

*Hydraulic conductivity is 10^{-3} m/s for aquifers (unconfined aquifer and confined aquifer) and preferential zone, while which is 10^{-5} m/s for the aquitard.

scenarios considering different locations of the preferential zone in the aquitard (see Fig. 1).

Case 1: The preferential zone was located in inland area, i.e., between −100 m and −90 m.

Case 2: The preferential zone was located in intertidal area, i.e., between −5 m and 5 m.

Case 3: The preferential zone was located in offshore area, i.e., between 30 m and 40 m.

3 RESULTS AND DISCUSSION

Figure 2 shows the freshwater-saltwater mixing zone in simulated system using salinity contours in four cases. For base case (Fig. 2a), the shape mixing zone (e.g., 0.9 salinity contour) is consistent with Michael et al. (2005). Figure 2b shows that an inland preferential zone (case 1) caused the freshwater-saltwater mixing zone in the confined aquifer to move landward, and induced seawater upcoming around the zone. Figure 2c shows that an intertidal preferential zone (case 2) caused the freshwater-saltwater mixing zone in the confined aquifer to move close to the preferential zone. Note that a high salinity zone existed in the intertidal zone.

Compared to the base case (Fig. 2a), Figure 2d indicates that an offshore preferential zone slightly affects on salinity distribution. However, it is expected to show a considerable effect on the water exchange through the preferential zone since the velocity vectors in the zone was intensive (Fig. 2d). In addition, given an

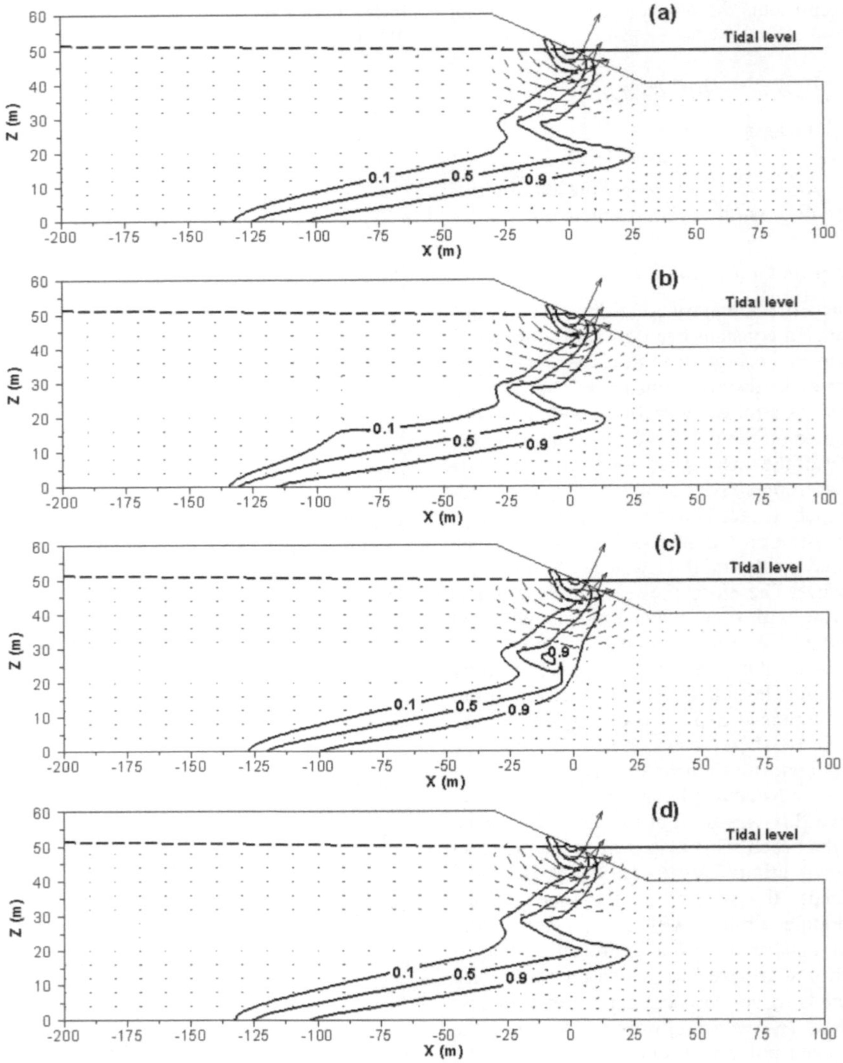

Figure 2. Salinity contours under four conditions: (a) no preferential zone (base case), (b) inland preferential zone (case 1), (c) intertidal preferential zone (case 2), and (d) offshore preferential zone (case 3). The three solid curves are contours of the pore-water/seawater salinity ratio (0.1, 0.5 and 0.9), denoting the freshwater-saltwater mixing zone.

offshore preferential zone closer to the freshwater-saltwater mixing zone, it is supposed to drive the mixing zone move seaward.

These simulation results demonstrate that the existence of a preferential zone has significant effects on the dynamics of the freshwater-saltwater mixing zone in the confined aquifer. One can regard a borehole or abandoned drinking well as a preferential zone in beaches/aquifers, for example, Shalev et al. (2009) conducted field measurements and three-dimensional numerical simulations to investigate the effects of long-screen borehole) on the depth of freshwater-saltwater mixing zone. They found that the in-well fluctuation magnitude of the mixing zone is an order of magnitude larger than that in the porous media of the actual aquifer. In addition, We expect that the preferential zone would have effects on solute and nutrients transport as well. Actually, some researches have indicated such effects of the preferential zone. For example, Atkinson et al. (1984) found that nutrients could enter shelf waters in southeastern United States continental shelf by upwelling at the shelf break. Moore et al. (2002) concluded that water exchange between the high permeability zone and the overlying ocean produced the cyclic temperature signal in Long Bay, North Carolina, and the exchange must occur through nearby breaches in the clay layer that allow water to flow through the permeable sand/shell layer. Santoro (2010) reviewed the importance of the freshwater-saltwater interface on both subsurface community composition and biogeochemical function of the microbial assemblage. He concluded that the salinization would alter microbial community composition for all functional groups involved in nitrogen cycling, and may lead to decreases in nitrification and coupled nitrification-denitrification, and increases in dissimilatory nitrate reduction to ammonium. In addition, such salinization is dependent on the central location and extent of the freshwater-saltwater mixing zone. Thus, the existence of a preferential zone must have potential effects on the salinization and microbial communities in coastal areas.

4 CONCLUSIONS

Preferential zones such as breached zones usually exist in confining layer of coastal aquifers or the continental shelf. This paper presented the effects of the preferential zone on the freshwater-saltwater mixing zone in a coastal multilayered aquifer system. The MARUN model was employed to conduct two-dimensional density-dependent numerical simulations considering the different locations of the preferential zone. The results show that the preferential zone existing in confining layer or aquitard has significant effects on the dynamics of the freshwater-saltwater mixing zone in underlying confined aquifer. Existing studies also indicated that the freshwater-saltwater mixing zone have a great potential on the solute/nutrient/contaminant transport in such a coastal environment. Future studies will be done to investigate the effects of the preferential zone on contaminant/nutrient transport in such a system.

REFERENCES

Ataie-Ashtiani, B., Volker, R.E. & Lockington, D.A. 1999. Tidal effects on sea water intrusion in unconfined aquifers. *J. Hydrol.* 216: 17–31.

Atkinson, L.P., O'Malley, P.G., Yoder, J. A. & Paffenhofer, G.A. 1984. The effect of summertime shelf break upwelling on nutrient flux in southeastern United States continental shelf waters. *J. Mar. Res.* 42: 969–993.

Boufadel, M.C. 2000. A mechanistic study of nonlinear solute transport in a groundwater-surface water system under steady state and transient hydraulic conditions. *Water Resour. Res.* 36 2549–2565.

Boufadel, M.C., Suidan, M.T. & Venosa, A.D. 1999. A numerical model for density-and-viscosity-dependent flows in two-dimensional variably-saturated porous media. *J. Contam. Hydrol.* 37: 1–20.

Brovelli, A., Mao, X. & Barry, D. 2007. Numerical modeling of tidal influence on density-dependent contaminant transport. *Water Resour. Res.* 43: W10426.

Falgàs, E., Ledo, J., Marcuello, A. & Queralt, P. 2009. Monitoring freshwater-seawater interface dynamics with audiomagnetotelluric data. *Near Surface Geophysics* 7: 391–399.

Galeati, G., Gambolati, G. & Neuman, S.P. 1992. Coupled and partially coupled Eulerian-Lagrangian model of freshwater-seawater mixing. *Water Resour. Res.* 28: 149–165.

Guo, H.P. & Jiao, J.J. 2007. Impact of coastal land reclamation on ground water level and the sea water interface. *Ground Water* 45(3): 352–367.

Kim, K.Y., Seong, H., Kim, T., Park, K.H., Woo, N.C., Park, Y.S., Koh, G.W. & Park, W.B. 2006. Tidal effects on variations of fresh-saltwater interface and groundwater flow in a multilayered coastal aquifer on a volcanic island (Jeju Island, Korea). *J. Hydrol.* 330: 525–542.

Li, H.L. & Boufadel, M.C. 2010. Long-term persistence of oil from the Exxon Valdez spill in two-layer beaches. *Nat. Geosci.* 3: 96–99.

Li, H.L., Boufadel, M.C. & Weaver, J. 2008. Tide-induced seawater-groundwater circulation in shallow beach aquifers. *J. Hydrol.* 352: 211–224.

Michael, H.A., Mulligan, A.E. & Harvey, C.F. 2005. Seasonal oscillations in water exchange between aquifers and the coastal ocean. *Nature* 436: 1145–1148.

Moore, W.S. 1999. The subterranean estuary: a reaction zone of ground water and sea water. *Mar. Chem.* 65: 111–126.

Moore, W.S., Krest, J., Taylor, G., Roggenstein, E., Joye, S. & Lee R. 2002. Thermal evidence of water exchange through a coastal aquifer: Implications for nutrient fluxes. *Geophys. Res. Lett.* 29(14): 1704.

Nielsen, P. 1990. Tidal dynamics of the water table in beaches. *Water Resour. Res.* 26: 2127–2134.

Post, V. & Abarca, E. 2010. Preface: Saltwater and freshwater interactions in coastal aquifers. *Hydrogeol. J.* 18(1): 1–4.

Robinson, C., Gibbes, B., Carey, H. & Li, L. 2007. Salt-freshwater dynamics in a subterranean estuary over a spring-neap tidal cycle. *J. Geophys. Res.* 112: C09007.

Santoro, A.E. 2010. Microbial nitrogen cycling at the saltwater-freshwater interface. *Hydrogeol. J.* 18(1): 187–202.

Shalev, E., Lazar, A., Wollman, S., Kington, S., Yechieli, Y. & Gvirtzman, H. 2009. Biased monitoring of fresh water-salt water mixing zone in coastal aquifers. *Ground Water* 47(1): 49–56.

van Genuchten, M.T. 1980. A closed-form equation for predicting the hydraulic conductivity of unsaturated soils. *Soil Sci. Soc. Am. J.* 44: 892–898.

Xia, Y.Q. & Li, H.L. 2009. The estimation of aquifer parameter using tidal effect in a coastal aquifer: A case study in Beihai Peninsula. *Earth Science Frontiers* 16(6): 276–281.

Xia, Y.Q., Li, H.L., Boufadel, M.C., Guo, Q.N. & Li, G.H. 2007. Tidal wave propagation in a coastal aquifer: Effects of leakages through its submarine outlet-capping and offshore roof. *J. Hydrol.* 337: 249–257.

Zheng, C.M. & Bennett, G.D. 2002. *Applied Contaminant Transport Modelling*. New York: John Wiley and Sons.

Zhou, X. & Wang, Y. 2009. Brief review on methods of estimation of the location of a fresh water-salt water interface with hydraulic heads or pressures in coastal zones. *Ground Water Monit. Rem.* 29(4): 77–84.

Modelling and Computation in Engineering – Zhu (ed)
© 2011 Taylor & Francis Group, London, ISBN 978-0-415-61516-7

Stability analysis of the built tunnel support affected by construction of overlapped tunnel based on numerical simulation

Jianqing Jia
School of Traffic and Transportation, Lan Zhou Jiao Tong University , Lanzhou, China; Key Lab for the Exploitation of Southwestern Resources & the Environmental Hazard Control Engineering, Chongqing, China

Hongtong Wang
Key Lab for the Exploitation of Southwestern Resources & the Environmental Hazard Control Engineering, Chongqing, China

ABSTRACT: Taking Da Ping tunnel as an example, support stability was simulated used 3D-σ finite element program. The simulation results showed that: (1) construction of the building tunnel has more influence to vertical displacement and vertical stress of the built tunnel support when the distance range from −20 m to 20 m, but the maximum values of vertical displacement and vertical stress are all small, from which we can know that the built tunnel is safety after the building tunnels have been constructed, (2) the support stress of built tunnel has been released after the building tunnel has been constructed, it maybe lead to the bulging phenomenon, but the maximum vertical displacement of the floor are all small which are separate 9.80 mm and 9.62 mm, that is to say, the bulging phenomenon will not occur in this project.

1 INTRODUCTION

We always encounter the situation of building new overlapped tunnels over or under the built tunnels with the improvement of urban traffic demand and resources exhaustion of urban underground space, and the distance between building and built tunnels is becoming smaller and smaller. So, we must face the problem how to ensure the stability of the built tunnels during construction of the building tunnels. The studying method for this question main is numerical simulation at present (CHEN 2003, FANG&HE 2007). However, these papers mainly studied surface settlement, and did not consider support stability of the built tunnel affected by the building overlapped tunnels (LIN 2006, FANG&ZHANG 2007). In this paper, taking Da Ping tunnel as an example, the support stability affected by the building tunnels is analyzed.

2 ENGINEERING BACKGROUND

2.1 *General situation of the engineering*

The engineering of Jia Hua-Jialing river is to be built to improve the traffic situation in Chongqing, and it is an important part of north-southern express artery. Da Ping tunnels are the controlling works of the express artery, and they composed of two parallel tunnels. The building tunnels I and J pass over them, and the distance is about 20 m. Tunnels distribution is shown as Fig. 1.

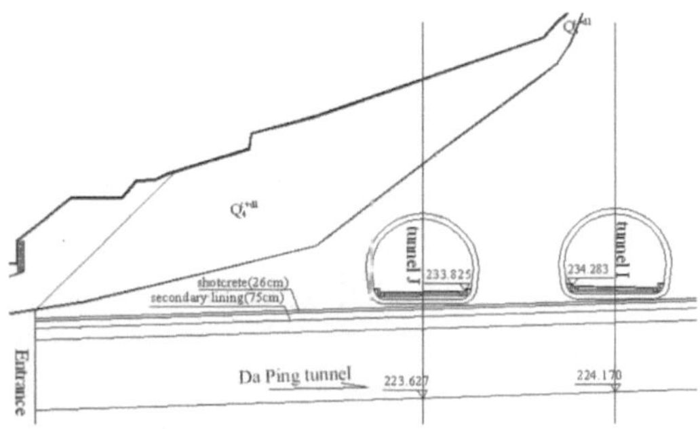

Figure 1. Tunnels distribution.

2.2 *Geology of surrounding rock mass*

The surrounding rock masses of Da Ping tunnels are Q_4^{c+dl}, J_{2s} and Q_4^{me}, and the lithology main are sandstone and sandy mudstone. The physical and mechanical parameters of surrounding rock masses and soil masses are listed in table 1 and table 2.

3 TUNNEL SUPPORTS

The primary supports of Da Ping tunnels are anchor and shotcrete, and the secondary lining is reinforced concrete. The EVA waterproofing material is constructed between primary support and the secondary lining. Supports of Da Ping tunnels are shown as Fig. 2 and Fig. 3.

Table 1. Physical and mechanical parameters of rock masses.

Lithology	Compressive strength / MPa	Elastic modulus / MPa	Passion ration	Friction angle / (o)	Internal cohesion / kPa
Andy mudstone	13.5	2127	0.37	29	570.0
Sandstone	37.0	5670	0.10	35	1000

Table 2. Physical and mechanical parameters of soil masses.

Soil features	Gravity / $KN.M^{-3}$	Internal cohesion /kPa	Friction angle /(°)
Earth-fill	20.0		
Alluvial loam	19.7	27.2	12.4
Alluvial gravel	20.0		

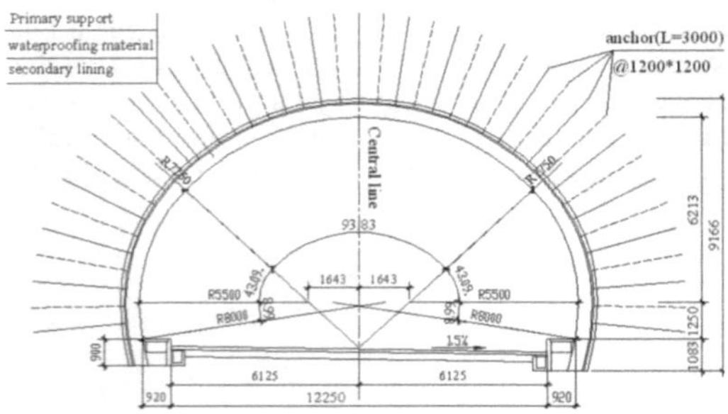

Figure 2. Support of Da Ping tunnel.

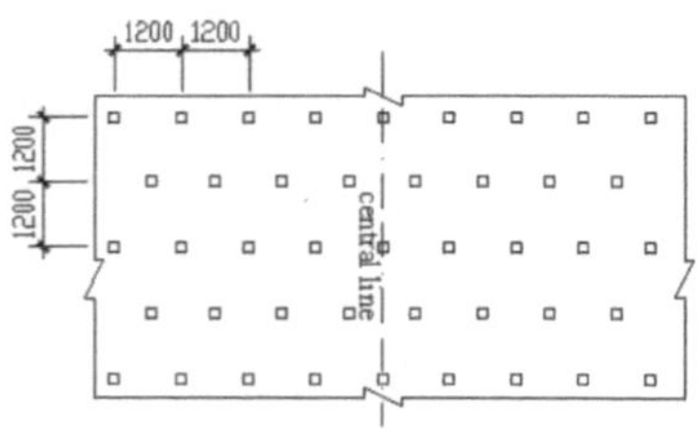

Figure 3. Arrangement of anchors.

4 NUMERICAL SIMULATION OF SUPPORT STABILITY OF DA PING TUNNELS

4.1 *Boundary conditions of numerical model*

The 3D-σfinite element program is used to simulate the stability of tunnel supports. The model length is 80 m, width is 65 m, and surface height is considered during building numerical model. The numerical simulation model is shown as Fig. 4, and tunnel supports are shown as Fig. 5 (In Fig. 5, the tunnel is numbered from 1 to 4 for analysis is easy, and 1 and 2 represent Da ping tunnels, and 3 and 4 represent the building tunnels I and J).

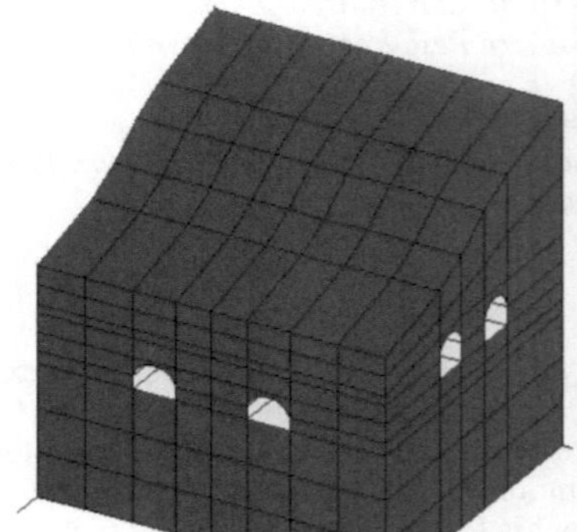
Figure 4. The numerical simulation model.

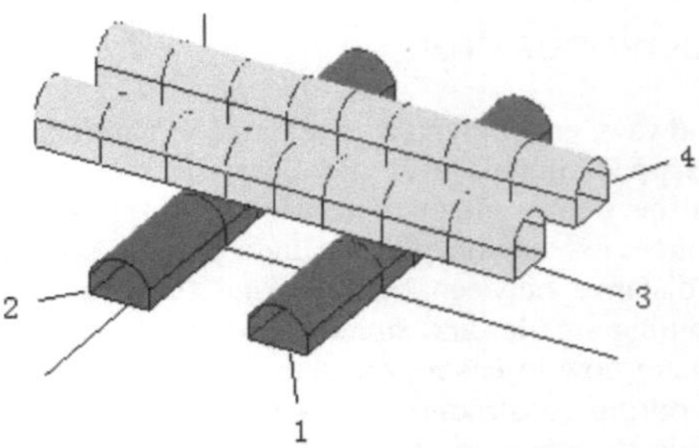

Figure 5. Tunnel support of numerical simulation model.

In the numerical model, the bolt element is used to simulate anchor, its elastic modulus E equal 210 GPa, and diameter equal 25 cm. The solid element is used to simulate primary support. The shell element is used to simulate secondary lining, and its elastic modulus is 25 GPa, thickness is 0.45 m, and Passion ration is 0.17.

The D-P criterion[1] was applied, and it will be

$$f = \alpha I_1 + \sqrt{J_2} - K = 0 \tag{1}$$

where $I_1 = \sigma_{ii} = \sigma_1 + \sigma_2 + \sigma_3 = \sigma_x + \sigma_y + \sigma_z$, it is the first stress invariant, J_2 is the second stress invariant

$$J_2 = \frac{1}{6}[(\sigma_1 - \sigma_2)^2 + (\sigma_2 - \sigma_3)^2 + (\sigma_3 - \sigma_1)^2]$$
$$= \frac{1}{6}[(\sigma_x - \sigma_y)^2 + (\sigma_y - \sigma_z)^2 + (\sigma_z - \sigma_x)^2] + 6(\tau_{xy}^2 + \tau_{yz}^2 + \tau_{zx}^2)]$$

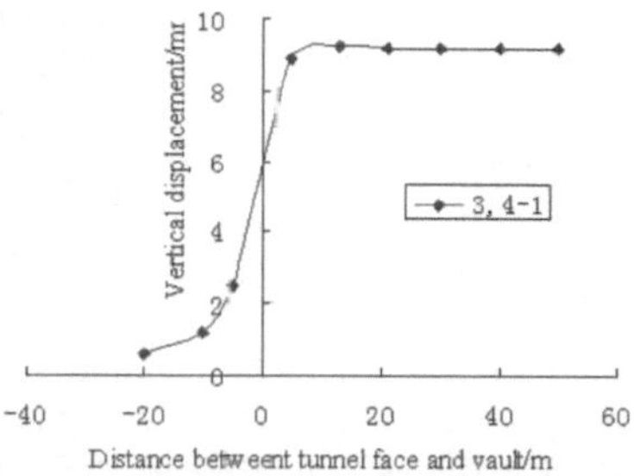

(a) Support vertical displacement of tunnel 1

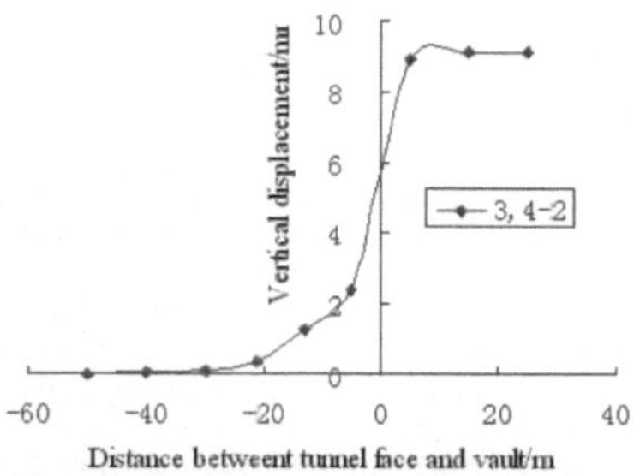

(b) Support vertical displacement of tunnel 2

Figure 6. Vertical displacement of the built tunnel support.

α and K are the experiment constants, they are related to the internal friction angle φ and the cohesive strength C. α and K will be expressed as follows:

$$\alpha = \frac{2\sin\varphi}{\sqrt{3}(3-\sin\varphi)} \tag{2}$$

$$K = \frac{6C\cos\varphi}{\sqrt{3}(3-\sin\varphi)} \tag{3}$$

4.2 *Numerical simulation results*

4.2.1 *Vertical displacement of tunnel support*

Vertical displacements of supports of tunnel 1 and 2 caused by construction of the building tunnel 3 and 4 are shown as Fig. 6 and Fig. 7. (In the figures, 3, 4-1 means that support vertical displacement of tunnel 1 caused by the building tunnel 3 and 4, and the other signs have the same meaning).

Fig. 6 shows that: (1) construction of building tunnels has little influence to the built tunnels when the distance between the tunnel face and the vault of built tunnel is larger than 20 m, after that, when the distance between the tunnel face and the vault of built tunnel is becoming small, the support vertical displacement increase very quickly, and the support vertical displacement is convergent when the tunnel face pass the vault of built tunnels which distance is about 15 m. (2) the support vertical displacements of tunnel 1 and 2 are small which are separate 9.23 mm and 9.15 mm, and the tunnel supports are stable.

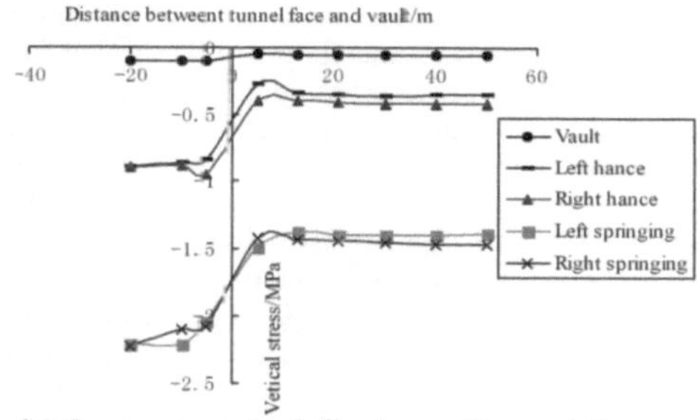

(a) Support vertical displace of tunnel 1

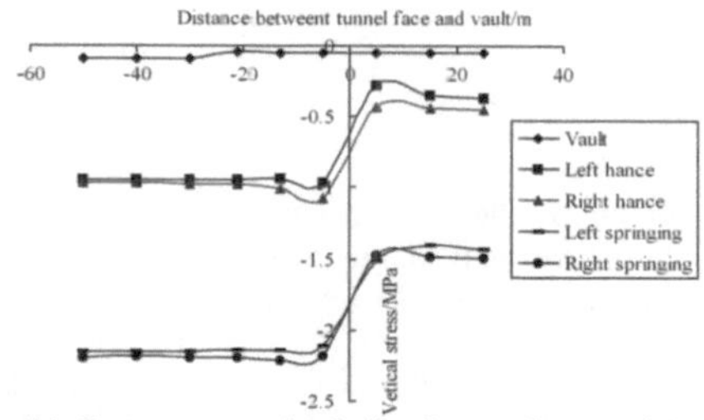

(b) Support vertical displace of tunnel 2

Figure 7. Vertical stress of the built tunnel support.

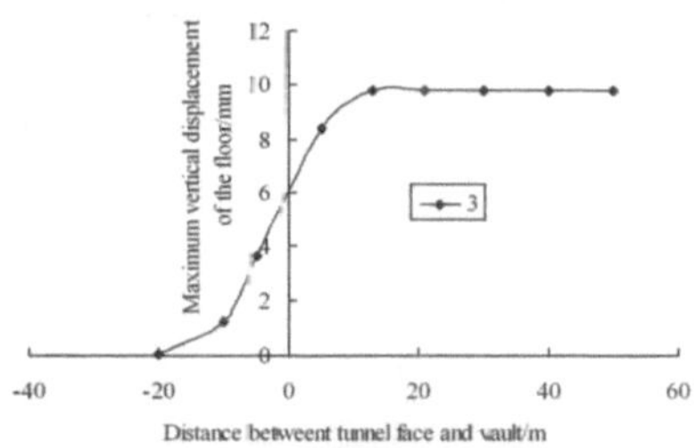

(a) Maximum vertical displacement of the floor of tunnel 3

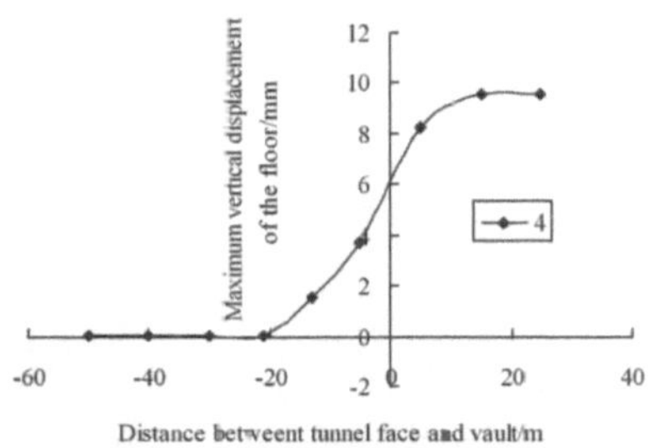

(b) Maximum vertical displacement of the floor of tunnel 4

Figure 8. Maximum vertical displacement of the building tunnel floor.

4.2.2 *Vertical stress of tunnel support*

Fig. 7 shows that: (1) the vault vertical stress is the largest one which is larger than hance and springing vertical stresses, (2) construction of building tunnels has little influence to the built tunnels when the distance between the tunnel face and the vault of built tunnel is larger than 20 m (from Fig. 7 we can know that the vertical stress has not change or is convergent), and when the distance between the tunnel face and the vault of built tunnel range from −5 m to 5 m, construction of building tunnels lead to the support vertical stress decrease quickly, that is to say, the support stress of built tunnel has been released after the building tunnel has been constructed, it maybe lead to the bulging phenomenon, but from Fig.8 we can know the maximum vertical displacement of the floor of the building tunnels are all small which are separate 9.80 mm and 9.62 mm, that is to say, the bulging phenomenon will not occur in this project, and the design plan of tunnel 3 and 4 which have no invert is reasonable.

5 CONCLUSIONS

(1) Construction of the building tunnels have some influence to vertical displacement and vertical stress of the built tunnel support when the distance range −20 m to 20 m, but the maximum values of vertical displacement and vertical stress are all small, from which we can know that the built tunnels are safety after the building tunnels have been constructed.

(2) The support stress of built tunnel has been released after the building tunnel has been constructed, it maybe lead to the bulging phenomenon, but the maximum vertical displacement of the floor are all small which are separate 9.80 mm and 9.62 mm, that is to say, the bulging phenomenon will not occur in this project.

ACKNOWLEDGMENT

Project (QL-08-16A) is supported by 'Qing Lan' Talent Engineering Funds by Lanzhou Jiaotong University; Project 50774106 is supported by National Natural Science Foundation of China; Project 2005CB221502 is supported by 973 Program.

REFERENCES

CHEN Xian-guo, WANG Xian-jun. 2003. 2D and 3D Finite Element Analyses of Overlapping Tunnels. Journal of southwest Jiaotong University, 38(6):379–385.

FANG Qian, ZHANG Dingli. 2007. Study on the Shallow Tunnel Construction Methods for a New Built Subway Station Passing through the Bottom of the Existing Subway. China Railway Science, 28(5):71–76.

FANG Yong, HE Chuan. 2007. Numerical analysis of effects of parallel shield tunneling on existent tunnel. Rock and Soil Mechanics, 28(7):1402–1406.

LIN Zhi, ZHU He-hua. 2006. Study of field monitoring on interaction between twin slurry shield tunnels in close space. Rock and Soil Mechanics, 27(7):1181–1186.

Modelling and Computation in Engineering – Zhu (ed)
© 2011 Taylor & Francis Group, London, ISBN 978-0-415-61516-7

Study on simulation test of heat transfer in unsaturated compacted red clay

Y.ZH. Tan
Three Gorges University Key Laboratory of Geological Hazards on Three Gorges Reservoir Area Ministry of Education, Yichang, China

L.W. Kong & AI.G. Guo
State Key Laboratory of Geomechanics and Geotechnical Engineering, Institute of Rock and Soil Mechanic, Chinese Academy of Sciences, Wuhan, China

ABSTRACT: A heat transfer test is performed for compacted test samples with three different initial moisture content values and three dry density values, with a temperature-driven water migration test instrument designed and produced by us, to simulate heat transfer in red clay subgrade soil in atmospheric environment. The test result indicates that the heat transfer process and final state are controlled by the initial dry density and initial moisture content of the test samples; the more compact the soil is and the higher the initial moisture content is, the shorter the time required to reach relative temperature balance will be; the temperature effect drives water migration, which, in return, affects temperature fluctuation; that is to say, temperature fluctuation and moisture content are always in an interactive state.

1 INTRODUCTION

The interaction between heat and soil is very common in geotechnical engineering, such as subterraneous storage of highly radioactive nuclear waste, directly buried high voltage cable engineering, subterraneous storage and transmission of thermal power, and road subgrade engineering, etc. The effect of atmospheric temperature on subgrade soil is not exhibited by thermal expansion/shrinkage phenomenon simply; instead, it is indirectly reflected in the form of moisture content change under temperature effect. The thermal expansion/shrinkage phenomenon caused by temperature fluctuation is not obvious. By now, a great deal of studies on the effect of temperature fluctuation on soil has been carried out, and rich payoffs have been obtained. Rose (1968) studied the pattern of moisture transfer under the action of temperature gradient, suction gradient, and gravity. Preece & Hitchcock (1970) simulated one-dimensional diffusion of heat and moisture in soil by test. Baladi & Ayers (1981) simulated transient moisture and heat transfer in infinite porous media by heating with a point source. Mitchell et al. (1981) simulated the field situation of directly buried high voltage cables by outdoor cable backfill test. Ewen & Thomas (1989) designed an apparatus composed of a horizontal glass tube and a heating rod mounted in the glass tube to simulate the process of temperature effect. It is seen that above tests employed heating rods to heat the soil directly, which may cause vaporization of moisture content in the contact part between heating rod and soil, especially under high temperature condition. Utilizing appropriate test means to study the heat transfer pattern in soil is of far reaching significance to further reveal the mechanism of action of moisture content on subgrade soil.

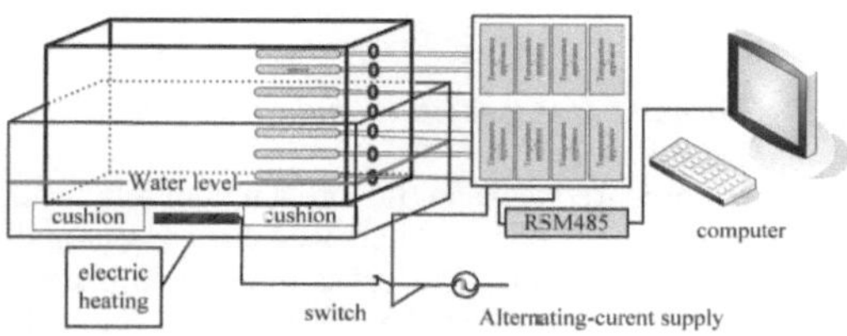

Figure 1. Schematic diagram of test instrument

2 PREPARATION FOR TEST

2.1 *Testing instrument*

The test instrument comprises heating and thermostatic control system, test sample forming and heat insulation system, and test data acquisition system. The working principle of each part is described as follows (Tan 2009). A general view of the test instrument is shown in Figure. 1.

2.2 *Preparation for Test*

2.2.1 *Basic properties of test samples*

The test samples are taken from Pile K27 on an expressway in Chenzhou, Hunan. The basic physical properties of the test samples are shown in Table 1.

Table 1. Physical property indexes of laterite soil.

Natural water Content%	Natural density g/cm^3	Specific gravity	Liquid limit%	Plastic limit%	Plasticity index	Free swelling%
30.9	1.80	2.65	61.0	35.0	26.0	28.5

2.2.2 *Preparation of test samples*

Choose reshaped and air-dried soil sample and sieve it using a sieve with mesh size = 5 mm, and calculate the mass of moisture to be added against the required moisture content. Then, spray water evenly on the soil sample with care with a spray can. Agitate the soil sample repeatedly in the preparation process, to make the moisture distributed evenly in the soil sample. Finally, encapsulate the prepared soil sample in an air-tight plastic bag and hold for 6 ~ 7 days; choose soil sample at 5 different positions in the bag and measure the moisture content in the 5 test samples in parallel with drying method. If the difference between moisture content values of the test samples taken from different positions is within the range of 0.3%, it indicate the holding time is enough; otherwise the soil sample must be taken out from the bag, agitated again, and then held, till the difference of moisture content meets above criterion.

2.2.3 *Forming of test samples and arrangement of sensors*

Load each prepared test sample into a rectangular box with inside dimensions as 41 cm×21 cm×31 cm and compact the test sample. The test sample is compacted in 7 layers, by controlling the dry density, and each compacted layer is in 4cm height. To ensure homogeneous density of the compacted test sample, the following measures are taken: (1) scrape the test sample to and fro with a ruler to achieve preliminary uniform distribution after the test sample is poured into the box; (2) utilize a thick steel plate (in 2.5 cm thickness) with area equal to the net area of the box, to ensure the force applied by the lifting jack on the dowel bar is evenly applied on the test sample; (3) place a small iron column (in 2 cm diameter and 4 cm height) at each corner in the box before each layer is compacted; when the steel plate is pressed to contact with the small iron columns, it indicates the planned cubic capacity is reached. That approach can effectively prevent inhomogeneous density of the test sample caused by eccentric compression. The surface of each layer is roughened after compaction, to ensure tight contact between layers.

After a layer of test sample is roughened and before the next layer of loose soil is poured into the box, a temperature sensor is pushed through a channel reserved in the box and fixed between the two layers of soil. Pay attention to prevent the sensor from displacement and protect the external shielding wire of the sensor in the compacting process, and verify the sensor performance is normal before the sensor is mounted.

2.3 *Test design*

(1) Tests for test samples with three different dry density values and three different initial moisture content values are performed. Dry density of test sample: 1.42 g/cm^3, 1.47 g/cm^3, 1.52 g/cm^3; corresponding degree of compaction: 90%, 93%, 96%; initial moisture content: 24.6%, 27.7%, 30.9 %.
(2) The test sample is heated to the set value of boundary temperature (48°C) in one operation, by conductive heating with water bath under the test sample box.
(3) The ambient temperature in the test room is controlled at 23 ~ 25°C.

3 TEST RESULT AND ANALYSIS

3.1 *Pattern of temperature fluctuation*

The patterns of internal temperature evolution in test samples at different degrees of compactness under the action of external constant temperature are shown in Figures 2 ~ 4.

3.1.1 *Initial moisture content (w = 24.6%)*

It is seen from Figure. 2(a) ~ (c), in the test sample with that moisture content, the time required to achieve temperature balance at the observing point (4 cm) at cold end is about 40 h; the time required to achieve temperature balance at the observing point (28 cm) at hot end is about 28 h. The times required to achieve temperature balance at the observing points from hot end to cold end increase incrementally. The temperature gradient between different points is obvious; however, the higher the degree of compactness is, the smaller the temperature gradient will be, because the higher the degree of compactness is, the lower the porosity in the test sample will be, and the heat-conducting property of soil grains is better than that of air.

In the initial stage of test (0 ~ 5d), the temperature is in a stable stage. As the migration of moisture increases, the effect on temperature distribution becomes more and more significant. It is seen from Figure. 2(d) ~ 2(f): the temperature decreases at different degrees in the time period (5 ~ 30d), and the temperature decrease in soil closer to the hot end is higher. That is a result of increased thermal resistance caused by reduction of moisture in the soil near the hot end through water migration. Though the moisture content increases and the thermal resistance decreases in the soil near the cold end, the temperature in the

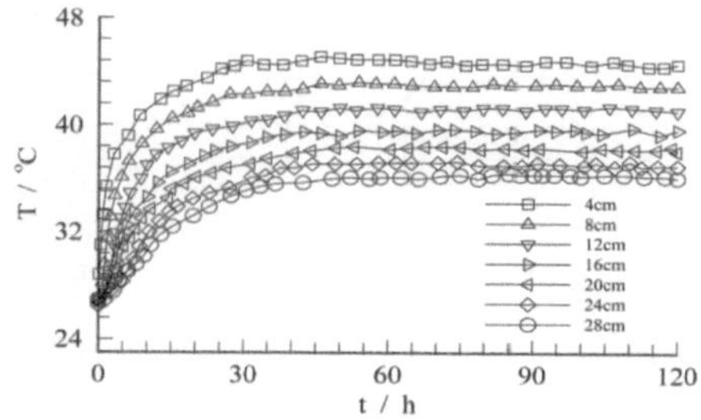

(a) Temperature Distribution in 120h (K=90 %)

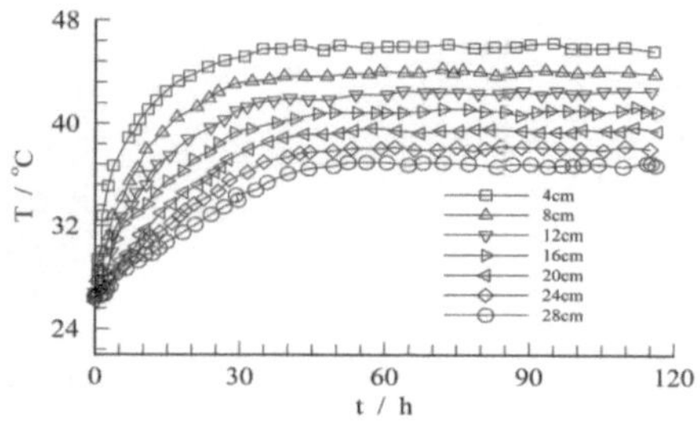

(b) Temperature Distribution in 120h (K=93 %)

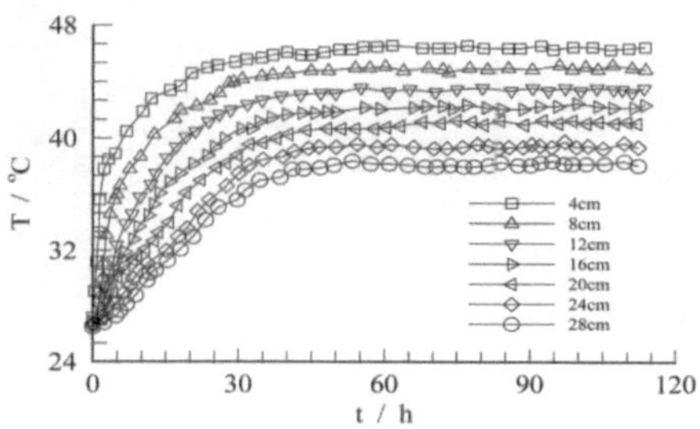

(c) Temperature Distribution in 120h (K=96 %)

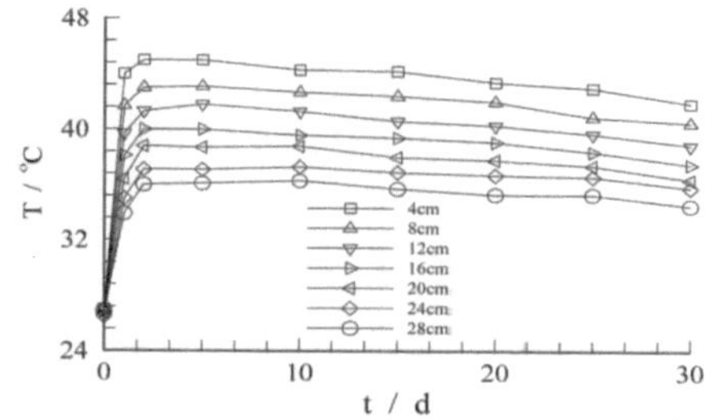

(d) Temperature Distribution in 30d (K=90 %)

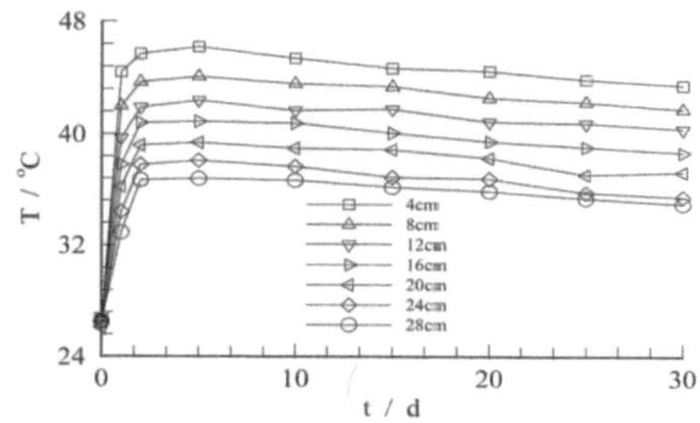

(e) Temperature Distribution in 30d (K=93 %)

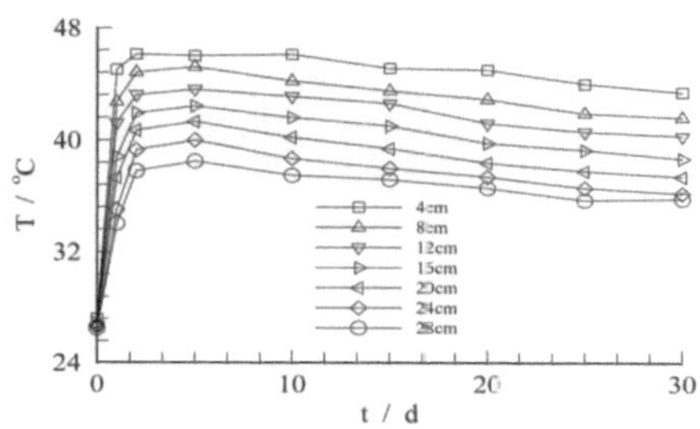

(f) Temperature Distribution in 30d (K=96 %)

Figure 2. Initial moisture content (w = 24.6%).

soil near the cold end is lower; therefore, the temperature gradient also increases in that soil, but is not as significant as the former.

3.1.2 *Initial moisture content (w = 27.7%)*

It is seen from Figure. 3(a) ~ (c): the time required to achieve temperature balance in test samples with that moisture content and different dry density values is about 30h. Compared to the case of test samples with initial moisture content (24.6 %), the time required to achieve temperature balance is less by 10h, but the time difference between test samples with different dry density values is not so obvious.

3.1.3 *Initial moisture content (w = 30.9%)*

As shown in Figure. 4, in test samples with that moisture content, the time required to achieve temperature balance is about 20 h. The temperature change with time is very stable in the entire test process, because no significant water migration occurs. The times required to achieve temperature balance in test samples with the same moisture content at different degrees of compactness are almost the same. However, the difference in time required to achieve temperature balance between test samples with different moisture content is obvious, because the heat conductivity of moisture is much higher than the heat conductivity of soil grains and the heat conductivity of air; therefore, the higher the moisture content is, the shorter the time required to achieve temperature balance will be.

3.2 *Analysis of heat transfer mechanism*

The main heat transfer approaches in soil are shown in Figure. 5. There are mainly three approaches: (1) heat conduction: at a temperature gradient, soil grains, liquid water, vapor, and air transfer heat by heat conduction; (2) heat convection: liquid water and vapor exchange heat with soil grains by heat convection as they flow through soil pores; (3) heat exchange during phase change: under the cyclic action of atmospheric temperature, evaporation and condensation

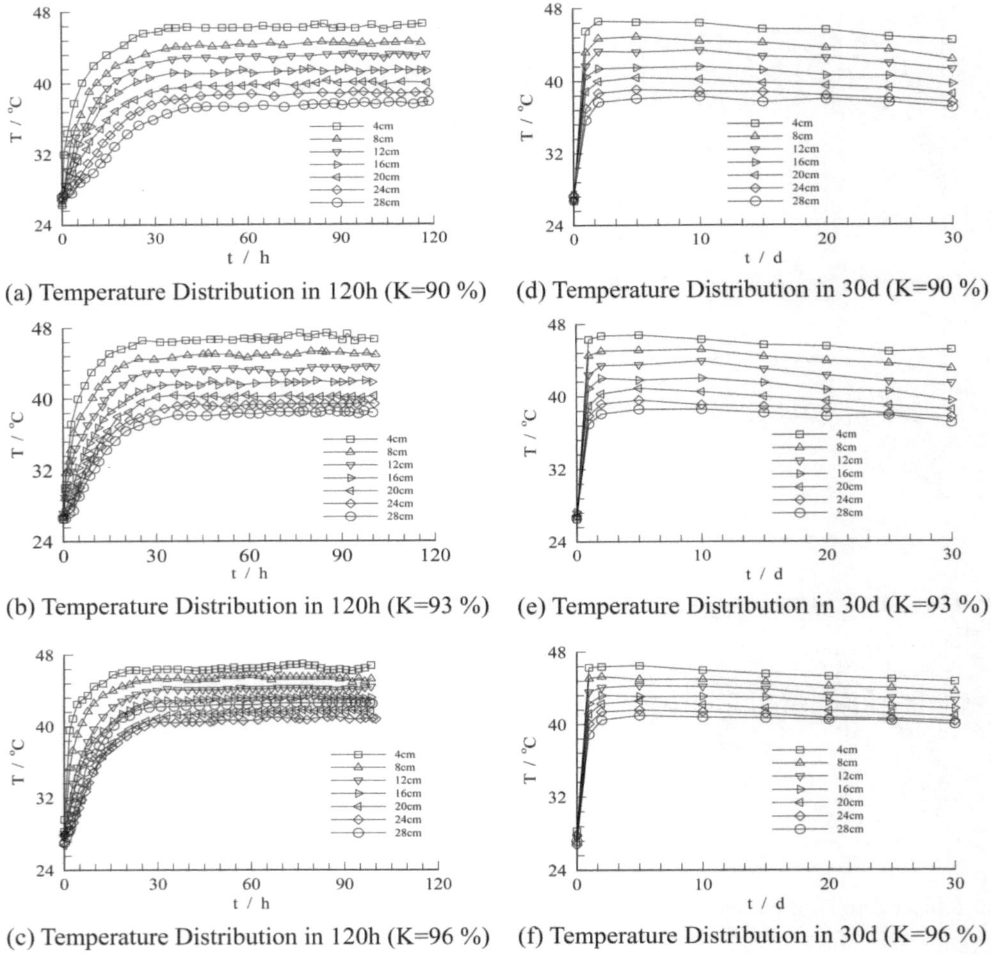

(a) Temperature Distribution in 120h (K=90 %) (d) Temperature Distribution in 30d (K=90 %)

(b) Temperature Distribution in 120h (K=93 %) (e) Temperature Distribution in 30d (K=93 %)

(c) Temperature Distribution in 120h (K=96 %) (f) Temperature Distribution in 30d (K=96 %)

Figure 3. Initial Moisture Content (w = 27.7%).

phenomena occur in soil and on the surface of soil, causing heat exchange during phase change.

The heat transfer rate in above approaches is related with factors such as mineral components, pore size, and moisture content of soil. In soil with low moisture content, the heat is mainly transferred by heat conduction through solid grains; as the moisture content in soil increases so that moisture pervades in small pores or water meniscus is formed among soil grains, the contact area between solid grains is virtually increased, and therefore provides wider heat transfer channels. When the moisture content is close to the saturated value, the heat transfer is faster in soil with higher moisture content at the same porosity, due to the fact that the heat-conducting performance of liquid water is higher than that of air. Heat convection caused by temperature gradient is only obvious in soil with large-size solid soil grains(Farouki 1981). Heat transfer caused by vapor convection or diffusion is negligible due to the low volumetric heat capacity. In addition, the heat conductivity of water vapor is much higher than that of pure air (De Vries 1958). Therefore, the heat transfer in unsaturated soil is mainly accomplished by heat conduction in the grain skeleton of soil and moisture evaporation and condensation.

It can be seen from the distribution pattern of moisture and temperature in the soil samples: there is close relationship between temperature fluctuation and moisture content variation, which is to say, temperature fluctuation causes moisture content variation, which, in turn, drives temperature redistribution in the soil; moisture migration and heat transfer in soil are closely coupled to each other.

4 CONCLUSION

Under the action of external constant temperature, the heat transfer process and temperature distribution pattern in enclosed soil samples are controlled by initial

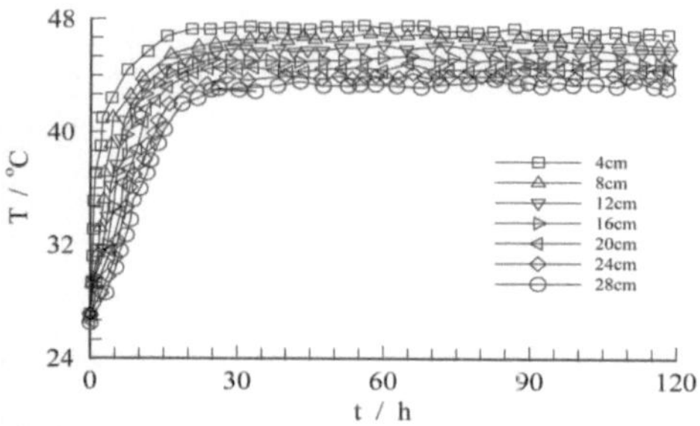

(a) Temperature Distribution in 120h (K=90 %)

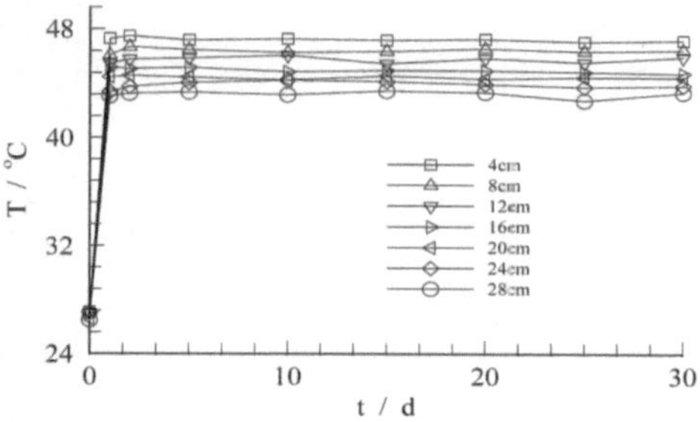

(b) Temperature Distribution in 30d (K=90 %)

Figure 4. Initial moisture content (w = 30.9%).

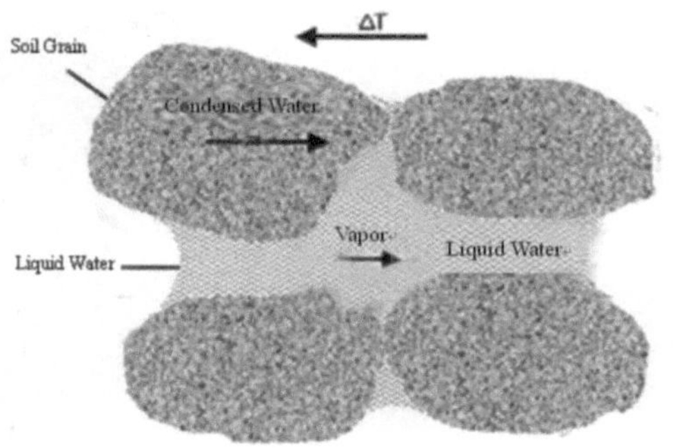

Figure 5. Schematic diagram of main heat transfer approaches in soil (Deru 2003).

dry density of soil and initial moisture content in soil. The times required to achieve temperature balance in test samples with the same moisture content at different degrees of compactness are almost the same. However, the difference in time required to achieve temperature balance between test samples with different moisture content is obvious, mainly because the heat conductivity of moisture is much higher than the heat conductivity of soil grains and the heat conductivity of air; therefore, the higher the moisture content is, the shorter the time required to achieve temperature balance will be.

ACKNOWLEDGEMENTS

We owe our heartfelt thanks to the critical project supported by National Natural Science Foundation of China in supporting the "research of disaster-causing mechanism of landslide, risk assessment and disaster reduction induced by rainstorm" (50839004)

REFERENCES

Baladi, J.Y. 1981. Transient heat and mass transfer in soil. *International Journal Heat Mass Transfer*, 24: 449–458.

De Vries, D. A. 1958.Simultaneous transfer of Heat and Moisture in Porous Media. *Transactions American Geophysical Union*, 39: 909–916.

Ewen, J., Thomas, H.R. 1989. Heating unsaturated medium sand. *Geotechnique*, 455–470.

Farouki, O.T. 1981.*Thermal Properties of Soils.* Hanover: Transaction Technology Publications.

Mitchell, J.K. 1981.Field testing of cable backfill system. *Underground cable thermal backfill*, 19–33.

Preece, R.J., Blowers, R.M. 1970. A numerical method for evaluating coupled heat and moisture difusion through porous media with varying physical properties. *Proceeding 1st International Conference on Numerical method in Thermal Problems*, 527–538.

Rose, C.W. 1968. Water transport in soil with daily temperature wave.I-Theory and experiment. *Australia Journal Soil Research,* 6: 31–44.

Rose, C.W. 1968. Water transport in soil with daily temperature wave.II - Analysis. *Australia Journal Soil Research,* 6: 45–57.

Tan, Y.ZH. 2009. *Study on engineering characteristics and moisture-heat coupling effect of compacted laterite soil.* Wuhan: Institute of Rock & Soil Mechanics, Chinese Academy of Sciences.

Modelling and Computation in Engineering – Zhu (ed)
© 2011 Taylor & Francis Group, London, ISBN 978-0-415-61516-7

Numerical simulation of free-space explosion based on LS-DYNA

Liping Liu, Yuanwen Yao & Yingmin Li
Civil Engineering College of Chongqing University, Chongqing, P.R. China

Kui Xia
Architectural design and research institute of Chongqing University, Chongqing, P.R. China

ABSTRACT: Numerical simulation is an important method for explosion studies. As the development of computer applied technology, explosion numerical simulation comes to be possible. In this paper, free space explosion has been implemented by LS-DYNA and the characteristics of blast incident wave pressure in free space explosion have been analyzed. The results show that: The LS-DYNA can effectively simulate the free space explosion, but it should be noted that reasonable parameter value such as element mesh, time step, and boundary simulation is the key to ensure the effectiveness of the simulation. Since the numerical simulation of a small scaled distance explosion may result in a relatively large discrete, a small scaled distance explosion should be cautiously analyzed by applying the method of numerical simulation. The time history of blast incident wave pressure in free space is shaped like pulse. The smaller the scaled distance is, the narrower the pulse is, the more obvious the double-peak effect of the time history is. The peak pressure of blast incident pressure time history in free space decreases with the increasing of the scaled distance.

1 INTRODUCTION

Explosion is a kind of chemical reaction process of high-speed, automatic transmission, with large amounts of heat released and gas produced at the same time. The explosion may cause damage such as vibration, combustion, and pressure blast wave, etc., and nuclear radiation and electromagnetic pulse may even exist during the process of nuclear weapon explosion. General, the main concern is the damage of the explosive blast wave. Although experiment is the best way to study the process and effect of explosion, there are many practical difficulties in such intensive research together with various uncertain factors because completing an explosion within such an extremely short period which is devastating per se may limit the experimental observation and measurement means; whereas, numerical simulation approach can be affected less by the environment, and it is easy to change the simulation conditions and parameters, which plays an important role in regularity study of performance and explosive experimental design. Numerical simulation has become an important means of explosion study.

Explosion is a type of highly nonlinear problem and has high requirement for computer performance and numerical simulation techniques. Therefore, the highly developed modern computer technology makes numerical simulation of explosion possible. Dynamic nonlinear finite element software, LS-DYNA, applying in two-dimensional and three-dimensional nonlinear structural high-speed collisions, explosions, metal forming and other nonlinear dynamic impact problems, is taken as a important tool for the explosive numerical simulation (Shi et al. 2005).

The numerical simulation of free-space explosion is achieved applying LS-DYNA in the paper. LS-DYNA explosion simulation model is proposed and the key technical issues in the numerical simulation of explosion are analyzed. The characteristics of blast incident wave pressure in free space have been concluded.

2 METHOD OF THE NUMERICAL SIMULATION OF EXPLOSION IN FREE SPACE BASED ON LS-DYNA

2.1 *Constitutive relationship of air and explosive*

The material constitutive relationship is the basis of numerical simulation. And in numerical simulation of free-space explosion, air and explosive materials are the two involving types. In the numerical simulation, the material model of air is assumed to be ideal gas. The relationship between Pressure p and Energy E0 is determined by the following equation.

$$P = (\gamma - 1)\rho E_0 \tag{1}$$

Where, P is air pressure, γ the gas adiabatic index, ρ the air density, E_0 the gas internal energy. In general, $\gamma = 1.4$, $\rho = 1.3 \times 10^{-3}$ g/cm^3, $E_0 = 2.5 \times 10^{-6}$ (10^5 MPa). (Ansys 1998a).

Material model of explosives is assumed to be JWL state equation, which can be used to calculate

Table 1. State equation parameters value of TNT explosive.

Sign	Parameters	Value
ρ	Density(g/cm^3)	1.6
D	Detonation Velocity (cm/μs)	0.70
PCJ	C-J Explosive Pressure (10^{5}MPa)	0.255
A	Material Constant (10^{5}MPa)	5.4094
B	Material Constant (10^{5}MPa)	0.093726
R1	Material Constant	4.5
R2	Material Constant	1.1
ω	Material Constant	0.35
E0	Initial Inner Energy(10^{5}MPa)	0.08
V0	Initial Relative Volume	1.0

the pressure converted by the explosive energy. And the relation between pressure and energy is (Ansys 1998b):

$$p = A(1-\frac{\omega}{R_1 V})e^{-R_1 V} + B(1-\frac{\omega}{R_2 V})e^{-R_2 V} + \frac{\omega}{V}E \quad (2)$$

Where, P is air pressure, V the relative volume of the explosive E the inner energy of the explosive, and A, B, R_1, R_2, ω are the material constants of explosives.

The release of explosive energy is controlled by combustion factor in LS-DYNA. Ignition time of each explosive unit is determined by the detonation velocity and the distance from the explosive centroid to the initiation point. The state equation parameters of TNT explosive used in the calculation of explosive loading are shown in Table 1 (Chen 2004).

2.2 *Explicit solution of nonlinear dynamic differential equation*

The most widely used central difference method of Explicit Solution of nonlinear dynamic differential equation and its motion equation is

$$\begin{cases} [M]\{\ddot{u}\}+[C]\{\dot{u}\}+[K]\{u_t\}=\{F_t\} \\ \{\ddot{u}\}=\frac{1}{\Delta t^2}(\{u_{t-\Delta t}\}-2\{u_t\}+\{u_{t+\Delta t}\}) \\ \{\dot{u}\}=\frac{1}{2\Delta t}(-\{u_{t-\Delta t}\}+\{u_{t+\Delta t}\}) \end{cases} \quad (3)$$

Where, [M] is the mass matrix; [C] the damping matrix; [K] the stiffness matrix; $\{u_t\}$ the nodal displacement vector; $\{\dot{u}\}$ the nodal velocity vector; $\{\ddot{u}\}$ the nodal acceleration vector; $\{F_t\}$ the nodal external load vector; Δt the time step. It can be deprived as follows from the formular above:

$$\begin{cases} ([M]\frac{1}{\Delta t^2}+\frac{1}{2\Delta t}[C])\{u_{t+\Delta t}\}=\{Q_t\} \\ \{Q_t\}=\{F_t\}-([K]-\frac{2}{\Delta t^2}[M])\{u_t\}-(\frac{2}{\Delta t^2}[M]\{u_t\}-\frac{1}{2\Delta t}[C])\{u_{t-\Delta t}\} \end{cases} \quad (4)$$

From the above it can be inferred that the recurrence formula is derived from the motion equations of time t. Then stiffness matrix will not appear in the left side of recurrence formula and the equations are non-coupled. Therefore, the mass matrix can be directly solved. When the mass matrix is diagonal and the damping can be ignored or damping matrix can be considered to be diagonal matrix, there is no need to solve any equations in time integration of motion equations and the integral process is simple and fast operated. Explicit integration method is conditional stability, the integration will be unstable when the integration step is larger than the critical value.

2.3 *Key issues of the numerical simulation*

2.3.1 *Integration time step*

Using central difference method to solve the ordinary differential equation is conditional stable and chronotropic time step increment solution is used in LS-DYNA. First, the maximum time step allowed by the stability condition in each element, namely the limit time step Δt_{ei}, should be calculated, and then the time step Δt should be taken as the minimum value of the limit time step of all the elements, that is, $\Delta t = \min\{\Delta t_1, \Delta t_2, \ldots\ldots \Delta t_N\}$, in which N is the number of elements. Limit time step Δt_{ei} for different element type is calculated by different method. The value of integration time step control parameters and artificial volume viscosity parameters, C0 and C1, are general taken as 1.5 and 0.6 in LS-DYNA.

2.3.2 *Artificial volume viscosity*

High-speed collision results in skip of pressure, density, acceleration and energy, which make it difficult to solve the dynamics differential equations. In LS-DYNA, artificial volume viscosity parameter, q, is used to consider this effect, which is controlled by keyword * CONTROL BULK VISCOSITY.

2.3.3 *Single-point Gaussian integration and hourglass control*

Single-point Gaussian integration reduce requirement of data operation and memory capacitance, but it may cause hourglass mode. In LS-DYNA, hourglass viscous damping method is used by keywords *HOURGLASS, *CONTROL_HOURGLASS.

2.3.4 *Non-reflecting boundary*

Non-reflecting boundary, also known as transmitting boundary or silent boundary, are boundary conditions mainly used in infinite or semi-infinite body to reduce the size of the calculated object. Non-reflecting boundary is accomplished to load additional viscous normal stress $\sigma = -\rho V_p V_{norm}$ and shear stress $\tau = -\rho V_s V_{\tan}$ in all boundary elements, where ρ is mass density, V_p is incident longitudinal wave velocity, V_{norm} is normal vector of V_p in the boundary, V_s is incident shear wave velocity, $V_{\tan}$ is tangential vector of V_s in the boundary.

2.4 *Numerical simulation models of free-space explosion*

Numerical simulation models of free-space explosion are shown in Figure 1. Figure 1a is to use the 1/8

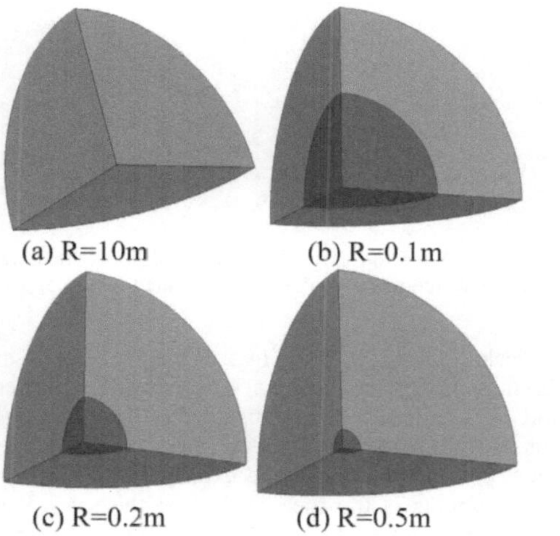

Figure 1. Numerical simulation models of free-space explosion.

sphere model with radius of 10 m (scaled distance $Z = 10\,m/kg^{1/3}$) to simulate free-space explosion. The part with the radius of 0-053 m (red or dark) are for the explosive (TNT equivalent is 1/8 kg, with a radius of 0.053 m), while the part with the radius of 0.053 m–10 cm (blue or light) are for the air. Explosive element is meshed into 0.005m, the air element with radius of 0.053–1 m is 0.01 m, the air element with radius of 1–10 m is 0.03 m, and there are a total of 98,901 elements.

In small scaled explosion condition, blast wave parameters in different points of same radius vary widely, significantly affected by element size. 3 models (R = 0.1 m, R = 0.2 m, R = 0.5 m) shown in Figure 1b, Figure 1c, Figure 1d are established to simulate small-scaled distance ($Z = 0.1\,m/kg^{1/3}$, $Z = 0.2\,m/kg^{1/3}$, $Z = 0.5\,m/kg^{1/3}$) explosions, where the two models (R = 0.1 m, R = 0.2 m) use the element size of 0.002 m, 0.005 m and 0.01 m; model (R = 0.5 m) uses the element size of 0.005 m and 0.01 m.

Non-reflection boundary is adopted in model boundaries to reflect the infinite medium. EULER material is used in air and explosives, both of which use 8-node 6-side solid element. The explosives centre is set for the initiation point to initiate at 0 time.

3 BLAST INCIDENT WAVE PRESSURE IN FREE-SPACE EXPLOSION

The above models are used to simulate the free-space explosions. The time-history of blast incident wave pressure in free-space explosion is analyzed. The peak value of blast incident wave pressure in free-space explosion is compared respectively with the empirical formula(Chen 2004 & Yang et al 2008) and blast wave parameters chart from Baker (Baker 1973). The characteristics of blast incident wave pressure in free space explosion are accomplished and the effectiveness of numerical simulation is verified.

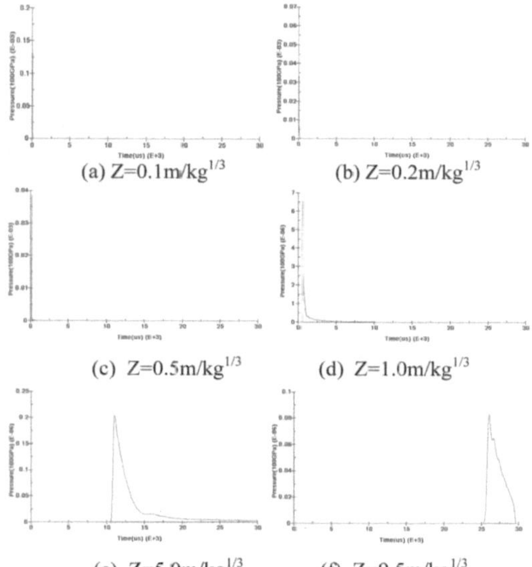

Figure 2. Time-histories of blast incident wave pressure in free-space explosion in different scaled distance of R = 10 m model.

3.1 *Time-history of blast incident wave pressure in free-space explosion*

The time-histories of blast incident wave pressure in free-space explosion in different scaled distance of R = 10 m model (Figure 1a) are shown in Figure 2. All time-histories are pulse curve and the smaller the scaled distance is, the narrower the pulse is. The peak values in different scaled distances arrive at different time, the smaller the scaled distance is, the sooner it arrives. The peak value is shown attenuation trend with the increase of the scaled distance.

It is necessary to increase small scaled-distance numerical examples since the time-history in small scaled-distance change too fast shown in Figure 2. Based on small scaled-distance numerical example models (R = 0.1 m, R = 0.2 m, R = 0.5 m) in last section, the time-histories of blast incident wave pressure in free-space explosion are shown in Figure 3, 4 and 5. There is double-peak effect in time-histories of small scaled-distance numerical examples, and the smaller the scaled distance is, more obvious the double-peak effect value is. The smaller the scaled distance is, the larger the discreetness of peak value is. Element size affects the calculation results significantly, in general, the smaller the size is, the closer to the actual value is.

3.2 *Peak value of blast incident wave pressure in free-space explosion*

Different researchers have proposed different empirical formulas of blast incident wave pressure in

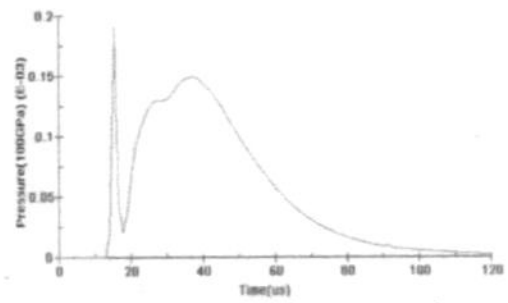

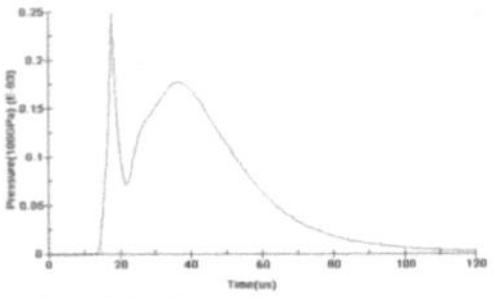

(a) 0.002m element size (b) 0.005m element size

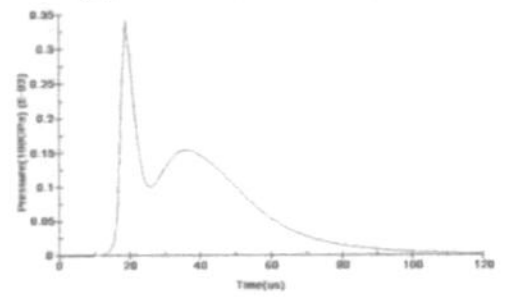

(c) 0.01m element size

Figure 3. Time-histories of blast incident wave pressure in $Z = 0.1\ m/kg^{1/3}$ of $R = 0.1$ m model.

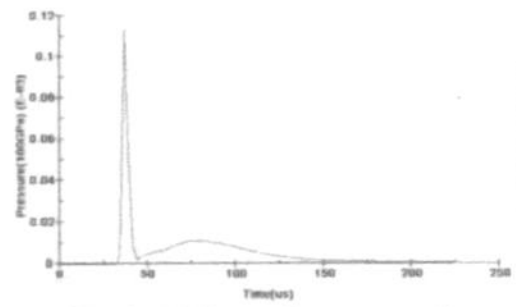

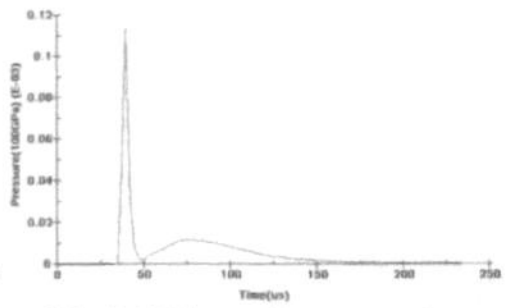

(a) 0.002m element size (b) 0.005m element size

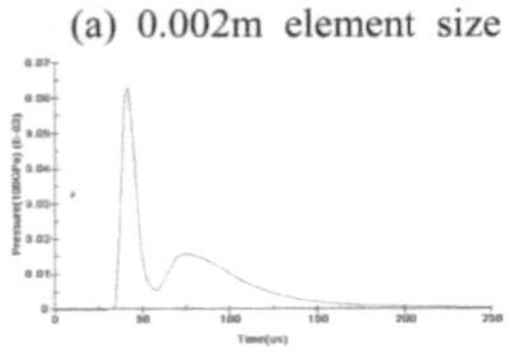

(c) 0.01m element size

Figure 4. Time-histories of blast incident wave pressure in $Z = 0.2\ m/kg^{1/3}$ of $R = 0.2$ m model.

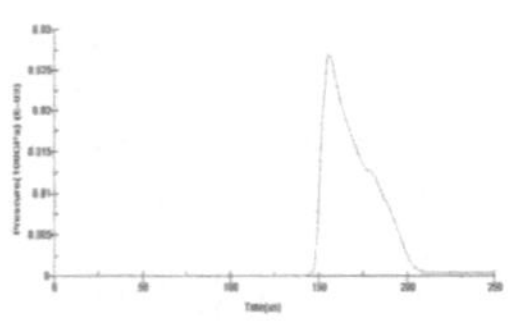

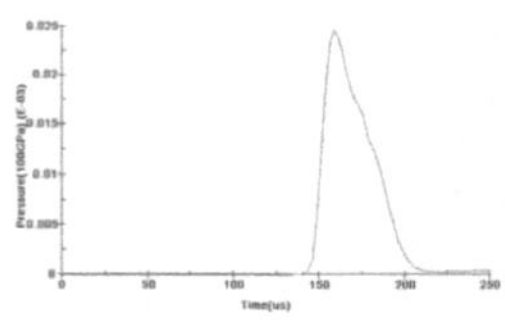

(a) 0.005m element size (b) 0.01m element size

Figure 5. Time-histories of blast incident wave pressure in $Z = 0.5\ m/kg^{1/3}$ of $R = 0.5$ m model.

free-space explosion, and WEBAKER presented the chart of the blast wave parameters (Baker 1973). Yang et al. 2008 concluded through finite element verification that the formula recommended by MA Sa-dovskyi can be used to describe the relation between peak value of blast incident wave pressure in free-space explosion and the scaled distance. In contrast, in a small scaled distance ($Z = 0.1$–$1.0\ m/kg^{1/3}$), there exists a big difference in peak value of blast incident wave pressure in free-space explosion among all the empirical formulas, and Chen 2004 gave the upper, median, and lower peak value of blast incident wave pressure in free-space explosion in small scaled-distance explosion condition.

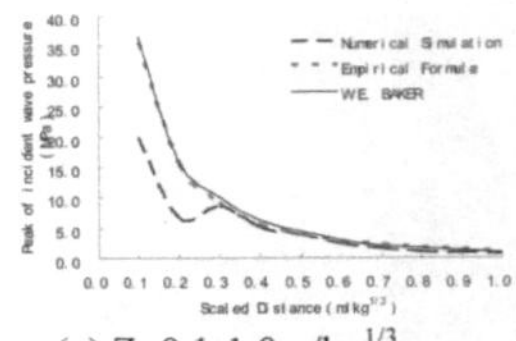

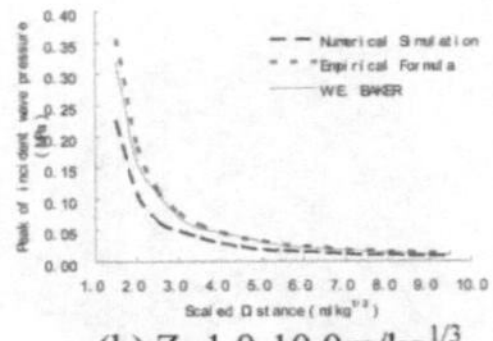

(a) Z=0.1-1.0m/kg$^{1/3}$ (b) Z=1.0-10.0m/kg$^{1/3}$

Figure 6. Variation of peak value of blast incident wave pressure in free-space explosion with scaled distance.

Peak value of blast incident wave pressure in different points on the same wave front is quite different, and the smaller the scaled distance is, the more obvious the difference is. The peak value of blast incident wave pressure in free-space explosion in all empirical formulas is the mean value on the same wave front, so in this section the peak value is also taken as the mean on the same wave-front. And the mean peak value of blast incident wave pressure in free-space explosion of different scaled distances are shown in Figure 6, and compared with WEBAKER Chart and the empirical formulas. For empirical formula, a small scaled distance ($Z < 1.0\ m/kg^{1/3}$) uses the median value provided by Chen Jianjie median (Chen 2004). MA Sa-dovskyi formula is adopted in the scaled distance $Z = 1.0$–$10\ m/kg^{1/3}$ (Yang et al 2008).

It can be clear seen from Figure 6 that, (1) the numerical simulation is verified to be effective since the value of numerical simulation shows good agreement empirical value; (2) when the scaled distance is taken as $Z < 0.3\ m/kg^{1/3}$, the value of numerical simulation differs from the empirical value greatly; when the scaled distance is $Z = 0.3$–$10\ m/kg^{1/3}$, the value of numerical simulation shows good agreement empirical value; in general, numerical simulation value is relatively smaller than the empirical value; (3) the peak value of blast incident wave pressure in free-space explosion decreases with the increase of the scaled distance, and attenuation velocity of small scaled distance is faster than that of large scaled distance.

4 CONCLUSION

In this paper, the numerical simulation of free space explosion has been implemented based on LS-DYNA. Both the key issues in numerical simulation and solutions are proposed, and the effectiveness of numerical simulation method is verified. Through numerical simulation method, the characteristic of time-history of blast incident wave pressure in free-space explosion is analyzed, and the peak value of blast incident wave pressure in free-space explosion is studied. The conclusions are as follows:

(1) Using LS-DYNA can effectively simulate the free-space explosive process, but it should be noted that reasonable value of parameters such as element mesh, time step, and boundary simulation is the key to ensure the effectiveness of the simulation.

(2) Since the numerical simulation of a small scaled distance may result in a relatively large discrete, the explosion of a small scaled distance should be cautiously analyzed by applying the method of numerical simulation.
(3) The time history time-history of blast incident wave pressure in free-space explosion is shown as pulse shape. The smaller the scaled distance is, the narrower the pulse is, and the more obvious the double-peak effect is.
(4) The peak value of blast incident wave pressure in free-space explosion decreases with the increase of the scaled distance and attenuation velocity of small scaled distance is faster than that of large scaled distance.

ACKNOWLEDGEMENT

The paper is supported by Natural Science Foundation Project of CQ CSTC (2009BB4202) and the National High Technology Research and Development Program of China (863 Program, 2007AA11Z125).

REFERENCES

ANSYS. 1998a. ANSYS/LS-DYNA User's Guide for Release 5.5.
ANSYS. 1998b. ANSYS/LS-DYNA Theoretical Manual.
Baker, W. E. 1973. Explosions in air. Austin, TX: University of Texas Press.
Chen, Jianjie. 2004. The research on antiknock quality of the inclosed structure which is under the internal action of blasting waves at a close distance. Nanjing: PLA Science and Engineering.
Shi, Dangyong & Li, Yuchun & Zhang, Shengmin, 2005. Basing on ANSYS/LS-DYNA8.1 to deal with the analysis of explicit dynamic. Beijing: Tsinghua university press.
Yang, Xin & Shi Shaoqing, 2008. The forecast and the Numerical simulation of the TNT peak value of explosive blast wave overpressure in the air. Explode 25(1):27–31.

Modelling and Computation in Engineering – Zhu (ed)
© 2011 Taylor & Francis Group, London, ISBN 978-0-415-61516-7

Dynamic response of harmonic plane compressional waves around a circular cavity in liquid-filled solid half-space

L.F. Jiang, S.X. Chen & Z. Han
State Key laboratory of Geomechanics and Geotechnical Engineering, Institute of Rock and Soil Mechanics, CAS, Wuhan, PR China

ABSTRACT: The dynamic response around a infinitely long cylindrical cavity of circular cross-section subjected to harmonic plane compressional waves in liquid-filled solid half-space is obtained by an complex function method based on the potential function and multi-polar coordinate. The steady state Biot's dynamic equations of liquid-filled solid in the absence of dissipation are uncoupled into Helmholtz equations via given potentials. Here, a circular cavity with large radius is used to replace the straight boundary of the liquid-filled solid half-space. The stresses and pore water pressures are obtained by using complex functions in multi-polar coordinates with certain boundary conditions of the solid matrix and the fluid. Then the variations of the coefficients of dynamic stress concentration and the pore water pressures concentration on boundaries of the cavity are discussed with different parameter conditions. The results of the given numerical example indicate that the method used in this paper is useful and efficient to the scattering and dynamic stress concentration of plane compressional waves in liquid-filled solid half-space.

1 INTRODUCTION

The scattering of harmonic plane compressional waves (i.e. Biot "fast" and "slow" waves) by an infinitely long cylindrical cavity of circular cross-section with various depths in a liquid-filled solid half-space is studied in this paper.

The two-dimensional problem of diffraction of elastic waves by cavities embedded in a poroelastic medium has been considered by many researchers: Zimmerman and Stern (1993) studied the diffraction of waves by a spherical cavity in an finite poroelastic soil medium by using the boundary element method (BEM). The problem of incident harmonic P and SV plane waves in tunnels in an infinite poroelastic saturated soil was studied by Katties et al. (2003) using the boundary element method. The scattering of harmonic plane waves by multiple elliptic cavities in a saturated soil medium was studied by Wang et al. (2005) using the potential function and the complex function. The dynamic interaction of time harmonic plane waves with a pair of parallel circular cylindrical cavities of infinite length buried in a boundless porous elastic fluid-saturated medium was investigated by Hasheminejad and Avazmohammadi (2007).

In the present study, an hybrid method combining the complex variable in multi-poplar coordinate system (Liu et al. 1982) and the convex approximation (Davis et al. 2001) is employed to solve the scattering and dynamic stress concentration of harmonic plane compressional waves by a unlined cylindrical cavity embedded in a saturated poroelastic medium of half-space. The behavior of the liquid-filled solid is governed by Biot's consolidation theory (Biot, M.A. 1941, 1956, 1962). Then, these equations are decoupled via introducing the potential functions and reduced to Helmholtz equations that the potential satisfy. Applying the boundary conditions of the liquid-filled solid, the solutions of dynamic stress concentration factor and pore pressure on the wall of the cavity impinged by an plane compressional waves normal to the axis of the cavity are presented and discussed.

2 GOVERNING EQUATIONS

Based on Biot's dynamic formulation, the field equations for a liquid-filled solid in the absence of dissipation are given by

$$N\nabla^2 u + grad\left[(D+N)e + Q\varepsilon\right] = \rho_{11}\ddot{u} + \rho_{12}\ddot{U} \tag{1}$$

$$grad\left[Qe + R\varepsilon\right] = \rho_{12}\ddot{u} + \rho_{22}\ddot{U} \tag{2}$$

Where u and U are, respectively, the displacements in the solid and liquid. The values of D, N, Q, R $\rho_{11}, \rho_{12}, \rho_{22}$ have been expressed by Deresiewicz H (1960)

The constitutive relations for the liquid-saturated porous solid are expressed as

$$\sigma_{ij} = 2N\varepsilon_{ij} + \delta_{ij}\left(De + Q\varepsilon\right) \tag{3}$$

$$\sigma = Qe + R\varepsilon \tag{4}$$

Where σ_{ij} is the stress in the solid; σ is the liquid stress per unit total area.

For the liquid-filled solid, we consider a Helmholtz resolution of each of the two displacement vectors in the form:

$$u = grad(\varphi) + curl(G) \tag{5}$$

$$U = grad(\psi) + curl(H) \tag{6}$$

Where φ and ψ to represent the scalar potentials of the solid and liquid, respectively; G and H to represent the vector potential of the solid and liquid, respectively;

$$G_{i,j} = 0, \quad H_{i,j} = 0. \tag{7}$$

Substituting from (5), (6) in (1), (2) yields a pair of quations:

$$grad\left(P\nabla^2\varphi + Q\nabla^2\psi - \rho_{11}\ddot{\varphi} - \rho_{12}\ddot{\psi}\right) + curl\left(\mu\nabla^2 G - \rho_{11}\ddot{G} - \rho_{12}\ddot{H}\right) = 0 \tag{8}$$

$$grad\left(Q\nabla^2\varphi + R\nabla^2\psi - \rho_{12}\ddot{\varphi} - \rho_{22}\ddot{\psi}\right) - curl\left(\rho_{12}\ddot{G} + \rho_{22}\ddot{H}\right) = 0 \tag{9}$$

Assuming that the scalar potentials φ and ψ have a harmonic time variation, i.e.

$$\varphi(x,y,t) = \varphi(x,y)e^{-i\omega t}, \psi(x,y,t) = \psi(x,y)e^{-i\omega t} \tag{10}$$

Substitution from (10) in (8), (9) yields upon elimination of $\nabla^2\psi$

$$\psi = (\rho_{22}Q - \rho_{12}R)^{-1}\left[(A/\omega^2)\nabla^2\varphi^2 + (\rho_{11}R - \rho_{12}Q)\varphi\right] \tag{11}$$

$$A\nabla^4\varphi + \omega^2 B\nabla^2\varphi + \omega^4 C\varphi = 0 \tag{12}$$

where

$$A = PR - Q^2, B = \rho_{11}R + \rho_{22}P - 2\rho_{12}Q, C = \rho_{11}\rho_{22} - \rho_{12}^2 \tag{13}$$

The solution of φ may be written as

$$\varphi = \varphi_f + \varphi_s \tag{14}$$

where

$$\left(\nabla^2 + k_f^2\right)\varphi_f = 0, \quad \left(\nabla^2 + k_S^2\right)\varphi_s = 0 \tag{15}$$

$$k_f^2 = \lambda_1^2\omega^2, \quad k_s^2 = \lambda_2^2\omega^2 \tag{16}$$

The wave corresponding to φ_f being the "fast" or P_f wave, propagating with a phase velocity λ_1^{-1}, that corresponded to φ_s being the "slow" or P_s wave whose phase velocity is λ_2^{-1}.

Finally, with the aid of (21), (22), we have the expression from (18)

$$\psi = \mu_1\varphi_f + \mu_2\varphi_s \tag{17}$$

where

$$\mu_1 = (\rho_{11}R - \rho_{12}Q - A\lambda_1^2)/(\rho_{22}Q - \rho_{12}R) \tag{18}$$

$$\mu_2 = \left(\rho_{11}R - \rho_{12}Q - A\lambda_2^2\right)/\left(\rho_{22}Q - \rho_{12}R\right) \tag{19}$$

Similarly,

$$H = -\frac{\rho_{12}}{\rho_{22}}G, \qquad \left(\nabla^2 + k_t^2\right)G = 0 \tag{20}$$

By introducing complex variables $z = x + iy$, $\bar{z} = x - iy$, Eqs. (15) and (20) can also be written as the following formula:

$$\frac{\partial^2\varphi_f}{\partial z\partial\bar{z}} = \left(\frac{ik_f}{2}\right)^2\varphi_f, \frac{\partial^2\varphi_s}{\partial z\partial\bar{z}} = \left(\frac{ik_s}{2}\right)^2\varphi_s, \frac{\partial^2 G}{\partial z\partial\bar{z}} = \left(\frac{ik_t}{2}\right)^2 G \tag{21}$$

Based on Liu et al. (1982), the general solutions of Eqs. (21) may be expressed in terms of Hankel function as follows

$$\varphi_f(z,\bar{z}) = \sum_{n=-\infty}^{n=\infty} a_n H_n(k_f|z|)\left(\frac{z}{|z|}\right)^n \tag{22}$$

$$\varphi_s(z,\bar{z}) = \sum_{n=-\infty}^{n=\infty} b_n H_n(k_s|z|)\left(\frac{z}{|z|}\right)^n \tag{23}$$

$$G(z,\bar{z}) = \sum_{n=-\infty}^{n=\infty} c_n H_n(k_t|z|)\left(\frac{z}{|z|}\right)^n \tag{24}$$

where $H_n^{(1)}(\ldots)$ is the Hankel function of the first kind of order n; a_n, b_n, c_n are arbitrary functions to be determined from the boundary conditions of the cavities.

3 EXPRESSIONS OF DISPLACEMENTS, STRESSES AND PORE PRESSURES

Complex variables $z = x + iy, \bar{z} = x - iy$ and polar coordinate system are introduced and the displacements, stresses and pore pressures are expressed as:

$$u_r + iu_\theta = 2\frac{\partial}{\partial z}\left(\varphi_f + \varphi_s - iG\right)\exp(-i\theta) \tag{25}$$

$$u_r - iu_\theta = 2\frac{\partial}{\partial z}\left(\varphi_f + \varphi_s + iG\right)\exp(i\theta) \tag{26}$$

$$w_r + iw_\theta = 2\frac{\partial}{\partial z}\left(\eta_1\varphi_f + \eta_2\varphi_s - i\alpha G\right)\exp(-i\theta) \tag{27}$$

$$w_r - iw_\theta = 2\frac{\partial}{\partial z}\left(\eta_1\varphi_f + \eta_2\varphi_s + i\alpha G\right)\exp(i\theta) \tag{28}$$

$$\sigma_r' + \sigma_\theta' = -2(\lambda + \mu)\left(k_f^2\varphi_f + k_s^2\varphi_s\right) \tag{29}$$

$$\sigma_r' + i\sigma_{r\theta}' = -(\lambda+\mu)\left(k_f^2\varphi_f + k_s^2\varphi_s\right) + 4\mu\frac{\partial^2}{\partial \bar{z}^2}\left(\varphi_f + \varphi_s - iG\right)\exp(-2i\theta) \tag{30}$$

$$\sigma_r' - i\sigma_{r\theta}' = -(\lambda+\mu)\left(k_f^2\varphi_f + k_s^2\varphi_s\right) + 4\mu\frac{\partial^2}{\partial z^2}\left(\varphi_f + \varphi_s + iG\right)\exp(2i\theta) \tag{31}$$

$$p = -A_f k_f^{\ 2}\varphi_f - A_s k_s^{\ 2}\varphi_s \tag{32}$$

$\sigma_r', \sigma_\theta', \sigma_{r\theta}'$ are the effective stress of the saturated porous elastic solid, respectively.

4 INCIDENT, REFLECTED, DIFFRACTION, AND TOTAL WAVES

4.1 *Incident wave*

In present paper, the incident harmonic plane compressional waves having the potential

$$\varphi^{(i)} = \varphi_f^{\ (i)} + \varphi_s^{\ (i)} \tag{33}$$

where

$$\varphi_f^{\ (i)} = \varphi_{f0}\exp[ik_f(x\sin\beta + y\cos\beta)]\exp(-i\omega t) \tag{34}$$

$$\varphi_s^{\ (i)} = \varphi_{s0}\exp[ik_s(x\sin\beta + y\cos\beta)]\exp(-i\omega t) \tag{35}$$

Here, φ_{f0} and φ_{s0} are the amplitudes of the fast and slow potentials; k_f and k_s are the wave numbers for the fast and slow waves; β is the angle of incidence and ω is the angular frequency.

4.2 *Reflected waves*

Deresiewicz (1960) shows the wave potentials generated at the boundaries. At the free surface ($y = 0$), the reflected longitudinal fast P_f, slow P_s and shear SV plane waves have the potentials $\varphi_f^{(r)}, \varphi_s^{(r)}$ and $G^{(r)}$, respectively. In virtue of complex variables $z = x + iy$, $\bar{z} = x - iy$ the reflected potentials $\varphi_f^{(r)}, \varphi_s^{(r)}$ and $G^{(r)}$ of the incident harmonic plane compressional wave can be expressed as the following formula:

$$\varphi_f^{\ (r)} = A_1\exp\left[\frac{ik_f}{2}\left(ze^{i(\pi/2-\beta_1)} + \bar{z}e^{-i(\pi/2-\beta_1)}\right)\right] \tag{36}$$

$$\varphi_s^{\ (r)} = A_2\exp\left[\frac{ik_s}{2}\left(ze^{i(\pi/2-\beta_2)} + \bar{z}e^{-i(\pi/2-\beta_2)}\right)\right] \tag{37}$$

$$G^{(r)} = A_3\exp\left[\frac{ik_t}{2}\left(ze^{i(\pi/2-\gamma)} + \bar{z}e^{-i(\pi/2-\gamma)}\right)\right] \tag{38}$$

$$\text{where } \beta = \beta_1,\ \lambda_1\sin\beta_1 = \lambda_2\sin\beta_2 = \lambda_3\sin\gamma \tag{39}$$

4.3 *Diffraction waves*

When a cavity is present, the incident wave and the reflected waves are scattered off the cavity surface. The total potentials of harmonic plane P_f, P_s and SV waves generated at the cavity are represented by $\varphi_{f1}^{(s)}, \varphi_{s1}^{(s)}$ and $G_1^{(s)}$. The cylinder vibrations are reflected off the half-space free surface generating new waves represented by $\varphi_{f2}^{(s)}, \varphi_{s2}^{(s)}$ and $G_2^{(s)}$.

Since the total potentials scattered at the saturated soil of half-space satisfy Helmholtz equation, the total scattering potentials are expressed

$$\varphi_f^{\ (s)} = \varphi_{f1}^{\ (s)} + \varphi_{f2}^{\ (s)} = \sum_{i=1}^{2}\sum_{n=-\infty}^{\infty} a_{in}H_n^{(1)}\left(k_f|z_{ij}|\right)\left(\frac{z_{ij}}{|z_{ij}|}\right)^n \tag{40}$$

$$\varphi_s^{\ (s)} = \varphi_{s1}^{\ (s)} + \varphi_{s2}^{\ (s)} = \sum_{i=1}^{2}\sum_{n=-\infty}^{\infty} b_{in}H_n^{(1)}\left(k_s|z_{ij}|\right)\left(\frac{z_{ij}}{|z_{ij}|}\right)^n \tag{41}$$

$$G^{(s)} = G_1^{\ (s)} + G_2^{\ (s)} = \sum_{i=1}^{2}\sum_{n=-\infty}^{\infty} c_{in}H_n^{(1)}\left(k_t|z_{ij}|\right)\left(\frac{z_{ij}}{|z_{ij}|}\right)^n \tag{42}$$

4.4 *Total waves*

The total potentials in the saturated soil are

$$\varphi_f^{\ (t)} = \varphi_f^{\ (i)} + \varphi_f^{\ (r)} + \varphi_{f1}^{\ (s)} + \varphi_{f2}^{\ (s)} \tag{43}$$

$$\varphi_s^{\ (t)} = \varphi_s^{\ (r)} + \varphi_{s1}^{\ (s)} + \varphi_{s2}^{\ (s)} \tag{44}$$

$$G^{(t)} = G^{(r)} + G_1^{\ (s)} + G_2^{\ (s)} \tag{45}$$

5 SOLUTIONS OF BOUNDARY PROBLEMS

5.1 *Stress boundary problems*

For stress boundary problems, the two cases are considered: permeable boundary and impermeable boundary. For permeable boundary, the pore pressure on the boundary of cavity is zero. For impermeable boundary, the normal displacement of fluid relative to the solid matrix is zero.

If the normal and tangential total stresses are given, the stresses on the boundary of cavity can be expressed as following formula by using Equates. (30), (31), (43), (44), (45)

$$\alpha_f\varphi_{fj}^{(t)} + \alpha_s\varphi_{sj}^{(t)} + 4\mu\frac{\partial^2}{\partial z^2}\left(\varphi_{fj}^{(t)} + \varphi_{sj}^{(t)} + iG_j^{(t)}\right)\exp(2i\theta) = f_{1j} - if_{2j} \tag{46}$$

$$\alpha_f\varphi_{fj}^{(t)} + \alpha_s\varphi_{sj}^{(t)} + 4\mu\frac{\partial^2}{\partial \bar{z}^2}\left(\varphi_{fj}^{(t)} + \varphi_{sj}^{(t)} - iG_j^{(t)}\right)\exp(-2i\theta) = f_{1j} + if_{2j} \tag{47}$$

where f_{1j} and f_{2j} are the normal and tangential total stresses and of jth boundary

$$\alpha_f = \alpha A_f k_f^{\ 2} - (\lambda+\mu)k_f^{\ 2} \quad , \quad \alpha_s = \alpha A_s k_s^{\ 2} - (\lambda+\mu)k_s^{\ 2} \tag{48}$$

For permeable boundary, pore pressure of the jth cavities is zero and Eq. (32) can be written as

$$P = -A_f k_f^{\,2} \varphi_{fj}^{(t)} - A_s k_s^{\,2} \varphi_{sj}^{(t)} = 0 \tag{49}$$

For impermeable boundary, the normal displacement of fluid to solid matrix of the jth cavities is zero and Eqs. (27) and (28) can be written as

$$w_{rj} = \frac{\partial}{\partial z}\left(\eta_1 \varphi_{fj} + \eta_2 \varphi_{sj} + i\alpha G_j^{(t)}\right)\exp(i\theta) + \frac{\partial}{\partial \bar{z}}\left(\eta_1 \bar{\varphi}_{fj}^{(t)} + \eta_2 \bar{\varphi}_{sj}^{(t)} - i\alpha \bar{G}_j^{(t)}\right)\exp(-i\theta) = 0 \tag{50}$$

5.2 *Displacement boundary problems*

For displacement boundary problems, the two cases are also considered: permeable boundary and impermeable boundary. For permeable boundary, the pore pressure on the boundary of cavity is zero. For impermeable boundary, the normal displacement of fluid relative to the solid matrix is zero.

If the normal and tangential total displacements are given, displacement on the boundary of cavity and free surface of half-space can be expressed as following formula by using Eqs. (25), (26)

$$u_r + iu_\theta = 2\frac{\partial}{\partial z}\left(\varphi_f + \varphi_s - iG\right)\exp(-i\theta) = g_1 + ig_2 \tag{51}$$

$$u_r - iu_\theta = 2\frac{\partial}{\partial z}\left(\varphi_f + \varphi_s + iG\right)\exp(i\theta) = g_1 - ig_2 \tag{52}$$

Eqs. (46), (47), (51), (52) and (49) are selected to solve the problem of permeable boundary and Eqs. (46), (47), (51), (52) and (50) are selected to solve the problem of impermeable boundary.

6 COMPARISON WITH PREVIOUS MODELS AND ANALYSIS OF NUMERICAL RESULTS

In engineering practice, the dynamic stress concentration factor and the pore water pressure concentration factor are most significant aspects in the study of wave propagation in the saturated porous elastic half-space. In this study, the dynamic stress concentration factor is defined as the ratio of tangential stress to the maximum amplitude of the effective stress at incident direction.

$$\sigma^* = \frac{\sigma_\theta'}{\sigma_0}$$

where

$$\sigma_\theta' = -2(\lambda + \mu)\left(k_f^{\,2}\varphi_f + k_s^{\,2}\varphi_s\right) - 2\alpha P,$$
$$\sigma_0 = -(\lambda + 2\mu)\left(k_f^{\,2}\varphi_{f0} + k_s^{\,2}\varphi_{s0}\right) - \alpha P$$

For the case of impermeable condition, the pore pressure concentration factor is defined as the ratio

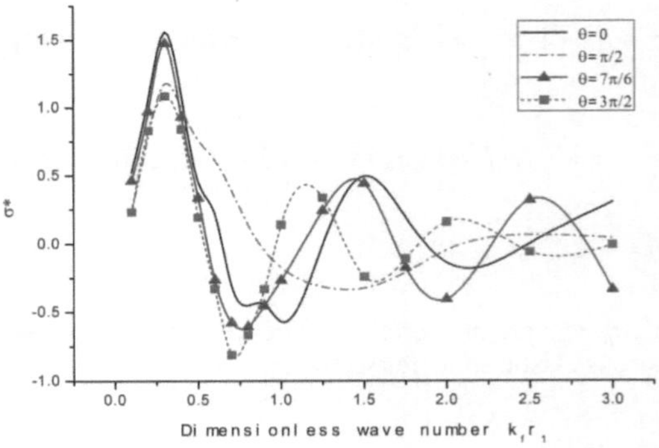

Figure 1. Distribution of the dynamic stress concentration factors around boundary in permeable condition.

of the pore pressure on the boundary of cavity to the maximum amplitude of pore pressure at the same point

$$P^* = \frac{P}{P_0}$$

where

$$P_0 = -A_f k_f^{\,2}\varphi_{f0} - A_s k_s^{\,2}\varphi_{s0}$$

7 NUMERICAL RESULTS

In the following analyses, the problem of an infinitely long cylindrical cavity of circular cross-section embedded in the saturated soil at half-space as illustrated in Figs. 1 is considered: For brevity, the cavity is only excited by harmonic plane fast compression component P_f.

$$\varphi_f^{(i)} = \varphi_{f0}\exp\left\{ik_f\left(x\cos\beta + y\sin\beta\right)\right\}\exp(-i\omega t).$$

The parameters are: $\beta = \pi/2$; $\rho_s = 2500\,kg/m^3$; $\rho_f = 1000\,kg/m^3$; $n = 0.3$; $\mu = 1.0 \times 10^7 Pa$; $\nu = 0.3$; $\alpha = 0.999$; $M = 1.0 \times 10^8 Pa$; $h/r_1 = 2, 5, 8$. Figs. 1–3 show the distributions of dynamic stress concentration factors and pore pressure along the boundary of cavity in permeable and impermeable conditions at incidence $\beta = 0$, depth $h/r_1 = 5$ with dimensionless wave number $0.5 \le k_t r_1 \le 3.0$.

8 CONCLUSIONS

An hybrid method combining the complex variable in multi-poplar coordinate system and the convex approximation is employed to solve the problem of scattering of harmonic plane compressional waves by a unlined cylindrical cavity embedded in a liquid-filled solid half-space. Figure 1–3 show the dynamic stress concentration factors and the pore pressure concentration factors on the cavity surface under different

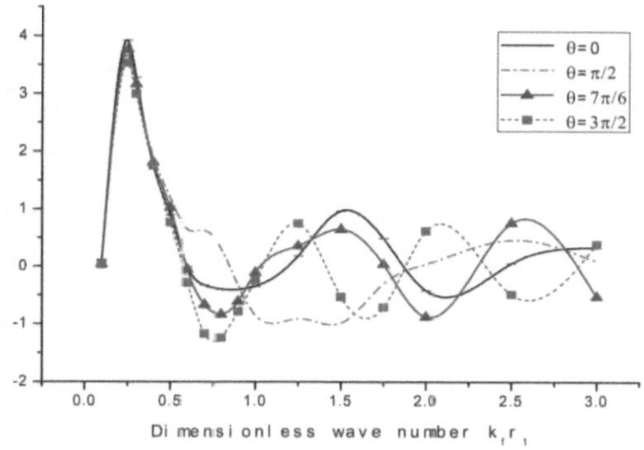

Figure 2. Distribution of the dynamic stress concentration factors around boundary of the cavity in impermeable condition.

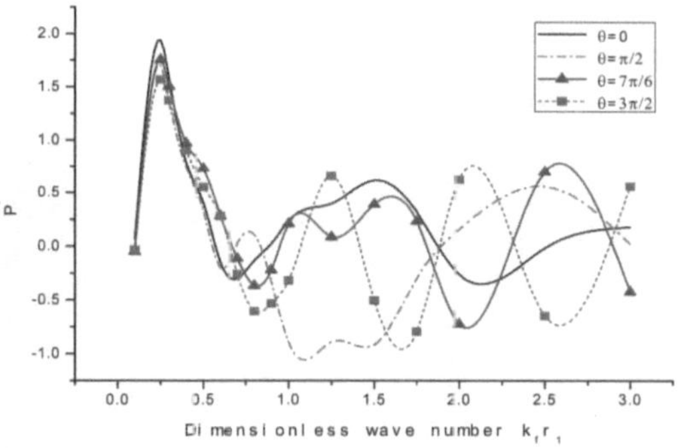

Figure 3. Distribution of the pore pressure concentration factors around boundary of the cavity in impermeable condition.

dimensionless wave number, and at high frequencies (i.e. $k_t r_1 > 1.0$) σ^* vary widely with θ and $k_t r_1$. At low frequencies (i.e. $k_f r_1 < 0.5$) the variation become uniform and the factors is almost most at $k_f r_1 = 0.3$. The dynamic stress concentration factors in permeable condition are smaller than those in impermeable condition. For the case of impermeable boundary condition, the concentration factors of pore pressures are less than those of dynamic stress. The dynamic stresses and pore pressures decrease with the dimensionless wave number increasing.

ACKNOWLEDGEMENTS

Research presented in this paper was supported by the State Key laboratory of Geo mechanics and Geotechnical Engineering (project numbers: SKLQ015 and SKLZ0803). The research was also financially supported by the Key Laboratory of Geotechnical Mechanics and Engineering of the Ministry of Water Resources (project number: G07-09) and Project supported by Funds of the Chinese Academy of Sciences for Key Topics in Innovation Engineering (Grant No. kzcx2-yw-150).

REFERENCES

Biot, M.A. 1962. Mechanics of deformation and acoustic porous media. J. Appl. Phy. 33 (4): 1482–1498.

Biot, M.A. 1956. Theory of propagation of elastic waves in a fluid-saturated porous solid. I. low frequecy rang. J.Acoust. Soc. Am, 28 (2), 168–178.

Biot, M.A. 1941. General theory of three dimensional consolidation. J. Appl. Phys., 12:155-164.

Davis CA, Lee VW, Bardet JP. 2001. Transverse reponse of underground cavities and pipes to incident SV waves. Earthquake Engineering and Structural Dynamics 30: 383–410.

Deresiewicz H. 1960. The effect of boundaries on wave propagation in a liquid-filled porous solid: I. Reflection of plane wave a free plane boundary (non-dissipative case). Bulletin of the Seismological Society of America. 50 (4): 599-607.

Hasheminejad SM, Avazmohammadi R.2007. Harmonic wave diffraction by two circular cavities in a poroelastic formation. Soil Dyn Earthq Eng. 27:29–41.

Kattis SE, Beskos DE, Cheng AHD. 2003. 2D dynamic response of unlined and lined tunnels in poroelastic soil to harmonic body waves. Earthq Eng Struct Dyn. 32(1):97–110.

Liu Diankui, Gai Bingzhang, Tao Guiyuan. 1982. Applications of the method of complex functions to dynamic stress concentrations. Wave Motion 4: 293–304

Wang JH, Zhou XL, Lu JF. 2005. Dynamic stress concentration around elliptic cavities in saturated poroelastic soil under harmonic plane waves. Int J Solids Struct. 42(14):4295–310.

Zimmerman C. 1993. Scattering of plane compressional waves by spherical inclusions in a poroelastic medium. J.Acoust. Soc.Am., 94(1):527-536

Zimmerman, C. and Stern, M. 1993. Boundary element solution of 3-D wave scatter problems in a poroelastic medium. Engineering Analysis with Boundary Elements. 12: 1591–1596.

Modelling and Computation in Engineering – Zhu (ed)
© 2011 Taylor & Francis Group, London, ISBN 978-0-415-61516-7

Unsaturated creep tests and empirical models of the sliding zone soils of the Qianjiangping landslide in Three Gorges

S.M. Wang & X.L. Lai
Key Laboratory of Geological Hazards in Three Gorges Reservoir Area of Education Ministry, China Three Gorges University, Yichang, Hubei, China

ABSTRACT: Creep of sliding zone soils may cause significant displacement in large-scale landslides in Three Gorges Reservoir Area. To investigate water effects on the creep behavior of the Qianjiangping Landslide, a series of unsaturated triaxial creep tests of the sliding zone soils were performed. Based on the analysis of the test results, a new stress intensity incorporating matric suction is defined and an unsaturated Singh-Mitchell creep model is developed. In addition, interrelation among parameters of the model, stress and time is analyzed quantitatively to modify the model. Predicted results are in good agreement with the experimental results, which indicates that the established unsaturated model can reasonably simulate the effects of water on the soil creep behavior of the Qianjiangping Landslide. This study provides a calculation model and parameters for the evaluation of long-term stability of landslides under the influence of rainfall and reservoir water fluctuation.

1 INTRODUCTION

A large number of large-scale landslides have been induced by the couple effects of rainfall and reservoir water fluctuation in Three Gorges Reservoir Area. As a typical one, the Qianjiangping Landslide caused enormous economic losses and social impact. Under the effects of rainfall and fluctuation of reservoir water, the deformation and long-term stability of these landslides depend on the time-dependent responses of the sliding zone soils significantly. Therefore, it is necessary to carry out creep tests of sliding zone soils of typical landslides in Three Gorges Reservoir Area to lay a mechanical foundation for the prediction of long-term stability of these landslides under the influence of water.

Time-dependent behavior of soils has been investigated in a number of uniaxial and triaxial tests. Based on studies of these creep test results, a number of famous empirical models were proposed, such as Singh-Mitchell creep model (Singh & Mitchell 1968), Mesri creep model (Mesri et al. 1981), Semple creep model (Semple 1973), etc. However, there is no creep model explicitly including effects of water reported to date. On the other hand, effects of suction of unsaturated soils on soil properties indicate the effects of water on them (Fredlund & Rahardjo 1997). Therefore, it is imperative to establish a suction-related creep model in order to seek the influence mechanism of water on soil creep behavior.

In this paper, a series of unsaturated drained triaxial creep tests of the sliding zone soils of the Qianjiangping Landslide were performed with the controlling of matric suction. And an empirical model explicitly considering the effects of matric suction was developed to study the effects of water on the creep behavior of landslide soil quantitatively.

Table 1. Physic-mechanical parameters of the sliding zone soils.

Specific Gravity	Water Content %	Density g/cm^3	Liquid Limit %	Plastic Limit %	C kPa	φ °
2.71	19	2.02	40.5	17	28.3	18

2 UNSATURATED DRAINED TRIAXIAL CREEP TESTS

2.1 *Basic properties of soil samples*

The Qianjiangping Landslide is located in Shaxi county, Zigui town, Yichang, Hubei province. It is on the south bank of Qinggan River, a branch of Yangtze River. Soil samples used for laboratory tests were taken from the sliding zone of the Qianjiangping Landslide. Then they were screened through a sieve with mesh size finer than 2 mm for the tests of laboratory basic index. The results are summarized in Table 1.

Besides, the unsaturated drained shear strength tests were performed to determine the unsaturated drained shear strength q_f of the sliding zone soils for the calculation of shear stress intensity of unsaturated creep tests (Table 2).

2.2 *Experimental apparatus and method*

The self-made *FSR*-6 unsaturated triaxial creep apparatus is employed to conduct the creep tests. It is

Table 2. Program of unsaturated creep tests.

No.	σ_3 kPa	u_a kPa	σ'_3 kPa	q_f kPa	q'_f kPa	q kPa
AC-1	250	50	200	262	261	31.4,70.6,101.9, 141.1,180.3
AC-2	300	100	200	297	350	51,111,171, 216,261
AC-3	350	150	200	331	370	65,128,191, 266,305
AC-4	400	200	200	365	380	70.6,129.4,188.2,247,305.8
AC-5	450	250	200	399	420	78.4,156.8,215.6,286.2

developed based on the principal of conventional triaxial creep apparatus and unsaturated triaxial apparatus, featuring both the loading system of the former and air pressure controlling system of the latter (Guan & Wang 2008). Hence, it is available to apply not only constant shear stress but constant air pressure. It is composed of several parts, such as confining pressure controlling system, matric suction controlling system, pore water pressure system, axial load system, cell pressure chamber, measurement and data collecting system, etc. (Figure 1).

The axial load was applied in a conventional way by applying dead weights in the axial direction. Specimens were cylindrical with length 120 mm and diameter 60 mm. Drained triaxial creep tests were performed on account of the drainage condition of the landslide in a long period of time. Test process is outlined as follows: (1) Preparing soil specimen. Remolded samples were prepared in the same way that in the unsaturated shear strength tests. Once finished, put them into saturator and make them saturated by using vacuum saturation. When the measured Skempton B-value exceeded 0.95, samples were considered saturated. (2) Suction equilibrium and consolidation. Prior to consolidation, an air pressure was applied to each specimen to make it reach the suction equilibrium. Then specimens were consolidated to the objective confining pressure as shown in Table 2. When the volume of drainage is less than 0.01 mm^3 within two hours, consolidation is complete. (3) Creep loading. The deviator stress was applied by the multi-stage loading method, which can avoid the inhomogeneity of specimens and obtain more test data in a single creep test. The constant deviator stress q applied in each level of creep is defined as the ratio of the drained shear strength q_f and the number of loading steps n, i.e., $q = sq_f/n$. Make $n = 5$ and specific values of q are summarized in Table 2. It should be noted that the shear strength of specimens were improved due to the step-by-step loading method, thus the drained shear strength q_f was redefined (Table 2) based on actual creep rupture in creep tests. If the axial displacement of specimens in each stage of creep was less than 0.01 mm within one day, then creep test in this stage was ended and the next level of q was applied until

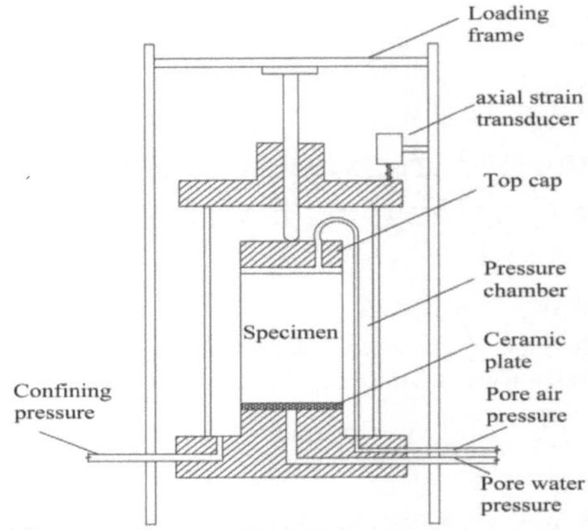

a Structure of unsaturated creep apparatus

b Double unsaturated creep apparatus

Figure 1. Unsaturated creep apparatus.

the creep rupture. Thus, the duration of each level of deviator stress was about one to two weeks.

2.3 *Test results*

It took nearly half a year for three creep apparatus to finish the unsaturated creep tests in Table 2, and obtained five groups of valid data. Due to the multi-stage loading method, test data are processed by the principle of Boltzmann superposition. A summary of test results is illustrated in Table 3, in which, ε_1 and ε_f are axial strain at the elapsed time of 60 min and at the end of creep test respectively. And t_f is elapsed time. Shear stress intensity D_r is defined as the ratio of the constant deviator stress q to the drained shear strength q_f, i.e., $D_r = q/q_f$. Figure 2 shows creep curves under matric suctions of 50, 100 and 150 kPa.

3 UNSATURATED CREEP MODEL

3.1 *Singh-Mitchell stress-strain-time function*

The total strain can be attributed to two components: (1) The instantaneous strain ε_i and (2) the time-dependent or creep strain ε_c. If a constant stress is

Table 3. Summary of creep test results.

No.	σ_3 kPa	u_a kPa	D_r	ε_1 %	ε_f %	t_f min
AC-1	250	50	0.120	0.29	0.59	8613
			0.270	1.78	2.46	9590
			0.391	1.95	3.12	13,957
			0.541	2.21	3.98	14,164
			0.691	2.51	5.08	11,594
AC-2	300	100	0.146	0.6	0.62	7607
			0.317	1.22	1.89	8961
			0.489	2.12	3.54	8456
			0.617	2.56	4.98	8836
			0.746	3.35	7.30	9162
AC-3	350	150	0.176	1.11	1.52	5687
			0.346	1.54	2.58	8455
			0.516	2.22	3.95	8576
			0.719	4.29	7.23	8450
			0.824	4.94	9.13	12,974
AC-4	400	200	0.186	1.56	2.10	9065
			0.340	1.83	2.83	10,389
			0.495	2.28	3.98	14,855
			0.650	2.63	5.11	15,193
			0.805	3.58	7.62	8860
AC-5	450	250	0.187	2.93	3.43	15,050
			0.373	3.63	4.35	18,170
			0.513	3.85	5.02	25,412
			0.681	4.34	6.18	21,031

applied under undrained condition, ε_c is defined as the undrained creep strain. If the stress is applied under drained condition, the ε_c represents the primary consolidation and secondary consolidation (or drained creep). There are many factors affecting the rheological properties of soil, such as overconsolidation ratio, aging and thixotropic effect. Without considering these factors, in general, ε_c can be approximated by a function of stress and time, i.e.,

$$\varepsilon_c = f_1(S) f_2(t) \tag{1}$$

Where $f_1(S)$ and $f_2(t)$ denote functions of stress and time respectively (Lin & Wang 1998).

Based on the observation of laboratory creep test results, Singh and Mitchell (1968) proposed a creep model which can well describe the creep behavior of soils. In the model, the exponential function and the power function have been adopted for $f_1(S)$ and $f_2(t)$ respectively, i.e.,

$$\dot{\varepsilon} = A\exp(\alpha D_r)(t_1/t)^m \tag{2}$$

Taking natural logarithm on both sides of Eq. (2), it can be rewritten as

$$\ln\dot{\varepsilon} = \ln A + \alpha D_r + m(\ln t - \ln t_1) \tag{3}$$

where $\dot{\varepsilon}$ denotes the axial strain rate at random time; Dr is shear stress intensity as defined before; t is the elapsed time of a creep test and t_1 a reference time, denoting a time at an early stage of creep test; $A$is

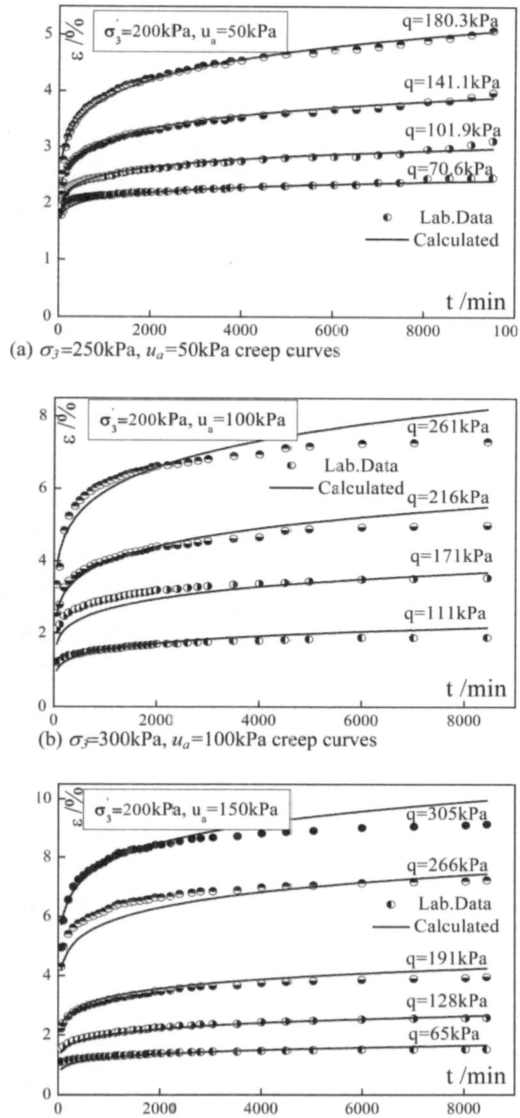

(a) σ_3=250kPa, u_a=50kPa creep curves

(b) σ_3=300kPa, u_a=100kPa creep curves

(c) σ_3=350kPa, u_a=150kPa creep curves

Figure 2. Unsaturated creep curves under net cell pressure of 200 kPa.

strain rate at $Dr = 0, t = t_1$; m is the inclination of $\ln\dot{\varepsilon}$ versus $\ln t$ plot and α the inclination of $\ln\dot{\varepsilon}$ versus Dr plot.

Integrations are performed on Eq. (2), while $m \neq 1$ and without considering initial strain, Eq. (2) can be rewritten as

$$\varepsilon = B\exp(\beta D_r)(t/t_1)^\lambda \tag{4}$$

where $B = At_1/(1-m)$, $\beta = \alpha$ and $\lambda = 1 - m$.

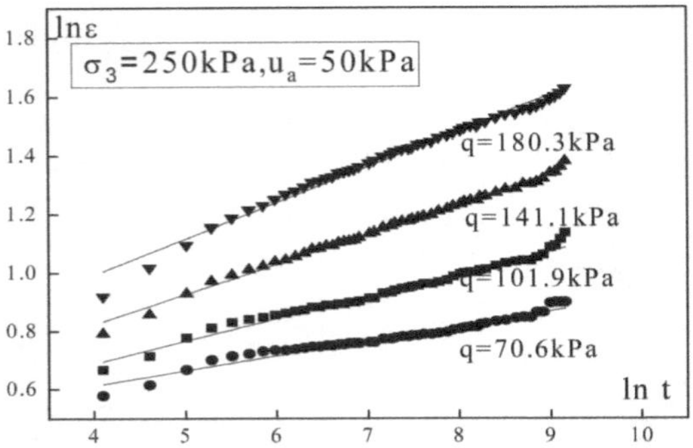

Figure 3. $\ln\varepsilon - \ln t$ curves.

3.2 *Unsaturated Singh-Mitchell stress-strain-time function*

Based on the observation of creep test results, it is found that there is a similarity between unsaturated creep curves and saturated curves. Thus, by using Singh-Mitchell creep model for reference, an attempt was made to develop an unsaturated Singh-Mitchell creep model of the sliding zone soils.

To include matric suction as a new stress variable in the unsaturated creep model, a new stress intensity was established. As can be seen from Table 3, creep strain decreases with the increasing of matric suction under the same stress intensity D_r. Thus, a new stress intensity D_R can be defined as follows,

$$D_R = \frac{P_a}{u_a} D_r \tag{5}$$

Substituting Eq. (5) into Eq. (4), an unsaturated creep model can be written as

$$\varepsilon = B\exp(\beta_{D_R}\frac{P_a}{u_a} D_r)(t/t_1)^{\lambda} \tag{6}$$

where P_a is the atmospheric pressure, i.e., 1 $P_a = 101.33$ kPa. β_{DR} is the corresponding parameter of β under the new stress intensity D_R. Taking natural logarithm on both sides of Eq. (6), it can be rewritten as

$$\ln\varepsilon = \ln B + \beta_{D_R} D_R + \lambda(\ln t - \ln t_1) \tag{7}$$

When soil is saturated, i.e., $u_a = P_a$, Eq. (6) is turned into Eq. (4) which is the saturated Singh-Mitchell model.

3.3 *Parameter fitting*

Creep curves under net confining pressure of 200 kPa and matric suction of 50 kPa were analyzed to describe developing process of the unsaturated creep model. Let $t_1 = 60$ min, and the corresponding axial strain ε_1 the initial strain. Firstly, plot $\ln\varepsilon$ versus $\ln t$ under different deviator stress as shown in Figure 3. Clearly, a parallel linear relationship exists between $\ln\varepsilon$ and $\ln t$. And values of the inclination of the linear relation λ are summarized in Table 4.

Table 4. λ values under different stress intensity.

q kPa	D_R kPa	λ	R^2	λ average
70.6	0.1810	0.0513	0.9513	0.0885
101.9	0.2615	0.0772	0.9792	
141.1	0.3620	0.1028	0.9931	
180.3	0.4626	0.1228	0.9852	

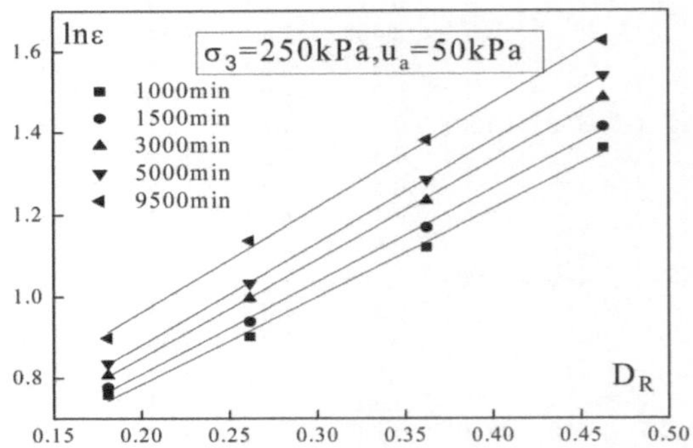

Figure 4. $D_R - \ln\varepsilon$ curves.

Table 5. B&β values at different times.

t min	β_{DR}	B	R^2	β_{DR} average	B average
1000	2.1508	1.1109	0.9967	2.3742	1.0404
1500	2.2648	1.0747	0.9985		
3000	2.4043	1.0229	0.9998		
5000	2.4943	0.9911	0.9999		
9500	2.5569	1.0023	0.9980		

Then, plot $\ln\varepsilon$ versus D_R at different elapsed times of 1000, 1500, 3000, 5000, 9500 min as shown in Figure 4. Clearly, a linear relationship exists between $\ln\varepsilon$ and D_R, with the slope β_{DR} and intercept $\ln B + \lambda(\ln t \ln t_1)$. Substituting λ and the elapsed time t into the intercept formulation, parameter B can then be obtained. Values of B and β_{DR} are summarized in Table 5. As parameters of Singh-Mitchell model are supposed to be independent of time and stress intensity, average values of these parameters are taken as their final values in the model.

Substituting the average values in Tables 4, 5 into Eq.(6), it can be rewritten as

$$\varepsilon = 1.0404\exp(2.3742\frac{P_a}{u_a} D_r)(t/t_1)^{0.0885} \tag{8}$$

Eq. (8) is the unsaturated Singh-Mitchell creep model of the sliding zone soils of the Qianjiangping Landslide under net confining pressure of 200 kPa

Table 6. Parameters of unsaturated Singh-Mitchell model.

σ_3' kPa	u_a kPa	λ	β_{D_R}	B
200	50	0.0885	2.3742	1.0404
	100	0.1199	6.2168	0.4185
	150	0.0998	6.9678	0.5882
	200	0.0828	5.7350	0.9903
	250	0.0622	3.8744	2.2439

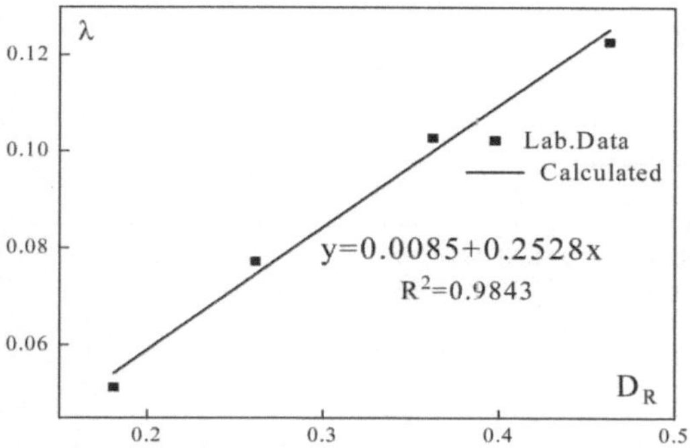

Figure 5. λ versus D_R plot.

and matric suction of 50 kPa. Models under other four matric suctions can be established by the same method.

4 PARAMETER ANALYSIS AND MODEL REVISION

Parameters of Singh-Mitchell creep model are supposed to be independent of deviatoric stress level and time. In other words, these parameters can be referred to as material constant. But data in Tables 4, 5 indicate that there are variations in the values of these parameters as the stress intensity and elapsed time change.

4.1 *Interrelation among parameters, stress intensity and time*

Data in Table 4 show that values of λ appear to be a function of the deviator stress level D_R. Plot λ versus D_R in Figure 5. Clearly, a linear relationship exists between λ and D_R, with correlation coefficient $R = 0.9843$. Data in Table 5 show that B and β_{DR} vary significantly with the elapsed time increasing. Figures 6, 7 indicate that B-$\ln t$, β_{DR}-$\ln t$ exhibit a linear relationship, with correlation coefficient both bigger than 0.95.

The above analysis shows that there is interrelation among the parameters, stress intensity and time, which can also be observed in other creep models under different suctions. Therefore, the established unsaturated Singh-Mitchell creep models need to be revised by taking into account these correlations to accurately describe the time-dependent stress-strain behavior of the sliding zone soils of the Qianjiangping Landslide.

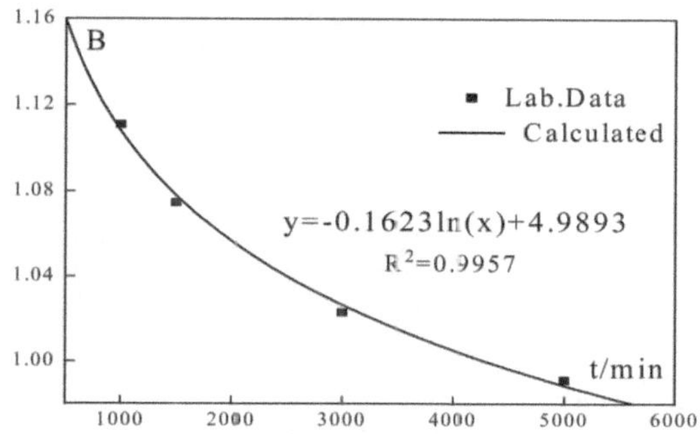

Figure 6. B versus t plot.

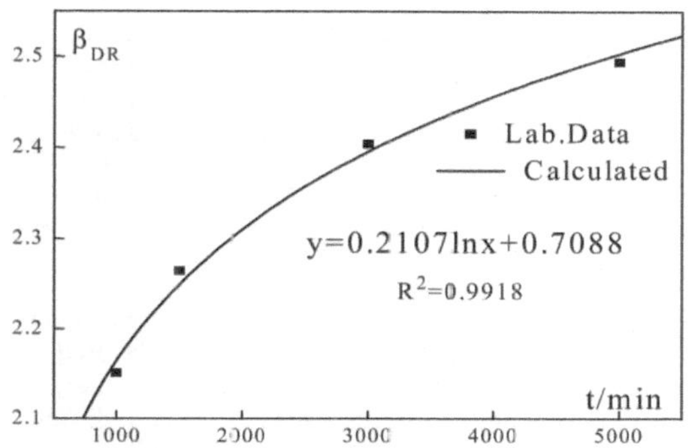

Figure 7. β_{DR} versus t plot.

4.2 *Revision of unsaturated Singh-Mitchell model and parameter fitting*

Relationships among R, β, λ and stress intensity as well as time are investigated because the parameter B is related to the of intercept of $D_R - \ln \varepsilon$ plot R and λ.

Let $R = a \ln t + b$, $\beta_{DR} = c \ln t + d$, $\lambda = eD_R + f$, where a, b, c, d, e, f are the slope and intercept of plots of R-$\ln t$, β_{DR}-$\ln t$, λ-D_R, respectively. Relationship between B and stress intensity as well as time can be obtained by functions of R and λ. Then substitute these functions into the unsaturated Singh-Mitchell model, Eq.(6) can be rewritten as:

$$\varepsilon = \exp(A \ln t D_R + B \ln t + C D_R + D)(\frac{t}{t_1})^{ED_R+F} \tag{9}$$

Where, $A = c - e$, $B = a - f$, $C = e \ln t_1 + d$, $D = b + f \ln t_1$, $E = e$, $F = f$.

There are six parameters A, B, C, D, E, F in Eq. (9) which are decided by another six parameters a, b, c, d, e, f. Firstly, through linear fitting of the test data, values of a, b, c, d, e, f can be obtained and so that of A, B, C, D, E, F. Take them as the initial values of A, B, C, D, E, F (Table 7) and obtain the final values (Table 8) by the nonlinear regression tool in Matlab.

4.3 *Verification of unsaturated Singh-Mitchell model*

Comparison of predicted results using the above models with test data were conducted for each laboratory

Table 7. Initial values of parameters of the model.

σ_3' kPa	u_a kPa	A	B	C	D	E	F
200	50	−0.042	0.0092	1.74343	0.2641	0.2527	0.0085
	100	−3.7861	0.2751	21.98121	−1.9164	3.7064	−0.1763
	150	−0.4433	0.0564	9.248102	−0.8016	0.3224	0.0537
	200	−0.0797	−0.0033	5.139227	0.1614	0.3163	0.0456
	250	−0.0143	−0.0015	2.878376	0.9586	0.2744	0.0066

Table 8. Final values of parameters of the model.

σ_3' kPa	u_a kPa	A	B	C	D	E	F
200	50	−0.0431	0.0081	1.7371	0.2578	0.2573	0.0131
	100	−3.8096	0.2522	21.8629	−2.0347	3.7899	−0.0928
	150	−0.4556	0.0439	9.1859	−0.8640	0.3656	0.0972
	200	0.0295	−0.105	5.3267	0.3201	0.4603	0.0964
	250	−0.0151	−0.0056	2.8631	0.9429	0.2889	0.0194

test, with comparison results shown in Figure 2. They show that these models can give reasonable predictions of the primary and secondary creep. But there is a difference when the stress intensity D_r is high. The calculated creep strain is a little lower than experimental data at the early stage of creep and a bit higher at the end of creep. This difference has also been mentioned in Ref. (Wang et al. 2005, Liu et al. 2008). Reasons for this difference need further investigation.

5 CONCLUSIONS

A set of unsaturated creep tests under a net cell pressure and five different matric suctions were conducted by using the self-developed unsaturated triaxial creep apparatus and a group unsaturated creep models were established. Based on the study in this paper, the following conclusions can be drawn:

1) By using Singh-Mitchell creep model for reference, a new stress intensity including matric suction was defined and thus the unsaturated Singh-Mitchell creep models of the sliding zone soils were developed. When soil saturated, this unsaturated creep model is turned into the saturated.
2) Through analysis of parameters of the models, it is found that they are not always independent of the stress intensity as well as time. In light of this, functions of parameters and stress intensity as well as time were established and taken to the unsaturated models to make them revised.
3) Comparison of predicted results using the above revised models with test data indicates that these models can give reasonable predictions of the primary and secondary creep. But there is a difference when the stress intensity D_r is high. The calculated creep strain is a little lower than experimental data at the early stage of creep and a bit higher at the end of creep. Reasons for this difference need further investigation.

ACKNOWLEDGEMENT

Research presented in this paper is part of projects supported by the projects by National Natural Science Foundation (No. 50879044, No. 50839004). This support is gratefully acknowledge.

REFERENCES

Fredlund D.G. & Rahardjo H. 1997. *Soil mechanics for unsaturated soil.* Chinese edition. Chen Z.Y., Zhang Z.M. & Chen Y.J. translators, China Beijing: China Architecture and Building Industry Press.

Guan L. & Wang S.M. 2008. Discussion on creep test methods of unsaturated soil. *Journal of Three Gorges University,* 30(2):32–34.

Lin H.D. & Wang C.C. 1998. Stress-strain-time function of clay. *Journal of Geotechnical and Geoenvironmental Engineering,* 124(4):289–296.

Lu P.ZH., Zeng J. & Sheng Q. 2008. Creep tests on soft clay and its empirical models. *Rock and Soil Mechanics,* 29(4):1041–1044.

Mesri, G., Rebres-Cordero E. & Shields, D.R. 1981. Shear stress-strain-time behavior of clays. *Geotechnique,* 31(4): 537–552.

Semple, R.M. 1973. *The effect of time-dependent properties of altered rock on the tunnel support requirements.* Ph. D. Thesis, University of Illinois, Urbana, Ill.

Singh, A. & Mitchell J.K. 1968. General stress-strain-time function for soils. *Journal of Soil Mechanics and Found Engineering Division,* ASCE, 94(1): 21–46.

Wang C., Zhang Y.L. & Liu H.W. 2005. A modified Singh-Mitchell's creep function of sliding zone soils of Xietan landslide in Three Gorges. *Rock and Soil Mechanics,* 26(3):415–418.

Modelling and Computation in Engineering – Zhu (ed)
© 2011 Taylor & Francis Group, London, ISBN 978-0-415-61516-7

Pounding effects of movable bearing on seismic behaviors of continuous girder bridges

W.L. Qiu, M. Jiang & L. Zhou
School of Civil Engineering, Dalian University of Technology, Dalian, P.R. China

ABSTRACT: The bearings are generally designed into two types: fixed and movable. The pier using fixed bearing bears most of seismic force caused by longitudinal earthquake. Only when the displacement of girder exceeds the gaps of the movable bearings, can the piers using movable bearings bear longitudinal seismic force besides the friction force. Using nonlinear time-history analysis method, considering the pounding of bearings and elastic-plastic nonlinearity, the seismic performance of continuous girder bridge under rear earthquake are studied. The influences of pounding stiffness, magnitude of gap, deformation of girder and intensity of earthquake on seismic responses are analyzed in detail. Based on the results of the study, the advices of reasonable gap and stiffness to reduce the seismic responses are given.

1 INTRODUCTION

The continuous girder bridges are largely used in bridge construction, and they are usually composed of 3 and more than 3 spans. Considering the longitudinal deformation of girder caused by temperature, shrinkage and creep of concrete, tension of tendons in girder and other factors, only one pier using fixed bearing (called fixed pier) can be adopted in one continuous girder bridge with normal pier height. So the fixed pier will bear very large force and be severely destroyed or collapse under strong earthquake. For continuous girder bridge with span of 20–40 m and pier height of about 5 m, the fixed pier can bear the moderate earthquake on its self. Under strong earthquake, large plastic displacement occurs in fixed pier. If the displacement of girder exceeds the gaps of movable bearing on the adjacent piers, the pier using movable bearing (called movable pier) can bear part of the longitudinal earthquake, which can avoid the fixed pier to be severely destroyed and make the whole bridge safer.

Many researchers have studied the pounding of bridge under earthquake. Considering some influencing factors such as pounding stiffness, gap and vibration period, Maleki (2004) studied the pounding effect between girder and side retainer of simply supported girder bridge. Teng (2007) studied the effects of span length, damping constant, pier height and characteristics of seismic wave on the transverse pounding of continuous bridge. Nie (2005) studied the seismic reduction of anti-seismic bolt of heteromorphic bridge. Wang (2008) studied the effects of three types of restrainers on seismic responses of continuous girder bridges. In this paper, considering longitudinal pounding of movable bearing, the seismic reduction of continuous girder bridge is studied in detail.

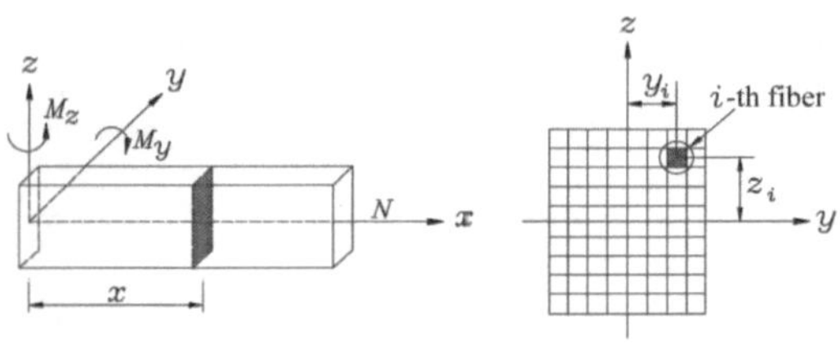

Figure 1. Fiber element model.

2 MODEL FOR NONLINEAR TIME-HISTORY ANALYSIS

2.1 *Elastic-plastic model of reinforced concrete pier*

The elastic-plastic fiber beam-column element model as shown in figure 1 is used to analyze the Elastic-plastic behavior of reinforced concrete pier (Enrico et al. 1996, Xie 2006). Firstly, the following assumptions are made to establish the fiber element model:

(1) Plane sections remain plane at any time.
(2) The slide between steel bar and concrete is ignored.

In the fiber model, the element is divided into some fibers along the longitudinal direction, if the number of the fibers is large enough, every fiber can be treated as only having axial strain and the curvature is ignored. The strain of every fiber is calculated by the axial strain and curvature of section. So the moment-curvature relationship of every section can be determined by the stress-strain relationship of the material and strain distribution on the section. For any section, the strain ε_i

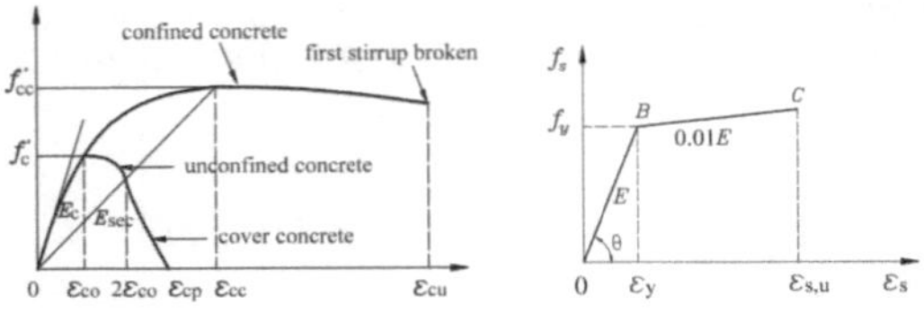

Figure 2. Stress-strain relationships of concrete and steel.

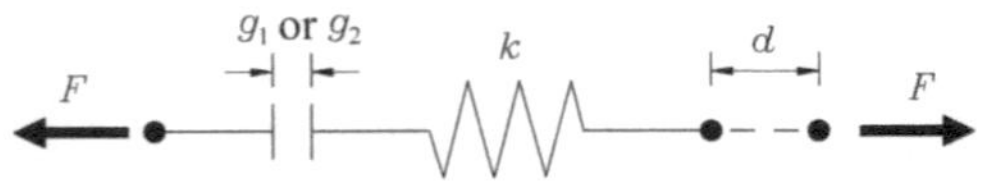

Figure 3. Pounding element of movable bearing.

of the ith fiber can be determined by the following expression

$$\varepsilon_i = [-z_i \quad -y_i \quad 1]\begin{Bmatrix} \varphi_y(x) \\ \varphi_z(x) \\ \varepsilon_x(x) \end{Bmatrix} \tag{1}$$

Where, x is the location of section on the element, $\varphi_y(x)$ is the curvature about axis y, $\varphi_z(x)$ is the curvature about axis z, $\varepsilon_x(x)$ is axial strain of element, y_i and z_i are coordinates of the ith fiber on the section. The axial force and moment of the section can be expressed as:

$$N = \sum_{i=1}^{n} (\sigma_{ci}A_{ci} + \sigma_{si}A_{si}) \tag{2}$$

$$M_y = \sum_{i=1}^{n} (\sigma_{ci}A_{ci} + \sigma_{si}A_{si})z_i \tag{3}$$

$$M_z = \sum_{i=1}^{n} (\sigma_{ci}A_{ci} + \sigma_{si}A_{si})y_i \tag{4}$$

Where σ_{ci} and σ_{si} are stresses of concrete and steel bar in the ith fiber respectively, which can be calculated by the stress-strain relationship of concrete and steel as shown in figure 2 (Chen & Duan 1999). A_{ci} and A_{si} are areas of concrete and steel bar in the ith fiber respectively. Using the above expressions, the moment-curvature relationship of every section can be determined.

2.2 *Pounding model of bearing*

In this paper, two pounding elements are used to simulate the gaps and poundings of one movable bearing in two directions when the bridge is subject to earthquake. One pounding element is shown in figure 3. In the model, k is the pounding stiffness of bearing, g_1 or g_2 is the gap of the bearing in one direction, d is the relative displacement of the girder and top of pier where the movable bearing located on. The pounding force can be determined by the following expression:

$$F = \begin{cases} k(d + g_i) & (d + g_i \le 0) \\ 0 & (d + g_i \le 0) \end{cases} \tag{5}$$

Additionally, a damping is used to consider energy dissipation of sliding friction of movable bearing.

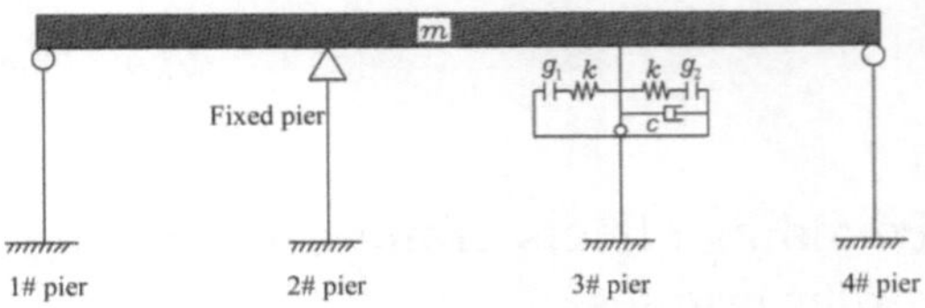

Figure 4. Analysis model of the bridge.

2.3 *Numerical example*

Taking a 3-span continuous girder bridge with span length of 30 m as numerical example, the seismic responses considering pounding of bearing are studied. The main girder is prestressed concrete box girder. The weight of each span girder is 12460 kN. Two circular section reinforced concrete columns with diameter of 1.5 m are adopted as pier. The cylinder compressive strength of concrete is 30 MPa, the tensile strength is ignored, and the initial Young's modulus 2.76e4 MPa. In the pier, 30 steel bars with diameter of 32 mm are used as longitudinal bars, and circular steel stirrups with diameter of 16 mm and space 100 mm are adopted. The yield strength of the steel bar is 335 MPa and the initial Young's modulus 2.08e5 MPa. Based on the design code of China, the yield curvature of the pier is 0.0025 rad/m, and the ultimate curvature is 0.028 rad/m.

Because the aim of this paper is to study the seismic reduction of movable pier, the El Centro seismic wave is adopted in the nonlinear time-history analysis. The peak earthquake acceleration is adjusted to be 0.2 g. The wave is input longitudinally. To make the study results representative, the heights of all piers take the value of 5 m. The analysis model of the bridge is shown in figure 4. The software MIDAS is used to analyze the seismic performance of the bridge. The influences of pounding stiffness, magnitude of gap, intensity of earthquake on seismic responds are analyzed in detail.

3 THE INFLUENCES OF POUNDING STIFFNESS OF BEARING

In order to determine the influences of pounding stiffness of bearing on the seismic responses of the bridge, gap g_1 and g_2 of the movable bearing on 3# pier take the same value of 25 mm, the stiffness of bearing k takes the value of $k1 = 1e7$ kN/m, $k2 = 1e6$ kN/m, $k3 = 1e5$ kN/m and $k4 = 0$ (no pounding) respectively. By using the nonlinear time-history analysis, time histories of moment, shear force and displacement of the fixed and movable piers and moment-curvature hysteretic loops of the piers are obtained. Some of them are shown in figures 5–8. Because the seismic responses after 5 s become smaller and smaller with the time increasing, the responses after 7 s are not given in the figure. The maximal responses are listed in table 1 for the purpose of comparison.

From figures 5, 6, we can find that the pounding of bearing reduces markedly the displacement and

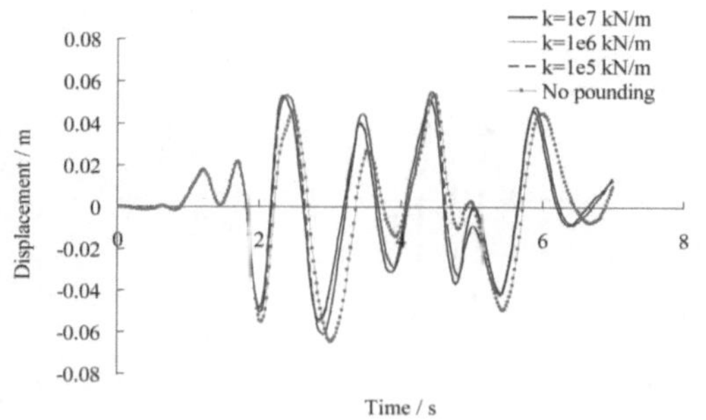

Figure 5. Displacement time-history of fixed pier top.

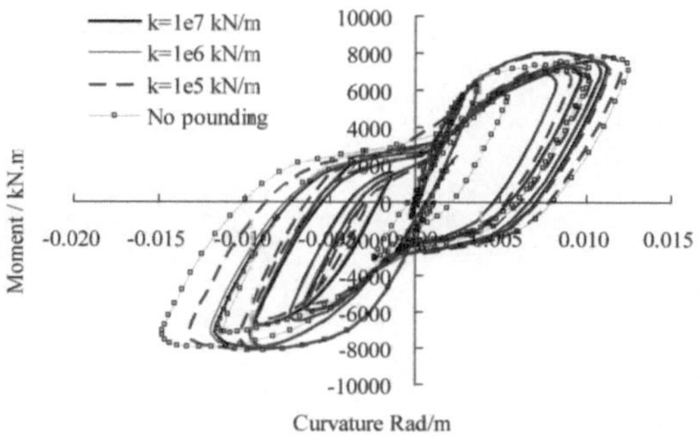

Figure 6. Moment-curvature relationship of fixed pier.

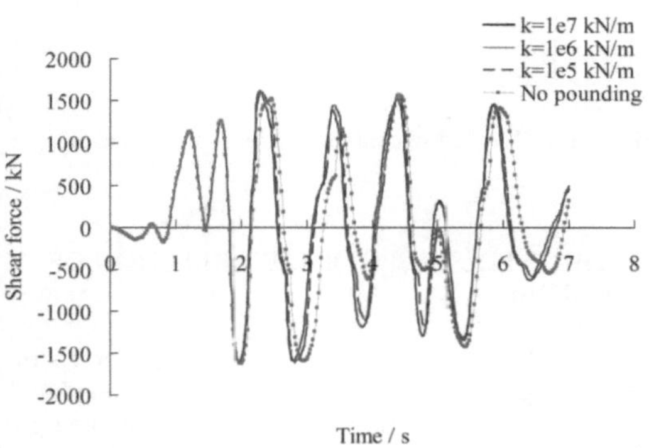

Figure 7. Shear force time-history of fixed pier.

curvature of the fixed pier. The maximal displacements of the fixed pier are nearly same for $k1$ and $k2$, their average value is about 84.7% of the displacement for no pounding. The maximal displacement for $k3$ is 94.7% of the value for no pounding. The maximal curvatures of the fixed pier are nearly same for $k1$ and $k2$, their average value is about 80.1% of the maximal curvature for no pounding. The maximal curvature for $k3$ is 89.2% of the value for no pounding. For the four cases, all of the maximal moments of the fixed pier bottom exceed the yield moment, the maximal shear forces considering pounding are nearly equal, the largest difference is less than 0.8%. The maximal pounding force for $k3$ is markedly smaller than the values of $k1$ and $k2$.

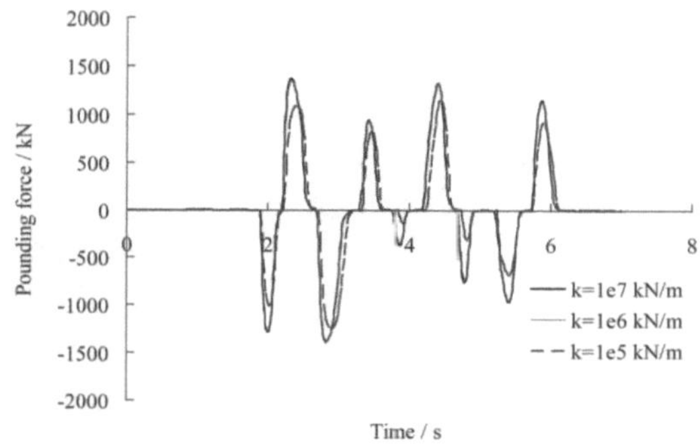

Figure 8. Pounding force of movable bearing.

Table 1. Maximal seismic responses.

k kN/m	D mm	φ Rad/m	S kN	F kN
$k1 = 1e7$	−53.4/52.3	0.0117	1620.1	1384.2
$k2 = 1e6$	−54.2/53.0	0.0120	1617.5	1375.8
$k3 = 1e5$	−61.1/55.2	0.0132	1613.4	1254.2
$k4 = 0$	−64.5/53.8	0.0148	1628.2	0.0

D is the displacement of fixed pier.
φ is the curvature of fixed pier.
S is the shear force of fixed pier.
F is pounding force of the movable bearing.

From the above comparison, we can see that, when the pounding stiffness of bearing takes the value about 1e6 kN/m, not only the desired seismic reduction effect can be obtained, but also the requirement of stiffness of bearing is reduced.

4 THE INFLUENCES OF BEARING GAPS

Considering the influences of the span length, temperature, shrinkage and creep of concrete and other factors, the total gap $G = g_1 + g_2$ is different for different bridges. In order to study the influences of the total gap on seismic responses, G takes the value of $G_1 = 30$ mm, $G2 = 50$ mm and $G3 = 70$ mm respectively. For every case, g_1 and g_2 are equal, and the stiffness k is 1e6 kN/m. By using the nonlinear time-history analysis, the three cases are analyzed. Some of them are shown in figures 9–11. The maximal responses are listed in table 2. For comparison, the results of no pounding are also given.

As shown in figures 9, 10, the total gap of bearing has large influence on displacement and curvature of the fixed pier. With decrease of the total gap, the effect of reduction is improved. The maximal displacements of the fixed pier for $G1$, $G2$ and $G3$ are respectively of 84.0%, 85.3% and 92.2% of the displacement for no pounding, and the maximal curvature of the fixed pier for $G1$, $G2$ and $G3$ are respectively of 72.9%, 81.1% and 87.2% of the value for no pounding. The

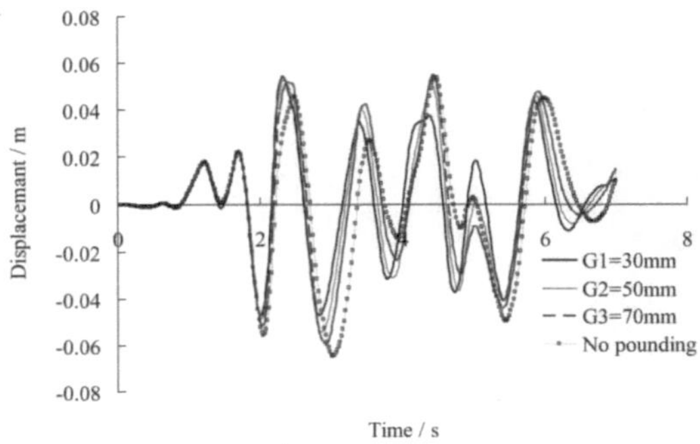

Figure 9. Displacement time-history of fixed pier top.

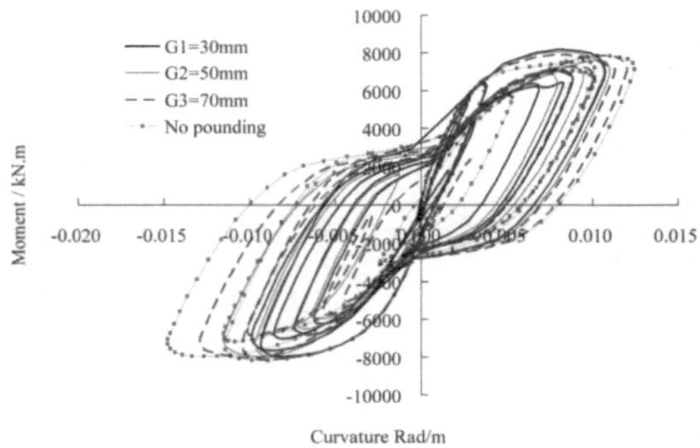

Figure 10. Moment-curvature relationship of fixed pier.

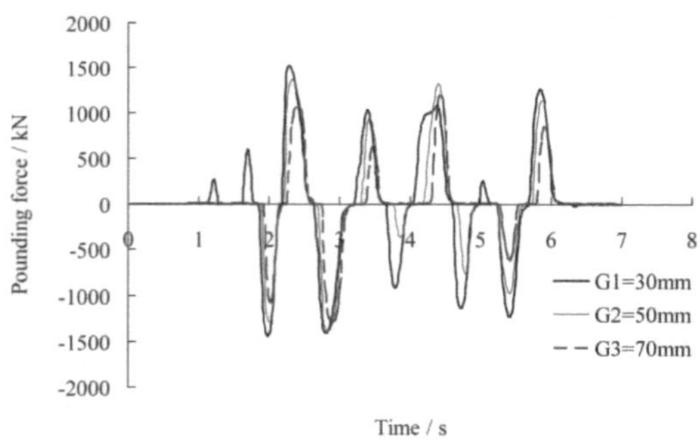

Figure 11. Pounding force of movable bearing.

Table 2. Maximal seismic responses.

G mm	D mm	φ Rad/m	S kN	F kN
$G1=30$	$-47.0/54.2$	0.0108	1588.6	1515.7
$G2=50$	$-54.2/53.0$	0.0120	1617.5	1383.9
$G3=70$	$-59.5/54.8$	0.0129	1620.6	1291.0
$G4=\infty$	$-64.5/53.8$	0.0148	1628.2	0.0

maximal shear forces considering pounding are nearly same with the difference of less than 2.5%. The maximal pounding forces for $G2$ and $G3$ are 91.3% and 85.2% of the values of $G1$ respectively.

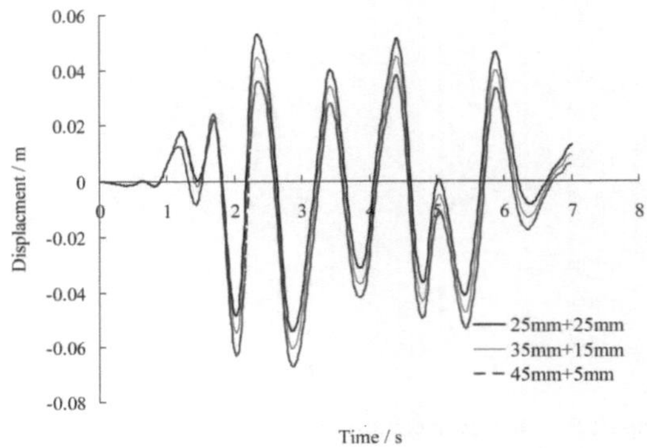

Figure 12. Displacement time-history of fixed pier top.

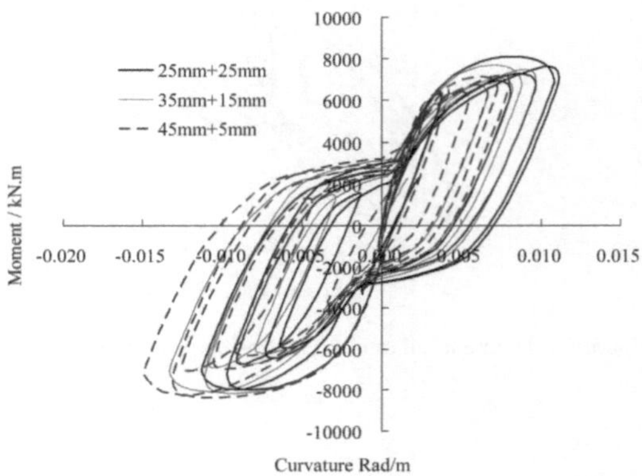

Figure 13. Moment-curvature relationship of fixed pier.

5 THE INFLUENCES OF DEFORMATION OF GIRDER

Influenced by temperature, shrinkage and creep of concrete, prestress and other factors, the length of girder changes at any time, and the gap g_1 and g_2 of movable bearing are not equal when the earthquake happen. Taking the total gap $G=50$ mm as sample, the influences of the changes of g_1 and g_2 on the seismic reduction of pounding are studied in this paper. Here, g_1 takes the different values of 25 mm, 35 mm, 45 mm, and g_2 takes the corresponding values of 25 mm, 15 mm and 5 mm respectively. The stiffness k is 1e6 kN/m. Some time-history of seismic responses are given in figures 12–14. The maximal responses are listed in table 3.

From figures 12,13, it is found that the deformation of girder largely affects on the seismic responses of the fixed pier. For the given value of total gap, the two gaps g_1 and g_2 are more same, the effects of seismic reduction are better. Not only too small gap or too large gap can not reduce the seismic responses, but also it maybe enlarges the responses and seriously destroys both fixed pier and movable pier.

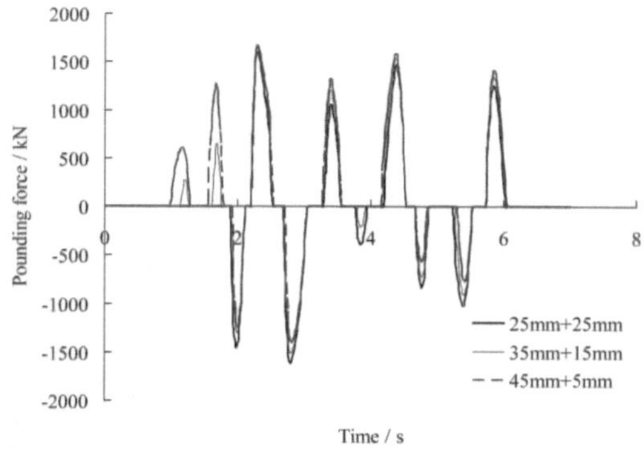

Figure 14. Pounding force of movable bearing.

Table 3. Maximal seismic responses.

g1/g2 mm	D mm	φ Rad/m	S kN	F kN
25/25	−54.2/53.0	0.0120	1617.5	1383.9
35/15	−60.4/44.9	0.0133	1633.9	1389.5
45/5	−67.1/38.0	0.0149	1674.7	1428.2

6 STUDY ON ANTI-SEISMIC CAPACITY

If the allowable coefficient of ductility is 6 (Fan 1997, Kowalsky 2000), the allowable curvature is 0.015 rad/m. In order to study the anti-seismic capacity of continuous girder bridge when considering pounding of bearing, by increasing the peak earthquake acceleration of El Centro wave, the anti-seismic capacity of the bridge can be obtain when the curvature of any pier reaches the allowable curvature. Taking the case of $k = 1e6$ kN/m and $g1 = g2 = 25$ mm as sample, the study result shows that, when the peak earthquake acceleration reaches 0.55 g, the curvature of fixed pier reaches the allowable curvature if considering pounding. But if not considering pounding, the peak earthquake acceleration is 0.42 g when the curvature of fixed pier reaches the allowable curvature. The ratio of the peak earthquake accelerations of the two cases are 1.31.

7 CONCLUSION

The study shows that reasonable design of movable bearings can efficiently reduce the seismic responses and damage of fixed pier. The effect of reduction is improved with increase of the stiffness of movable bearing and decrease of total gap of movable bearing. Too small stiffness has little effect of reduction, but the effect is improved very little when the stiffness larger than a certain value. The minimal gap of movable bearing must satisfy the need of deformation of girder caused by temperature, shrinkage and creep of concrete, tension of tendons and other factors. In order to obtain better reduction effects, the two gaps of bearing should take the nearly equal value. Too small gap maybe enlarge the responses and seriously destroy both fixed pier and movable pier. So we advise that the gaps of the movable bearing can be adjusted to lessen the total gap and to make the two side gap nearly equal after deformation of girder caused by shrinkage and creep of concrete occurs.

Because the movable bearing is subject to large pounding force, the capacities of bearing horizontal force of fixed bearing and movable bearing is advised to be equal. And the capacities should be large than the flexural capacity of the piers.

REFERENCES

Chen W.F. & Duan L. 1999. *Bridge Engineering Handbook*, CRC press LLC.

Deng Y.L., Peng T.B. & Li J.Z. 2007. Pounding model of bridge structures and parameter analysis under transverse earthquakes. *Journal of Vibration and Shock* 26(9): 104–106.

Enrico S, Filip C. & Filippou. 1996. Fiber beam-column model for nonlinear analysis of RC Frames: Part I. Formulation. *Earthquake Engineering and Structural Dynamics* 1996, 25: 711–725.

Enrico S, Filip C. & Filippou. 1996. Fiber beam-column model for nonlinear Analysis of RC Frames: Part II. Application. *Earthquake Engineering and Structural Dynamics* 25: 727–742.

Fan, L.C. 1997. *Seismic design of bridge*. Shanghai: Tongji University Press.

Kowalsky, M. J. 2000. Deformation limit states for circular reinforced concrete bridge column. *Journal of Structural Engineering, ASCE* 26 (8):869–878.

Maleki S. 2004. Effect of side retainers on seismic response of bridges with elastomeric bearings. *Journal of bridge engineering* 9(1): 95–100.

Nie L.Y., Li J.Z. & Fan L.C. 2005. Selection of pounding analysis parameters and its effects on structure under earthquake. *Engineering Mechanics* 22(5): 142–146.

Wang J.W., Li J.Z. & Fan L.C. 2008. Effects of Restrainers on Seismic Responses of Continuous Beam Bridges, *Journal of the China railway society* 30(3):71–77.

Xie, X. 2006. *Seismic response and earthquake resistant design of bridges*, Beijing: China Communications Press.

Modelling and Computation in Engineering – Zhu (ed)

ISBN 978-0-415-61516-7

Research on transient state of ultrasonic feeding

Liang Li
College of science, Liaoning University of Technology, Jinzhou, Liaoning, China

Qing He
Institute of Vibration Engineering, Liaoning University of Technology, Jinzhou, Liaoning, China

ABSTRACT: In this paper, a dynamical model describing the motion of the specimen of the ultrasonic feeding is used to study the transient state of the ultrasonic feeding. By solving the dynamical equation, the specimen's normal velocity of the ultrasonic feeding can be calculated. The results indicate that the time for the specimen to reach the steady state is much greater than that for the oscillator, and the time for specimen to stop is also much greater than that for the oscillator. In addition, the motion of the specimen in the transient state embraces the low frequency oscillation and the high frequency oscillation.

1 INTRODUCTION

In recent years, with the development of the piezoelectric drive, especially the development of the ultrasonic motor, the ultrasonic feeding has been rapidly developed. Nowadays, the ultrasonic feeding is gaining ever increasing attention due to its advantages such as running without noise, fast response of startup and stopping, light-duty and diversified structures etc. (Mracek & Wallaschek 2005, Yang & Li 2003, He & Wolfgang 2003). The motion of the specimen of the ultrasonic feeding embraces the steady state behaviour and the transient state behaviour. The steady state behaviour is the steady motion of the specimen with the actuation of the steady oscillation of the ultrasonic oscillator, and the transient state behaviour is the motion of the specimen under the conditions of the startup and the stopping of the ultrasonic feeding. Although the ultrasonic feeding mainly works in the steady state, the transient state is very important to control the motion of the specimen, as well as to the design of the ultrasonic feeding.

In this paper, a dynamical model describing the motion of the specimen of the ultrasonic feeding is used to study the transient behaviour of the specimen.

2 DESCRIPTION OF DYNAMICAL MODEL

The normal kinematical equation of the ultrasonic oscillator has the following form

$$y_0 = f(t) \tag{1}$$

The normal dynamical equation of the specimen with the actuation of the ultrasonic oscillator can be given by (Li & He 2009)

$$m\frac{\mathrm{d}^2 y'}{\mathrm{d}t^2} = F_c - mg - m\frac{\mathrm{d}^2 y_0}{\mathrm{d}t^2} - \gamma\left(\frac{\mathrm{d}y_0}{\mathrm{d}t} + \frac{\mathrm{d}y'}{\mathrm{d}t}\right) \tag{2}$$

Where, y' denotes the ordinate of the specimen relative to the oscillator's surface (noninertial system), m is the mass of the specimen, t is the time, γ is the specimen damping coefficient, and F_c is the normal contact force which has the following form (Li & He 2009)

$$F_c(y') = \frac{\pi^2 E_{\mathrm{equ}} a r^2}{4\sqrt{2\pi}\cdot b\sigma} \cdot \int_{y'}^{\infty} \frac{(l-y')}{l} \cdot \exp\left\{-\frac{[l-2(Ra_o + Ra_s)]^2}{2\sigma^2}\right\} \cdot \mathrm{d}l \tag{3}$$

Where, a and b are the upper radius and the lower radius of the micro peaks on the rough surfaces respectively, r is the radius of the cylindrical specimen, σ is the model parameter describing the contact surfaces. Ra_o and Ra_s are the roughness of the oscillator and the specimen respectively. E_{equ} is equivalent Young modulus and has the following form (Li & He 2009)

$$E_{\mathrm{equ}} = \frac{E_s E_o (Ra_o + Ra_s)}{E_s Ra_o + E_o Ra_s} \tag{4}$$

Where, E_s and E_o are the Young moduli of the specimen and the oscillator respectively.

The ordinate of the specimen in the inertial system is given by

$$y = y_0 + y' \tag{5}$$

And then, the normal velocity of the specimen has the following form

$$v = \frac{\mathrm{d}y}{\mathrm{d}t} = \frac{\mathrm{d}y_0}{\mathrm{d}t} + \frac{\mathrm{d}y'}{\mathrm{d}t} \tag{6}$$

If the kinematical equations of the oscillator are given, the normal displacement of the specimen relative to the oscillator's surface y' can be calculated by solving Equation (2), and the velocity of the specimen can be calculated by the equation (6).

In this paper, it is assumed that the kinematical equation of the ultrasonic oscillator in the startup state is

$$y_0 = [A - A\exp(-\beta t)]\cos(\omega t + \phi) \quad (7)$$

The kinematical equation of the ultrasonic oscillator in the steady state is

$$y_0 = A\cos(\omega t + \phi) \quad (8)$$

And the kinematical equation of the ultrasonic oscillator in the stopping state is

$$y_0 = A\exp(-\beta t)\cos(\omega t + \phi) \quad (9)$$

Where, A, ω, ϕ is the steady amplitude, the angular frequency and the initial phase of the oscillator's oscillation respectively, β is the oscillator damping coefficient.

3 RESULTS AND DISCUSSION

The material of the oscillator under consideration is copper and its surface is much bigger than the specimens'. The specimens studied in this paper are made in steel. All the specimens are cylinder and their heights are all 5×10^{-3} m and their radii are all 3×10^{-3} m. In theoretical analysis, the upper radius a and the lower radius b of the micro peak are 5×10^{-12} m and 8×10^{-8} m respectively.

In the initial state, the oscillator doesn't move, and the specimen keeps motionless on the surface of the oscillator.

Figure 1 is the normal velocity curve of the ultrasonic oscillator. From 0s to 5 s, the amplitude of the oscillator is increasing and the oscillation gradually reaches the steady state, whereas, from 30 s to 35 s, the amplitude is decreasing and the oscillation gradually stop. That is, in the case of the Figure1, the time for the oscillator to reach the steady state is about 5 s, and the time for the oscillator to stop is also about 5 s.

Figure 2 shows the calculated normal velocity curve of the specimen with the actuation of the oscillator of figure 1. Figure 2 indicates that the time for the specimen to reach the steady state is over 15 s, and the time for the specimen to stop is over 10 s.

Comparing figure1 and figure 2, the conclusion can be drawn that the time for the specimen to reach the steady state is about three times of that for the oscillator, and the time for the specimen to stop is about two times of that of the oscillator, this is, it takes much more time for the specimen, than for the oscillator, to reach the steady state and stopping state in the same conditions.

Figure 3 and Figure 4 show the specimen's calculated normal velocity in the startup state. From these two figures, it can be seen that, in the startup state, the motion of the specimen embraces the low frequency

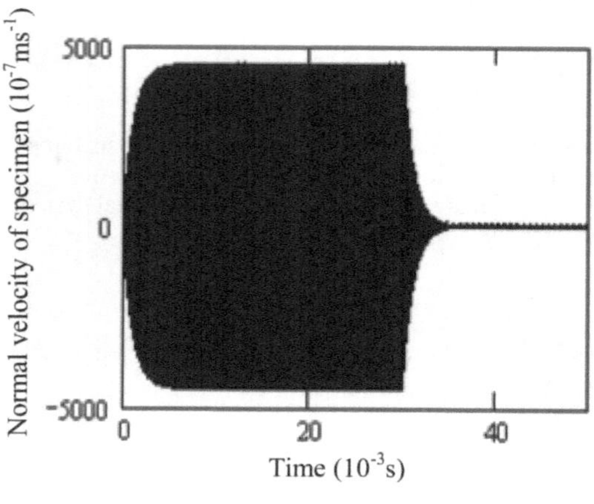

Figure 1. The normal velocity curve of the ultrasonic oscillator. $A = 3 \times 10^{-6}$ m, $\omega = 1.5 \times 10^5$s^{-1}.

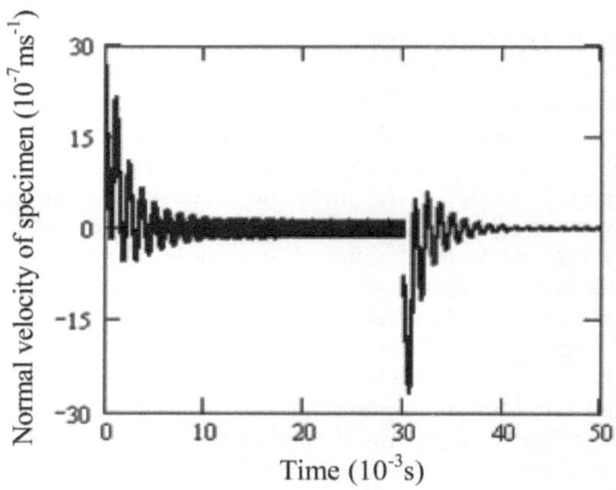

Figure 2. The calculated normal velocity curve of the specimen with the actuation of the ultrasonic oscillation. The normal velocity curve of the oscillator is given in Figure 1.

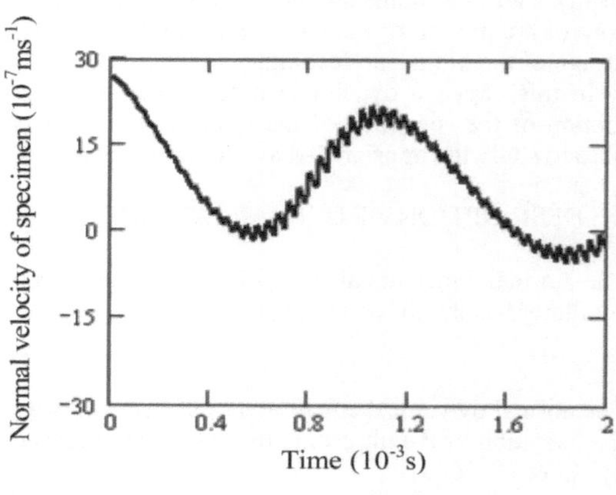

Figure 3. The part of the curve in the figure 2 showing the specimen's normal velocity in the startup state.

oscillation and the high frequency oscillation, and the low frequency oscillation gradually decreases in amplitude and dies out, whereas the amplitude of the high frequency oscillation increases with time. In addition, it is quite apparent that the frequency of the high frequency oscillation approaches the frequency of the oscillator.

Figure 5 shows the steady behaviour of the specimen. The figure indicates that, in the steady state, the low frequency oscillation dies out, and the high frequency oscillation of the specimen approaches the simple harmonic oscillation. In the SI units, the specimen's amplitude of the steady state is three orders of magnitude lower than that of the oscillator's, and the frequency of the specimen, in the steady state, is equal to that of the oscillator.

Figure 6 and Figure 7 show the specimen's calculated normal velocity curves in the stopping state. According to the two figures, the motion of the specimen in the stopping state also embraces the low frequency oscillation and the high frequency oscillation, and the amplitude of the low frequency oscillation suddenly gets greater just when the oscillator's amplitude begins to decrease, and then decreases with the oscillator's amplitude approaching zero. In addition, the frequency of the low frequency oscillation nearly equals to that in the startup state, and the frequency of the high frequency oscillation is also equal to the frequency of the oscillator. Both the low frequency oscillation and the high frequency oscillation all gradually decrease in amplitude and die out at last.

4 CONCLUSIONS

The specimen's motion with the actuation of the ultrasonic oscillator embraces the steady state behaviour and the transient state behaviour, and the transient state behaviour is the motion of the specimen under the conditions of the startup and the stopping of the ultrasonic feeding. In this paper, the dynamical model is used to study the specimen's transient state behaviour. By

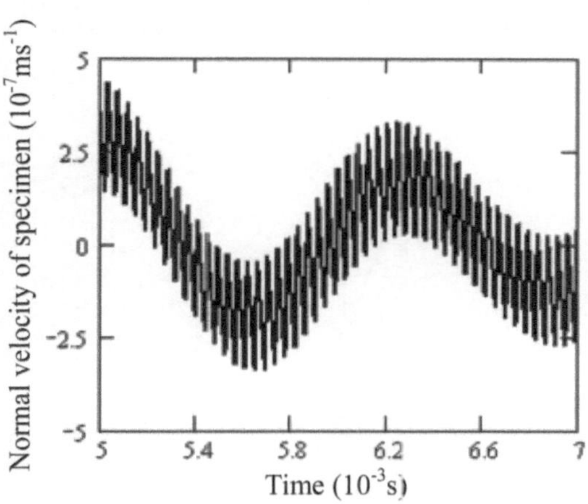

Figure 4. The part of the curve in the figure 2 showing the specimen's normal velocity in the startup state.

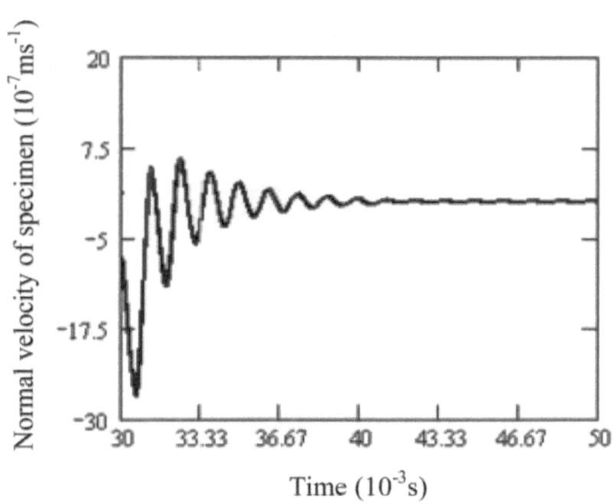

Figure 6. The part of the curve in the figure 2 showing the specimen's velocity in the stopping state.

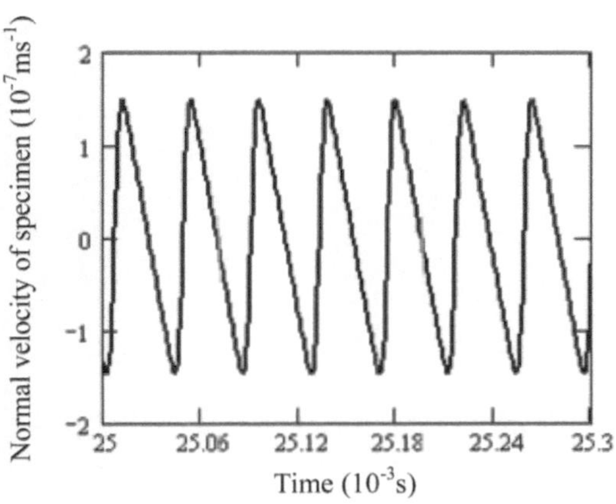

Figure 5. The part of the curve in the figure 2 showing the specimen's steady behaviour.

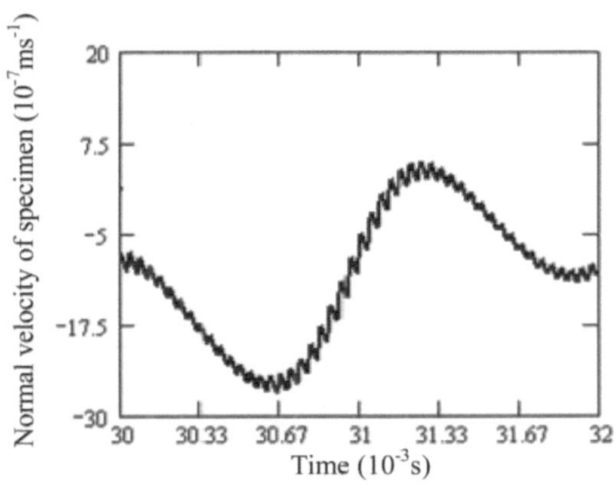

Figure 7. The part of the curve in the figure 2 showing the specimen's normal velocity in the stopping state.

solving the dynamical equation of the specimen, the normal velocity curves of the specimen in both the steady state and the transient state are obtained. From these curves, the conclusions can be drawn that:

(1) The time for the specimen to reach the steady state is much greater than that for oscillator, and the time for specimen to stop is also much greater than that for oscillator.
(2) The motion of the specimen in the transient state embraces the low frequency oscillation and the high frequency oscillation.
(3) The frequency of the high frequency oscillation in both the startup state and the stopping state equals to the frequency of the oscillator. The high frequency oscillation increases in amplitude with the time in the startup state, whereas decreases in the amplitude with the time in the stopping state.
(4) The amplitude of the specimen's low frequency oscillation in the startup state decreases with the time. Whereas, in the stopping state, the amplitude of the low frequency oscillation suddenly becomes much greater just when the oscillator's amplitude begins to decreases, and then decreases with the time. The frequency of the low frequency oscillation in the startup state is equal to that in the stopping state.

ACKNOWLEDGMENT

This paper is supported by national natural science foundation of China (50675094).

REFERENCES

He, Q. & Wolfgang, S. 2003. Modal analysis of a ring type linear ultrasonic traveling wave motor. *Piezoelectrics & Acoustooptics* 25(6): 514–516, 520.

Li, L. & He, Q. 2009. Modelling of normal contact between specimen and oscillator, *Proceeding of 2009 International Conference on Simulation and Optimation*, Peking, China: 52–54.

Mracek, M. & Wallaschek, J. 2005. A system for powder transport based on piezoelectrically excited ultrasonic progressive waves. *Materials Chemistry and Physics* 90: 378–380.

Yang Y. & Li X. 2003. Experimental and analytical study of ultrasonic micro powder feeding. *J. Phys. D: Appl. Phys.* 36: 134–1354.

Zhang, G.L. Guo, H. & Zhao, C. S. 2002. Experimental Study on the Feeding Ability of Ultrasonic Powder feeding device. *Piezoelectrics & Acoustooptics* 24(6): 489–491.

Modelling and Computation in Engineering – Zhu (ed)
© 2011 Taylor & Francis Group, London, ISBN 978-0-415-61516-7

Crack identification of multi-layer plane frames based on wavelet transform of rotation mode

De-Qing Guan, Nan Jiang & Yun-Tao Dong
School of Civil Engineering and Architecture, Changsha University of Science and Technology, Changsha, China

ABSTRACT: By means of the wavelet analysis theory the damage identification of plane frame was studied. Based on the finite element anâlysis of modal parameters, the method using the wavelet analysis of rotation mode was proposed for two-story plane frames crack identification, and the hear wavelet transform was applied to analyse the rotation mode of plane frame, so the location of crack of the plane frame can be identified by the maximum of wavelet coefficients. Some useful conclusions for the plane frame have been drawn and this method may be useful in damage identification and diagnosis in structures.

1 INTRODUCTION

Framework is the basic structure type, which in many engineering sructures are extensively used. As the impact of earthquake and wind load, the internal structure of materials will change in the period of structural service and resulte in a variety of damage. These damage threatening people's life and property all the time. Therefore, the research of damage identification of framework structure not only has important theoretical significance, but also has a useful reference value for engineering applications. According to structural dynamics theory, the structural damage will cause the change of structural modal parameters, such as natural frequencies, normal vibration mode, stiffness and damping.

At present, the main method about damage identification of frame structure is as follows: wavelet analysis (Liu, Gurgenci, Veidt 2004), methods basedon testing frequency and mode (Nikolakopoulos, Katsareas 1997), neural network analysis (Elkordy, Chang, Lee 1994). Wavelet Analysis is regarded as "mathematical microscope". Its quality of localization of a space (on low scale signal's wavelet transform depended on local quality at certain point) can be used to analyze the locality and degree of signal's singularity. In recent years, damage identificaton based on wavelet analysis has been applied for the damage detection of frames. C.S. Huang. etc. (Huang, Su 2007) used continuous wavelet transform for analysing acceleration response of three-steel frames to identify structural damage. Hou Zect (Hou, Noori, Amand 2000) applied wavelet decomposition to the health monitoring of frame structure with some cracks caused by the discontinuous load and accumulated fatigue damage. A.V. Ovanesova, etc. (Ovanesova, Suárez 2004) used the discrete wavelet transform for detecting dynamic parameters of the support pillar and beam-column joints where the frame contain cracks. A. Hera. etc. (Adriana Hera, Zhikun Hou 2004) to study the damage identification of four-story steel frame, firstly, obtained acceleration response of each node by finite element analysis, and then decompose the signal by db4 wavelet to identify the location of damage. Wang Jun etc. analyzed the response signal of defective a frame model under pulse excitation and use wavelet packet decomposition to identify the damage of structure. Lihong Quan based on vibration test about frame model of reinforced concrete used wavelet transform for decomposing signal of seismic response as different frequency for analyzing local features of signal, and then identified the location of structural damage. Guan Deqing and Jiang Xin (Guan Deqing & Jiang Xin 2007) proposed that continuous wavelet transform of rotation mode could identify cracks of continuous beam, and found that wavelet transform of rotation mode is more effective to identify the damage than the fundamental mode of vibration. On this basis, through analyzing dynamic parameters of plane frame by the finite-element analysis. This paper proposed that wavelet transform of rotation mode could identify structural damage. Take two-plane frame for example, we respectively studied cracks in the beam and column. Firstly, mode of frame is established by the finite element analysis and then hear wavelet is applied to wavelet transform of rotation mode for location of crack of the plane frame can be identified by the of maximum wavelet coefficients. Calculation results show that the method is effective and this paper had the important significance for damage detection of the building construction.

2 THEORETICAL BASIS

2.1 *Singularity identification theory of wavelet*

Set function $\psi(t) \in L^1(R) \cap L^2(R)$ and $\int_{-\infty}^{+\infty} \psi(t) d_t = 0$, where $\psi(t)$ is the basic wavelet or the mother wavelet.

Translating and shifting mother wavelet can obtain Eqn. 1.

$$\psi_{a,b}(t) = \left(\sqrt{|a|}\right)^{-1/2} \psi\left(\frac{t-b}{a}\right), \quad a,b \in R, \quad a \neq 0 \tag{1}$$

Where $\psi_{a,b}(t)$ is the wavelet function, a is the scale factor that reflect the signal frequency information, b is the shift factor that reflect the signal time information.

In which mother wavelet should meet admissible condition.

$$c_\psi = \int_{-\infty}^{+\infty} \frac{|\hat{\psi}(\omega)|^2}{|\omega|} d\omega < +\infty \tag{2}$$

So, based on Eqn. 2, wavelet function requires not only has a definite concussion (it contains the characteristics of a certain frequency), but also has quality of space localization (it constants equal to 0 or converges to 0 quickly within a fixed range).

Assume that $\psi(t)$ is the basic wavelet, $\psi_{a,b}(t)$ is the continuous wavelet function. For arbitrary function or signal $f(t) \in L^2(R)$, continuous wavelet transform can be expressed by Eqn. 3.

$$WT_f(a,b) = \langle f(t), \psi_{a,b}(t) \rangle = |a|^{-1/2} \int_{-\infty}^{+\infty} f(t) \psi^*\left(\frac{t-b}{a}\right) dt \tag{3}$$

Where $a \neq 0$, b, t are continuous variables, $\psi^*(t)$ denotes the complex-conjugate of $\psi(t)$.

Continuous wavelet transform can be written as The following Eqn. 4

$$WT_f(a,b) = \left(\sqrt{|a|}\right)^{-1/2} \int_{-\infty}^{\infty} f(t) \psi^*\left(\frac{t-b}{a}\right) dt = |a|^{1/2} f * \psi_{|a|}(b) \tag{4}$$

In which $\bar{\psi}_{|a|}(t) = |a|^{-1}\psi^*(-t/a)$, so wavelet transform can also be considered as the signal and the filter convolution operation. From engineering significance of angle $\bar{\psi}_{|a|}(t)$ can be considered as the high pass filter.

Generally, singular signal can divided into two cases: (i) Mutation in the signal amplitude at one point will lead to signal non-continuous. This type of mutation is called as discontinuity point of the first kind. (ii) Signal is very smooth in appearance and signal amplitude has no mutation, whereas First-order differential of a signal has suffered a mutation and non-continuous. This type of mutation is called as discontinuity point of the second kind.

Assume that $\theta(t)$ is a smooth function $\int_{-\infty}^{+\infty} \theta(S)ds = 1$, and $\theta(s)$ denotes $(1 + x^2)^{-1'}$s infinitesimal of higher order. At the same time we know that $\theta_s(t) = \theta(1/s)/s$ and the wavelet function $\psi(t)$ denotes the first derivative of $\theta(t)(\psi(t) = d\theta(t)/dt)$, so $f(t)$ can be expressed by continuous wavelet transform as Eqn. 5.

$$Wf(s,u) = s^{1/2}\left(f * \bar{\psi}_s\right)(u) = s^{1/2} \frac{d}{du}\left(f * \bar{\theta}_s\right)(u) \tag{5}$$

The wavelet transformation module maximum $|Wf(s,u)|$ the first derivative maximum of "f" polished by θ_s, which just correspond to point mutations of the signal "f". Analyzing the signal on scales and considering the wavelet function as the first or second derivative of a smooth function, the absolute value of the wavelet coefficients are a bit large in point mutations of the signal. So, we can detect the position of singular points by the module maximum. Detecting singular signals, the wavelet base should satisfy the compact support and enough vanishing moment at a certain interval. In this paper, the strain mode of a double-tower cable stayed bridge is analyzed using continuous wavelet transform by Mexh wavelet (Mexican Hat wavelet) and the function satisfies the wavelet admissible condition and vanishing moment.

2.2 *Rotation mode theory*

For a multi-degree of freedom system with damping, the motion equations of system can be expressed as:

$$[M]\{\ddot{x}\} + [C]\{\dot{x}\} + [K]\{x\} = \{f(t)\} \tag{6}$$

where, [K] stands for the stiffness matrix, $[M]$ stands for the mass matrix, $[C]$ stands for the damping matrix, $\{f(t)\}$ stands for the load vector.Reporter: $\{x\} = \{X\}e^{j\omega t}$, the law of without damping free vibration corresponding to the above system can be expressed as:

$$[K]\{\Phi_i\} = \omega_i^2 [M]\{\Phi_i\} \tag{7}$$

Where, ω_i stands for vibration frequency of the i-order model, $\{\Phi_i\}$ stands for the i-order vibration mode vector (feature vector).This paper applies finite element method and modal analysis with the Lanczos method directly generating a set Lanczos vectors of mutually orthogonal, reducing the motion equations, and then obtain the characteristics solution of the original system equation by solving the Eigenvalue problems of the motion equations reduced. In this way, the basic vibration and rotation modes of framework can be obtained by the finite element analysis for plane frame structure. Usually, when analyzing the flexural rigidity of components with transverse cracks, the bending rigidity of the cracks of corresponding section can be replaced by the stiffness of the section height after loss.This article applies the finite element analysis for the plane frame structure, to deal with the bending rigidity of the crack corresponding section with this approach. In the crack section there is $EI(v^+) \neq EI(v^-)$, but the structure still satisfies the deformation conditions and internal force balance condition.

In model analysis, the basic modes reflect the relative deformation of different parts of the structure, such as the mutation node of geometry in a layer of frame (beam-column joints), regardless of the existence of cracks, the relative deformation is the largest, its singularity of the corresponding to wavelet coefficients is the most obvious features, so it is inconvenience to identify structural damage by the basic modes. In damage (cracks) area of the structure, the relative variation of the rotation is very significant, so it is very easy to reflect the injury in the vicinity of the beam-column connections with the rotation mode. Therefore, we propose the wavelet transform of rotation mode to identify the location of structural cracks, when analyzing structural damage of the framework with the wavelet method.

3 COMPUTATIONAL ANALYSIS

According to this principle, problems about two frame's crack identification were researched in this paper. The framework was for the rigid beam-column joints, and the support was consolidation of indeterminate plane frame. The various bars which frames consist of were uniform bar. Assume that the length of each parts is $100\,mm$, section size was $20 \times 50\,mm$, materials for the steel Q345, elastic modul $E = 2.07 \times 10^{11}\,N/m^2$, density $\rho = 7800\,kg/m^3$, Poisson's ration $\mu = 0.3$. The model of frame was shown in Figure 1. Finite element method was used to analyse the dynamic characteristics of the frame that each bar is divided into 100 equal units. By the A-B-C-D-E-F-B-E order all units of frame was numbered from 1 to 600 in the overall coordinate system. Lanczos method was applied to calculate five conditions of framework. Condition 1: BC column has a crack; Condition 2: AB and BC columns, each has a crack; Condition 3: CD beam and DE column, each has a crack; Condition 4: BE beam has a crack; Condition 5: BE and CD beams, each has a crack. Hear wavelet was used to transform the rotation mode. By the maximum of wavelet coefficients to identify the location of cracks in the five working conditions, and recognition results were compared.

3.1 *Condition I*

There was a crack that assumed on BC column from B node 1/5 location. The crack damage element is 120. Hear wavelet was used as mother wavelet to continuous transform the rotation mode of injuries framework that get the wavelet coefficients are shown in Figure 2. A significant curve break point was been found in wavelet coefficients, just the location of the injury.

3.2 *Condition II*

There was two cracks that assumed on AB column from B node 3/5 location and BC column from B node 1/2 location. The crack damage element is 160 and 50. Hear wavelet was used as mother wavelet to continuous transform the rotation mode of injuries framework that get the wavelet coefficients are shown in Figure 3. A significant curve break point was been found in wavelet coefficients, just the location of the injury.

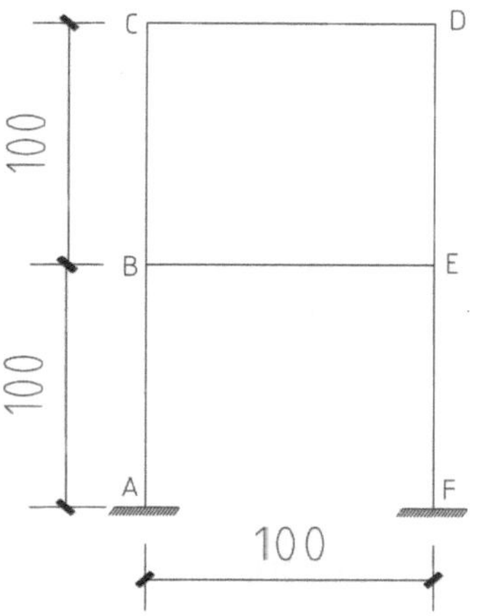

Figure 1. Calculating diagram (Unit: mm).

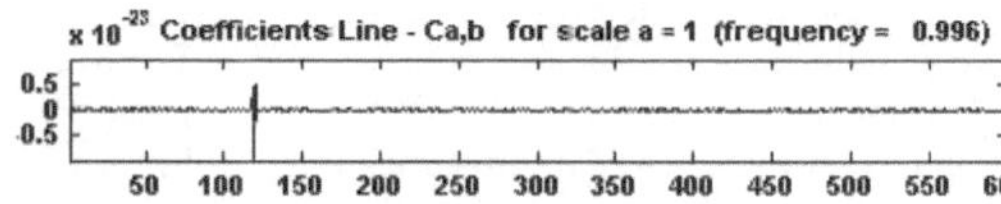

Figure 2. The wavelet coefficients of rotation mode.

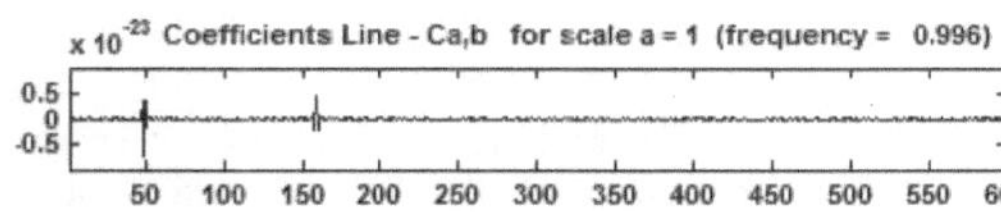

Figure 3. The wavelet coefficients of strain mode.

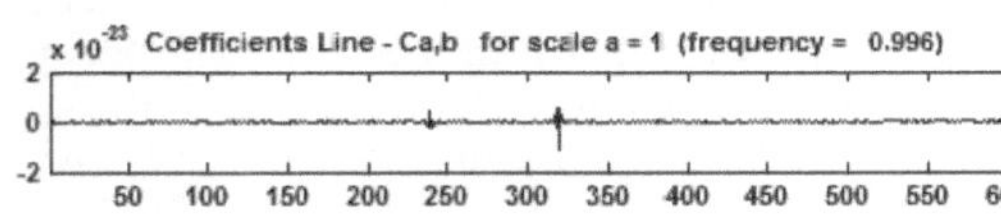

Figure 4. The wavelet coefficients of strain mode.

3.3 *Condition III*

There was two cracks that assumed on CD beam from C node 2/5 location and DE column from D node 1/5 location. The crack damage element is 240 and 320. Hear wavelet was used as mother wavelet to continuous transform the rotation mode of injuries framework that get the wavelet coefficients are shown in Figure 4. A significant curve break point was been found in wavelet coefficients, just the location of the injury.

3.4 *Condition IV*

There was two cracks that assumed on CD beam from C node 3/10 location and BE beam from B node 1/10 location. The crack damage element is 330 and 510. Hear wavelet was used as mother wavelet to continuous transform the rotation mode of injuries framework that get the wavelet coefficients are shown in Figure 5.

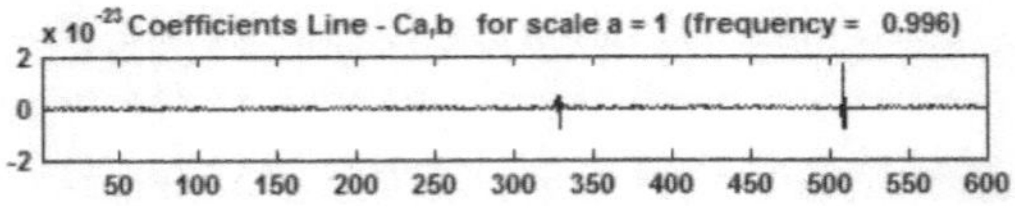

Figure 5. The wavelet coefficients of rotation mode.

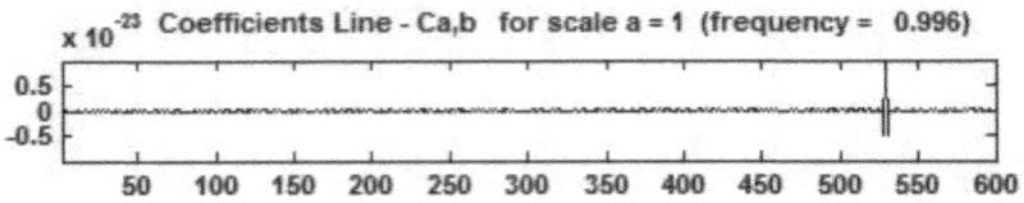

Figure 6. The wavelet coefficients of rotation mode.

A significant curve break point was been found in wavelet coefficients, just the location of the injury.

3.5 *Condition V*

There was a crack that assumed on BE beam from B node 1/4 location. The crack damage element is 525. Hear wavelet was used as mother wavelet to continuous transform the rotation mode of injuries framework that get the wavelet coefficients are shown in Figure 6. A significant curve break point was been found in wavelet coefficients, just the location of the injury.

4 CONCLUSIONS

The plane frame of the crack was analysed in this paper. By the finite element analysis we studied the rotation mode of plane frame and deduced the following results:

(I) Based on the wavelet analysis of rotation mode the damage could be identified. We could obtain the rotation mode of damaged using the finite element method, and then the position of the crack could be identified by modulus maximum of the wavelet coefficients. The numerical analysis showed that the effectiveness of this method.

(II) First of all, we could obtain the vibration characteristics of damaged plane frame using the finite element method and calculate the rotation mode by Lipschitz exponent,and then we analyzed the rotation mode of the damaged plane frame. The method above could be used for damaged beams identification.

(III) When crack is identified by the wavelet analysis of rotation mode, the method could avoid other singularity caused by change of geometry and was very clear for the crack identification. So, in practice engineering, this method may be useful for damage identification of plane frame structures.

ACKNOWLEDGEMENT

The authors gratefully acknowledge the research support for this work provided by National Natural Science Foundation of China (No. 50578018).

REFERENCES

Adriana Hera, Zhikun Hou (2004). Application of Wavelet Approach for ASCE Structural Health Monitoring Benchmark Studies. Journal of Engineering Mechanics. 130(1):96~104

Guan Deqing & Jiang Xin (2007). Crack Identification of Timoshenko Beam by Means of the Wavelet Analysis. Vibration and Shock. 26(5):67–70.

Guan Deqing & Shi Licheng (2010). Research on damage of single tower cable-stayed bridge identification by wavelet analysis of the curvature mode. Journal of Architecture and Civil Engineering. 27(1):21–25.

Hou Z, Noori M, Amand R St. (2000). Wavelet-based Approach for Structural Damage Detection. Journal of Engineering Mechanics [J], Journal of Engineering Mechanics. 126(7):677~683

Huang, C.S., W.C. Su. (2007). Identification of Modal Parameters of a Time Invariant Linear System by Continuous Wavelet Transformation. Mechanical Systems and Signal Processing. 21(4):1642~1664

Liew, K.M., Q. Wang (1998). Application of Wavelet Theory for Crack Identification in Structures. Journal of Engineering Mechanics. ASCE, 124(2):152–157.

Liu, D., H. Gurgenci, M. Veidt (2004). In Situ Damage Detection in Frame Structures through Coupled Response Measurements. Mechanical Systems and Signal Processing, 18(3):573~585

Ming-Hung Hsu (2005). Vibration Analysis of Edge-cracked Beam on Elastic Foundation with Axial Loading Using the Differential Quadrature Method. Computer Methods in Applied Mechanics and Engineering, 194(1):1–17

Nikolakopoulos, P.G., D.E. Katsareas, C.A. (1997). Papadopoulos. Crack identification in frame structures. Computers & Structures. 64(14):389~406

Ovanesova, A.V., L.E. Suárez. (2004). Applications of Wavelet Transforms to Damage Detection in Frame Structures. Engineering Structures. 26(1):39~49

Quan Wang, Xiaomin Deng (1999). Damage Detection with Spatial Wavelets. International Journal of Solids and Structures. 36:3443–3468

Sun Yankui. (2005). Wavelet Analysis and Its Applications. China Machine Press.

Sun Zengshou, Han Jianggang & Ren Weixin (2005). Structural Damage Identification Based on Curvature Mode and Wavelet Transform. Journal of Vibration, Measurement & Diagnosis. 25(4):263–267.

Modelling and Computation in Engineering – Zhu (ed)
© 2011 Taylor & Francis Group, London, ISBN 978-0-415-61516-7

Research on lateral shearing deformation of asphalt pavement under heavy axle load

Yunsheng Zhu, Xiaojing Kong & Kaifeng Wang
Wuhan University of Technology, Wuhan, China

ABSTRACT: In order to analysis how lateral shear flowing deformation of asphalt layers influences the asphalt pavement rutting under heavy axle load, taken the visco-elastic properties of the asphalt concrete, the non-uniform heavy contacting pressure of tire under different axle load and the frictional condition between pavement layers into consideration, the distribution regularity of lateral shearing stress and strain in a typical asphalt pavement was analyzed by means of nonlinear finite element method. The shear yield zone in the asphalt layers is judged by the secondary developed program based on the finite element software COSMOS/M, the rutting mechanism of asphalt pavement is discussed by means of the shear yield zone. The results reveal that: the lateral shear flowing deformation of the middle surface layer in asphalt pavement structure is the master cause of pavement rutting and the middle surface layer is a key layer for the asphalt pavement structure to resist rutting.

1 INTRODUCTION

Asphalt pavement is at a complex stress condition under the heavy axle load, because the wheel rolls repeatedly when the temperature is high, the lateral shearing deformation of the asphalt mixture of pavement layers produces. Epps et al. 1998 have conducted the long-term observational study to the experimental road of US's AASHTO and the West Track, discovered that the lateral shearing deformation is the primary course for the asphalt pavement rutting when the strength of the base course is enough. Nielsen 1966 studies the characteristics of the asphalt mixture deformation by means of a full scale rutting tester (Danish Asphalt Rutting Tester), the researching result reveals that the rutting is almost caused by the shearing flow of the asphalt mixture, the structure and abrasion rutting can be neglected in the total permanent deformation. Myers et al. 1998 & Xiaodi 2003 have studied the shearing stress in the asphalt pavement by means of the actual tire load, and found that the great shearing stress flow in the asphalt pavement is an important cause for the rutting. Shearing stress in asphalt pavement usually exceeds the shearing strength of the asphalt mixture when inner temperature of asphalt pavement is relatively high, the unrecoverable plastic flowing deformation and the shearing yield zone in the asphalt pavement produces. Taken the different heavy axle-load and non-uniform contacting pressure into consideration, the distribution regularity of lateral shearing strain in the asphalt pavement was analyzed by means of the nonlinear finite element method.

Table 1. Pavement structure of FEM.

Pavement Layer name and number		Thickness
Upper Surface Layer(AK-16A)	I	4 cm
Middle Surface Layer(AC-20I)	II	5 cm
Bottom Surface Layer(AC-25I)	III	6 cm
Base Course (6% Cement Stabilized Aggregate)	IV	18 cm
Base Course (5% Cement Stabilized Aggregate)	V	18 cm
Subbase (Cement Stabilized Sand)	VI	20 cm

2 FINITE ELEMENT MODEL OF TYPICAL PAVEMENT STRUCTURE AND RELATED CALCULATING PARAMETERS

2.1 *Selection of typical pavement structure*

In order to analyze the developing regularity of shearing flow deformation in asphalt pavement, a typical pavement structure was selected to calculated the deformation response by means of finite element software COSMOS/M. The structure composition and the layer number are shown in Table 1.

2.2 *Selection of heavy axle load and contacting pressure distribution*

A lot of relative studies reveals that the more obvious the non-uniform distribution of tire contacting pressure is showing, the more different the mechanical response of pavement structure will be. Therefore, the tire contacting pressure can not be simplified to circular uniform load as usually under heavy axle load which produces heavy contacting stress on the pavement structure. So, the rectangle contacting pressure shown in Figure 1 was adopted according to Guoping 2004. The rectangle contacting area was divided into three parts, two fringe areas is 20% of the width and the central area is 60% of the width. The contacting pressure of each area was calculated according to the standard vehicle of type JN-150. The tire width is 26.3 cm, spacing between two wheels is 34.6 cm. In order to analyze how the heavy axle-load takes effect on mechanical response of pavement structure, six wheel load levels of 25,50,80,100,120, 150 KN were calculated, which contains the contacting pressure and size shown in Table 2. Since the actual tire width varies with wheel load, it was assumed to be a constant for simplicity, just the length of contacting area varies with the axle load.

2.3 *Constitutive model and parameters of materials in pavement layer*

Taken into account the asphalt mixture viscoelastic properties, a time hardening creep model was adopted for the asphalt mixture layers in calculation:

$$\dot{\varepsilon}_{cr} = c_1 \cdot \sigma^{c_2} \cdot t^{c_3} \cdot e^{-c_4/t} \qquad (1)$$

Where: $\dot{\varepsilon}_{cr}$: creep rate; σ: creep stress; t: loading time; C_1, C_2, C_3, C_4 : regression coefficient;

The material parameters is shown in Table 3.

2.4 *Establishment of finite element model of the pavement structure*

According to the selected surface structure, calculating model of finite element analysis is shown in Figure 2, where y is the direction of driving, x direction is the road cross-section, z direction is perpendicular to pavement layer, coordinate origin is located in the two-wheel gap center, H is the thickness of total pavement layers. Analyzing scope of x, y-axis direction is 5 m, subgrade depth is also 5 m.

In order to accurately express the contacting area and contacting pressure of axle-load as assumed above, the three-dimensional geometric models were established after the model is divided into six blocks which were meshed individually according to the fringe area and central area of non-uniform contacting pressure area, the calculating load was set in accordance with Table 2. The element type is 8-node hexahedral element. The assumption of boundary condition is: no Z direction displacement at the bottom of subgrade, no X direction displacement at the left and right sides of and no Y direction displacement at the front and back of the calculating model. According to Xueliang et al. 2006, a slipping coefficient 0.5 of half-sliding

Table 3. Calculating parameter of pavement layer.

Layer	E (MPa)	ν	C_1	C_2	C_3	C (Kpa)	Φ (°)
I	1200	0.30	2.5E-7	0.95	−0.61	178	38
II	1600	0.30	8.2E-8	1.08	−0.75	259	35
III	1450	0.30	2.2E-8	1.24	−0.88	268	34
IV	1900	0.20	–	–	–	–	–
V	1550	0.20	–	–	–	–	–
VI	1350	0.25	–	–	–	–	–
Subgrade	45	0.45	–	–	–	–	–

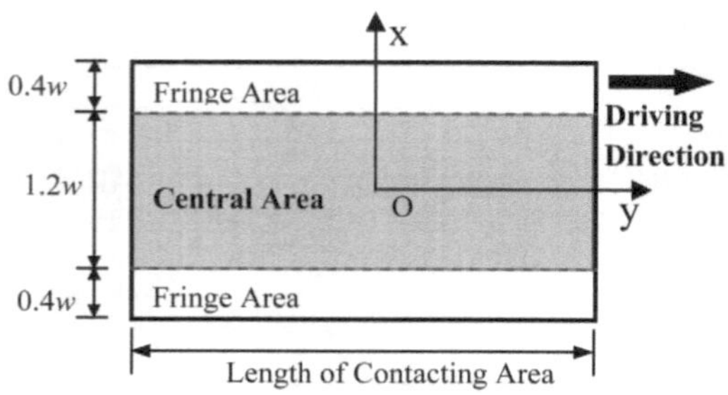

Figure 1. Sketch map of contacting

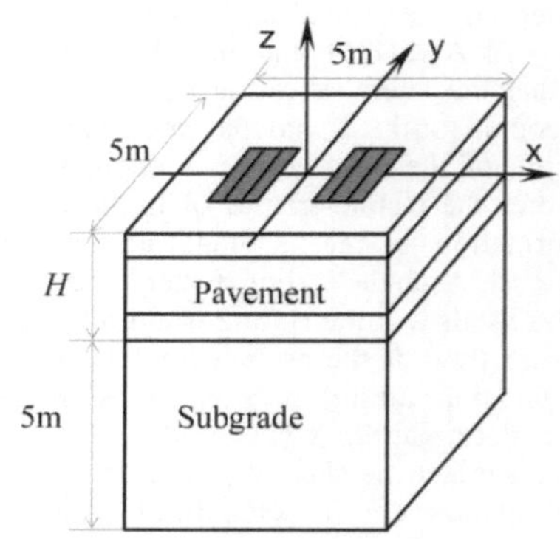

Figure 2. Calculating structure of FEM mode.

Table 2. Calculating result of tire and contacting size.

Weight of a Single Wheel (KN)	25	50	80	100	120	150
Air Pressure of Tire (MPa)	0.7	0.75	0.9	1.0	1.1	1.2
Contacting Pressure of Fringe Area (MPa)	0.47	0.60	0.81	0.94	1.08	1.26
Contacting Pressure of Central Area (MPa)	0.51	0.63	0.77	0.87	0.96	1.11
Width of Contacting Area (cm)	26.3	26.3	26.3	26.3	26.3	26.3
Length of Contacting Area (cm)	19.6	31.1	38.4	41.6	44.1	47.6

state between the surface layer and subgrade was was applied to simulate the contacting state of these two layers, the other layers were assumed to be completely continuous.

3 SHEARING DEFORMATION OF ASPHALT SURFACE

3.1 *Lateral shearing strain variation with pavement depth*

According to the calculating result, the shearing strain value can be gained in a vertical plane which the peak point of shearing stress gets through. Figure 3 shows the curve between lateral shearing strain and depth of pavement structure under different wheel load level. The maximum lateral shearing strain and vertical compressive strain varies with the wheel load level. Figure 4 shows the varying curve.

Figure 3 and Figure 4 shows that:

(1) The lateral shearing strain along the pavement depth increases gradually, and then decreases rapidly after it reaches peak value, the attenuation rate increases as the wheel load augment, the position of the peak value of the maximum shearing strain moves down with the wheel load augment. When the wheel load is in a low level, the shearing strain is small and relatively uniform in the whole asphalt layer, changes little from top to bottom. In heavy axle load conditions, the asphalt surface layer is in a higher level of shear strain, maximum shear strain appears in the middle asphalt layer, the shearing deformation occurs in this layer and the asphalt mixture is more easily coming into shearing plastic state. So, the rutting occurs in the middle asphalt layer because of large lateral shearing stress flow.

(2) The maximum lateral shearing strain increases gradually with the wheel load augment, 16.4%, 23.3%, 27.4% 31.8%, 40% increment can be gained when the wheel load increases. When the wheel load is large, the shear strain increases significantly larger. Therefore, heavy axle load conditions can result in sharply enlargement of the lateral shear deformation, which benefits the occurrence of transverse instability of structural rutting.

(3) With respect to increment of the maximum lateral shearing strain, the increment of the maximum vertical compressive strain is much smaller in the asphalt layer. When the wheel load increases gradually from 25 KN to 150 KN, maximum vertical compressive strain increases as follows: 3.1%, 4.8%, 6.2%, 7.2%, 10.6%. While the Wheel load is small, the maximum vertical compressive strain is much larger than the maximum lateral shearing strain in asphalt layers, pavement rutting is caused mainly by vertical deformation of the pavement structure. However, while the wheel load increases gradually, the increment of the maximum shearing strain is much larger than the maximum vertical compressive strain, and approaching to it. So, under the condition of heavy axle load, the lateral shearing strain develops rapidly, which results in the uplift deformation out of the wheel tracking. The rut depth increases rapidly under the common action of lateral shearing deformation and vertical compressive deformation of the pavement structure.

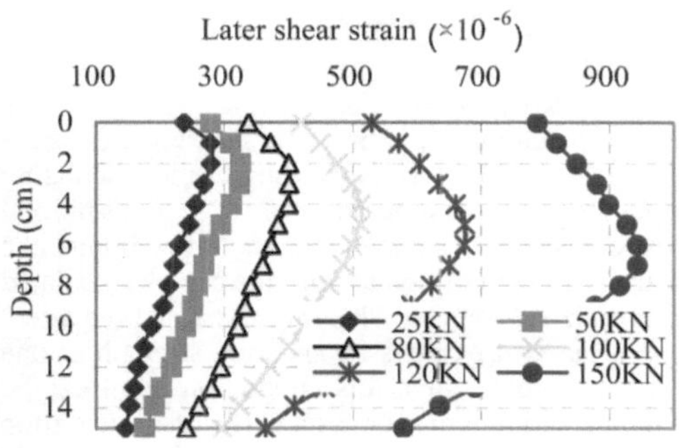

Figure 3. Lateral shearing strain curve along the depth of pavement structure.

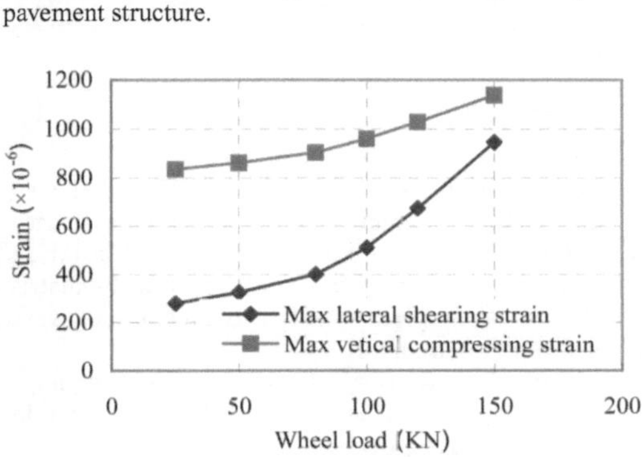

Figure 4. Curve of Maximum strain varying with wheel load.

3.2 *Relationship between parameters of pavement and lateral shear deformation in asphalt layers*

The regularity of the maximum lateral shearing strain of the asphalt layers in analyzed under the condition of different resilient modulus of surface course and base course and thickness of surface course. Figure 5 to 7 shows the varying curve of the maximum lateral strain under the different parameters of the pavement structure.

Figure 5 to Figure 7 shows that:

(1) The maximum lateral shearing strain is sensitive to the resilient modulus of the surface course under the different wheel axle load. The maximum lateral shearing strain decays rapidly as the modulus increases, the decaying amplitude of shearing strain is much larger under the heavy wheel axle load especially. This decaying amplitude is tend to stable when the resilient modulus exceeds 2000 Mpa. Therefore, appropriate improvement of the resilient modulus will contribute to reduce the lateral shearing deformation and enhance the

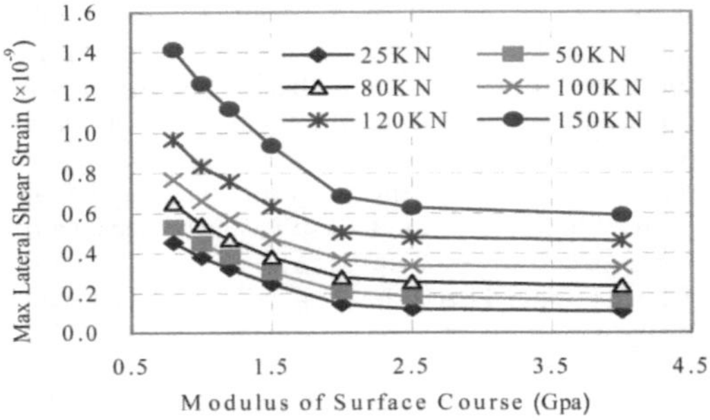

Figure 5. The curve of maximum lateral shearing strain varying with the resilient modulus of the surface course.

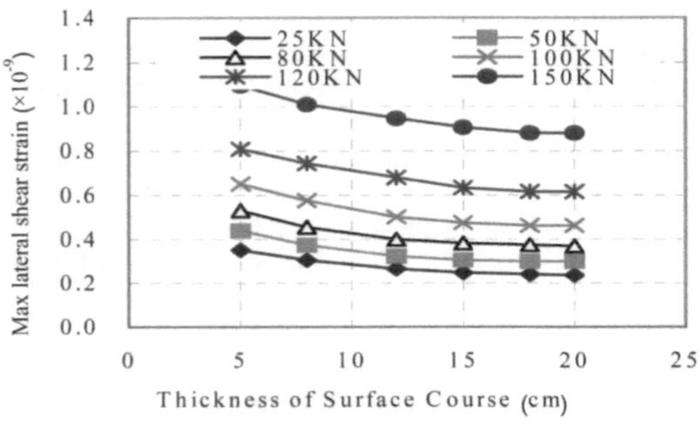

Figure 6. The curve of maximum lateral shearing strain varying with the thickness of the surface course.

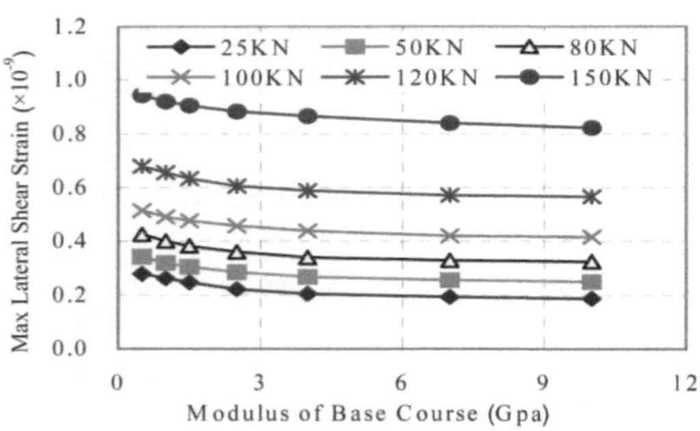

Figure 7. The curve of maximum lateral shearing strain varying with the resilient modulus of the base course.

anti-rutting performance of pavement structure. Whereas, this improvement was not effective after the resilient modulus is over 2000 Mpa.

(2) The maximum lateral shearing strain decays gradually with increase of the surface course, but the decaying amplitude is relative small. Especially when the thickness exceeds 15cm, the decaying rate is even smaller. Although the maximum lateral shearing strain can be reduced with the increase of surface course thickness, the vertical compressive strain increases rapidly, which is attribute to enhance the vertical compressive deformation for the pavement structure. So, the increase of asphalt

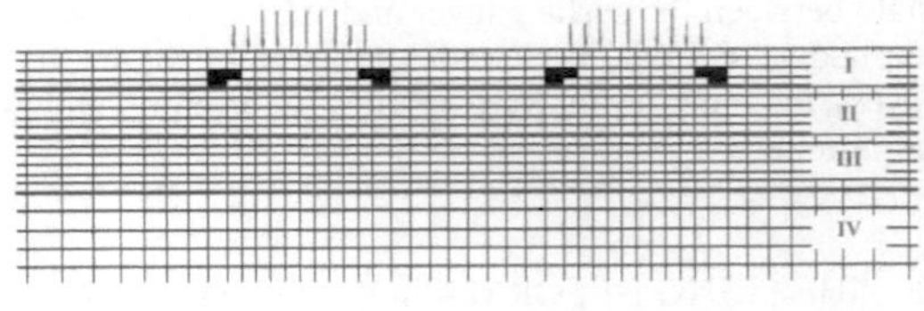

Figure 8. Shearing yield zone distribution under 50 KN wheel load.

surface course thickness is benefit to the shearing resistance performance of the asphalt pavement structure, but too large thickness of asphalt surface layer will exacerbate the Rut of the asphalt pavement structure.

(3) The maximum lateral shearing strain is less sensitive to the resilient modulus of base course, it decays at a low rate with the increase of the base course resilient modulus. On the contrary, the vertical compressive strain of the pavement structure increases significantly. So, just taken the deformation of the asphalt pavement into consideration, the improving function of the lateral shearing deformation in the asphalt pavement structure is not obvious for the asphalt pavement under heavy axle load only depending on increase of the resilient modulus of base course, which is benefit to the pavement rutting.

3.3 *Distribution of shear yield zone in asphalt layer*

A secondary developing program is gained by means of COMMAND LANGUAGE in Cosmos/M software, which can automatically draw the shearing yield zone in asphalt surface layer. According to the calculating output file*.out, the program reads each unit's normal stress and shearing stress at the integration points, the shearing strength of asphalt mixture can be calculated by formula 2. When two shearing stress in three directions of an unit exceeds its shearing strength at one integration point, then, this integration point can be thought entering into plastic state. If 6 of 8 integration points in one unit is at plastic state, this unit is a plastic unit. The program records the number of this unit and saves it into another data file, all plastic unit can be drawn in Cosmos/M software.

$$\tau_f = \sigma \cdot \tan \varphi + c \tag{2}$$

Where: τ_f: shearing strength of asphalt mixtures (Kpa); σ: normal stress of calculating point; ϕ: internal friction angle of asphalt mixture(°); C: asphalt mixture cohesion (Kpa); Figure 8 to Figure 12 show the distribution of shearing yield zone in asphalt surface layers under different axle load

From the chart, the characteristics of shearing yielding zone in asphalt surface layers under different axle load are shown as below:

(1) The area of shearing yield zone in asphalt layers expands gradually with the increase of wheel load.

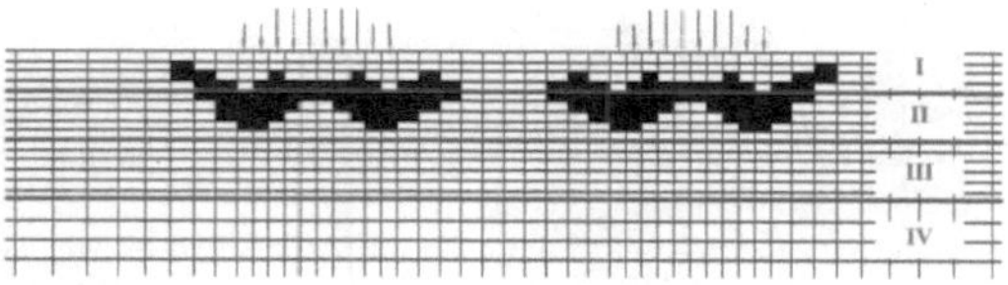

Figure 9. Shearing yield zone distribution under 80 KN wheel load.

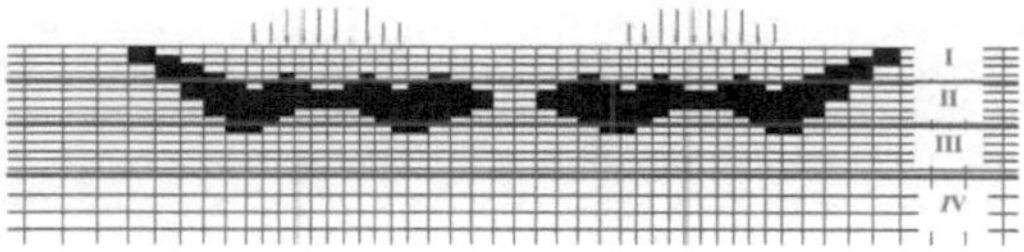

Figure 10. Shearing yield zone distribution under 100 KN wheel load.

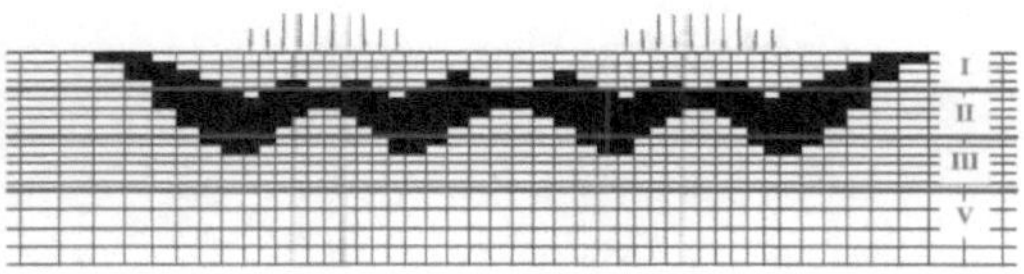

Figure 11. Shearing yield zone distribution under 120 KN wheel load.

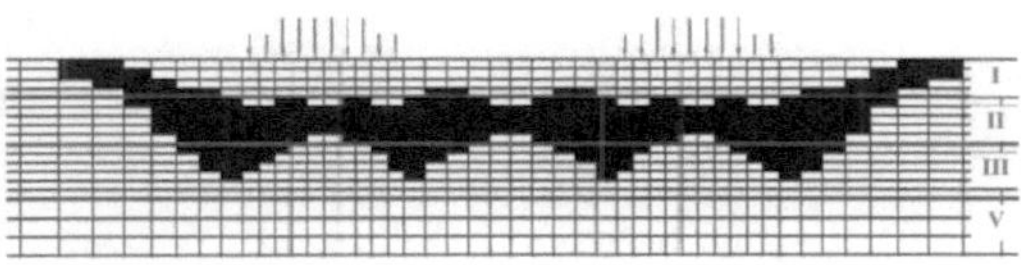

Figure 12. Shearing yield zone distribution under 150 KN wheel load.

When the wheel load is in a low level, the distributing range of shearing yielding zone is small. No shearing yield zone appears under 25 KN wheel load, and just a little area of shearing yield zone appears in the upper surface layer under 50 KN wheel load. When the wheel load reaches 80 KN, the area of shearing yielding zone expands with a large amplitude, and then, it enlarges slowly with the increase of axle load.

(2) The position of shearing yield zone is basically located in the asphalt surface course nearby the peak point of maximum shearing stress. With the increase of wheel load, the shear yield zone expands gradually from the peak point of maximum shear stress to its periphery and distributes along the outside edge of wheel load, the shear yield area of this position is biggest, area under the center of wheel load and gap between two wheels is relative small.

(3) When the wheel load changes between the 80~120 KN, the shear yield zone in asphalt surface course are mainly located in the middle surface layer. Large area of shear yield zone appears in the bottom layer only when the wheel load reaches 150KN. That is to say, the shear flowing deformation in asphalt layers starts from the middle surface layer and does not start from the bottom surface layer. In other words, the middle surface layer of asphalt pavement plays an important role in the anti-rutting or shear resistance. This conclusion is in accordance with the research of Shisheng 1999, & Fenella 1996, which reveals that the rutting of asphalt pavement usually occurs below the top of pavement surface 5~10 cm.

(4) When the wheel load is small, the shear yield zone can not reach the upper surface layer, which is basically not continuous. With the increase of wheel load, the shear yield zone extends gradually to the top of pavement and forms a continuous area. When the wheel load reaches 100 KN, the shear yield zone expands to the upper surface layer. When the wheel load exceeds 120 KN, the shear yield zone of outer edge of wheel load extends to the top of pavement, and then, the continuous shearing surface forms through the top of the pavement structure. The shear flowing deformation increases rapidly because of the formation of continuous shearing surface, at the mean time, the asphalt mixture outside the wheel load uplifts gradually to the top of pavement. The vertical deformation and the uplift deformation outside the wheel load of the asphalt mixture under the contacting area of tire constitute the macro deformation of the asphalt pavement.

4 CONCLUSIONS

The lateral shear deformation of typical asphalt pavement structures is calculated by means of finite element, the following conclusions can be drawn:

(1) Under the condition of the heavy axle load and high contacting pressure of tire, the lateral shear deformation develops sharply with the increase of the wheel load, the maximum lateral shear strain is located in the middle surface layer, the middle surface layer is a key layer for the asphalt pavement structure to resist rutting.

(2) Under the condition of the heavy axle load and high contacting pressure of tire, the increasing rate of the lateral shear strain is much larger than the vertical compressive strain, the lateral shear flowing deformation of asphalt mixture in asphalt surface course is the primary cause of the pavement rutting.

(3) Under the condition of the heavy axle load and high contacting pressure of tire, appropriate improvement of the resilient modulus of asphalt mixture will contribute to reduce the lateral shearing deformation and enhance the anti-rutting performance of pavement structure.

(4) Under the condition of the heavy axle load and high contacting pressure of tire, the shear yield zone expands from the middle surface layer to the top of asphalt pavement with the increase of the wheel load, the rutting of asphalt pavement develops progressively with the formation of shear yield zone.

REFERENCES

Epps J A, Monismith C L, Seeds S Betal. 1998. WesTrack performance interim findings. *Journal of the Association of Asphalt Paving Technologists*. Vol.67, Association of Asphalt Paving Technologists, St. Paul, Minnesota.

Fenella, M.L. 1996. Permanent deformation of asphalt concrete pavements: A nonlinear viscoelastic approach to mix analyses and design, *PHD Dissertation, University of California, Berkeley.*

Guoping Qian. 2004. Research of Complex Working State and Mechanic Response of Asphalt Pavement Under Heavy-duty Conditions *[PHD Dissertation]. Shanghai: Tongji University.*

Myers, L.A. and Roque, R. et al. 1998. Mechanisms of Surface-Initiated Longitudinal Wheel Path Cracks in High-Type Bituminous Pavements[C]. *Proceedings of the AAPT.*

Nielsen, J.P. 1966. The Modulus of Deformation of Asphalt Concrete. *Naval Civil Engineering Lab Port Hueneme Calif NCEL-TN-822.*

Shisheng Wu. 1999. Report of horizontal shear stress between asphalt surface and the base course with cement stabilized. *Xi'an Highway Institute.*

Xiaodi Hu. 2003. Measuring Tire Ground Pressure And Stress Response Analysis of Asphalt Pavement *[PHD Dissertation]. Shanghai: Tongji University.*

Xueliang, Zhang, Wei-gang & Hong-jie Liang. 2006. The Mechanical Response Analysis of Asphalt Pavement in Different Interface Condition Between Layers. *Journal of Shenyang Jianzhu University (Natural Science), Vol.22 No.4.*

Modelling and Computation in Engineering – Zhu (ed)
© 2011 Taylor & Francis Group, London, ISBN 978-0-415-61516-7

Theoretical explanation on the characteristics of earthquake induced landslides by computation on model slopes

Yu-Hua Lang
China Three Gorges University, Yichang, China

Hiroyuki Nakamura
Tokyo University of Agriculture and Technology, Tokyo, Japan

ABSTRACT: The characteristics of earthquake induced landslides are explained theoretically by calculation on model slopes using a limit equilibrium method of an application of dynamic programming which can search the critical slip surface under giving conditions. The influences of such factors as slope angle, slope type, the depth of surface layer, cohesions of surface layer and bedrock, horizontal and vertical seismic forces, are analyzed quantitatively.

1 INTRODUCTION

Earthquake induced landslides have lots of special characteristics compared to rainfall induced landslides. For instance, they often occurred in step slope with a shape of convex. They occurred in shallow surface layer and concentrated along earthquake faults frequently etc. Though there are many reports and references on these kinds of characteristics, such as Yamakuchi 1982, Jibson & Keefer 1988, and Towhata 2008, the reason for the formation of these characteristics are not explained clearly and quantitatively. The correlation among topographical features, distributional features, soil parameters and coefficient of earthquake are important tasks for study or discussion. Therefore this paper integrated and analyzed these characteristics according to references including authors' study firstly. Then computed the variation of safety factor and slip surface on model slopes using a limit equilibrium method of an application of dynamic programming which can search the critical slip surface under giving conditions. The influences of several factors are analyzed quantitatively to explain the formation of these characteristics.

2 METHOD, CALCULATING FACTORS AND CONDITIONS

The possible factors which may affect the characteristics of earthquake induced landslides were selected and put into the model slope. The critical slip surface and factor of safety were calculated under different combinations of factors in order to explain the formation of characteristics.

The stability analysis method used in this paper is a limit equilibrium method of an application of dynamic programming. The method can auto-search the critical slip surface under giving geography and soil mechanics conditions (Nakamura et al. 1987, Lang et al. 2002). The necessary data were slope shape, soil strength parameters, and coefficient of earthquake.

The affecting factors of earthquake induced landslides were discussed in much view of points. For example, Keefer (1984) analyzed 27 landslides then supposed 5 kinds of affecting factors as following: slope angel, height of slope, weathering, soil, and erosion conditions. Li (1986) reviewed earthquake induced landslides in China and proposed 3 factors to judge the slope stability under earthquake as slope angel, rock type, and underground water level. Yoshimatsu (1997) analyzed the correlation among factors and suggested that following factors have good relation with earthquake induced landslide: width of landslide, length of landslide, shape of slope, and height difference. But these papers did not discuss the correlation among factors quantitatively.

This paper take the characteristics of earthquake induced landslides into account then select 7 factors of slope to do theoretical computation as section shape, depth of surface layer, slope length, slope angel, cohesion , friction angle, and coefficient of earthquake. The correlation among these factors and slip surface and safety factor were calculated in order to explain the characteristics quantitatively.

Figure 1 shows the model slopes and calculation factors. Table 1 shows the factors, calculating range, and cases.

The earthquake induced landslides often occur in step slopes. The calculating cases for slope angle are set to 30° and 45° 2 cases.

There are lots of reports (Yoshimatsu 1990) about the features that the convex slope many easy slide during earthquake. There is no clear explanation

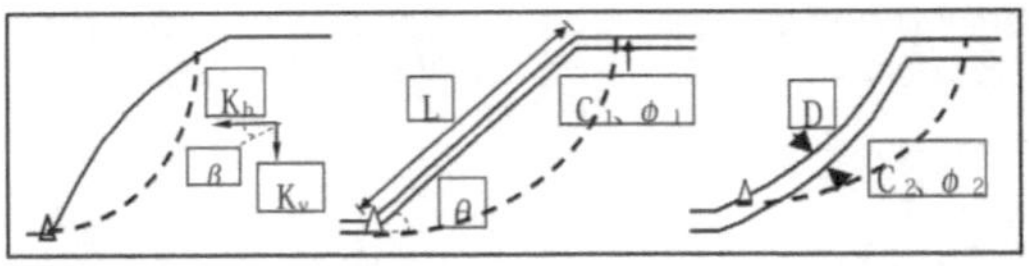

Figure 1. Model slopes and calculation factors.

Table 1. Affecting factors, calculating range, and cases.

No.	Factors	Symbol	Range	Case
1	slope angle	θ	30°, 45°	2
2	shape of section	Q,Z,P	convex, plan, concave	3
3	depth of surface layer	D	0, 2, 4 m	3
4	slope length	L	10, 50 m	2
5	cohesion	C	0∼50 kPa	11
6	friction angle	φ	30°, 35°	2
7	seismic coefficient	K_v, K_h	0∼1.0	6

about the phenomena. Three shapes of section (convex,plan,concave) were used to check the influence of geography.

As for the depth of surface layer, there are three cases of 0 m, 2 m, and 4 m. The 0 m case is set for the homogeneous stratum, and 2 m, 4 m cases are set for the stratum with different weathering layer in surface. For estimating of scale influence, slope length were set to 10 m and 50 m 2 cases.

The cohesion of soil may be the most influencing factor according to research (Lang & Nakamura 1996). And friction angle has great influence on the factor of safety. This paper set cohesion to 0–50 kPa 11 cases continually and friction angle to 30°, 35° two cases.

Though there are lots of references reports that seismic force will affect stability of slopes (Kobayasi 1981, Yeats 1997) there are rarely research on the different influence of horizontal seismic force and vertical seismic force to the slip surface and factor of safety. This paper changed the horizontal seismic coefficient and vertical seismic coefficient from 0 to 1.0 as 6 cases to calculate the critical slip surface and factor of safety under different conditions.

The characteristics of earthquake induced landslides will be explained by the theoretical computation on model slopes from following section.

3 SLOPE ANGLE AND LANDSLIDE

The earthquake induced landslide disseminate in step slopes especially when slope angel bigger than 30°. This paper will try to analyze the effecting of slope angle quantitatively.

Figure 2 shows the influence grade of slope angle on factor of safety. The horizontal axis is seismic coefficient and the vertical axis (1-F_{45}/F_{30}) is the decreasing coefficient of safety factor when slope angle increased

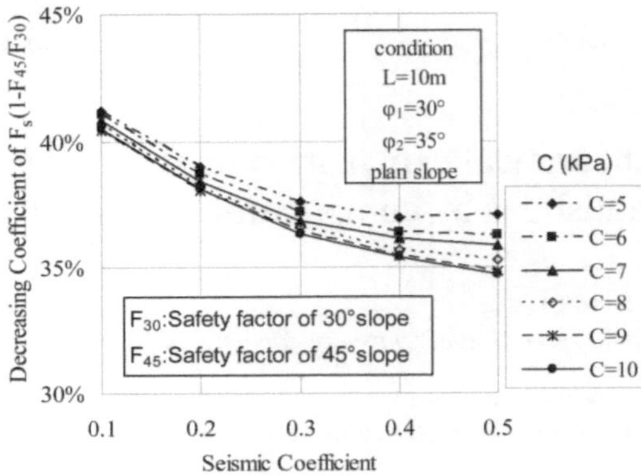

Figure 2. The influence grade of slope angle on factor of safety.

from 30° to 45°. F_{45} is the safety factor when slope angel equal to 45° and F_{30} is the safety factor of 30° slope. When slope angle increase 15 degrees from 30° to 45°, the factor of safety will decrease about 34%–42%. The factor of safety may decrease 2.5% when slope angle increase 1.0 degree namely.

4 SLOPE SHAPE AND LANDSLIDE

The influencing of landform especially the convex shape of slope had been explained from the view point of amplification effect of seismic waves (Yasue et al. 1981). This paper ignores the amplification effect of seismic waves and considers the influence by slope shape only then discusses the variation of slip surface and safety factor. The landform of slope in section can be divided to convex, plan, and concave three types. It is a presumption that the shorter the slope length, the smaller the safety factor of convex slope, the earthquake induced landslides will easily occur in convex slope. According to the model computation on the relation between section shape and position of slip surface, the exit of slip surface locate in the change point of inclination in bottom of slope for the convex and plain slopes of homogenous soil. But the exit of slip surface locates above the change point of inclination for concave slope. The slip cliff occur from upper part of convex slope while it occur above the change point of inclination in top of concave and plain slopes (see slip surface in Figure 1).

The seismic coefficients are changed in each type of slope in order to investigate their influence on the position of slip surface. There is a slight tendency that along with the increase of seismic force, the depth of slip surface will become bigger. The average change of the depth is smaller than 5%. And the change of slip surface of the convex slope is smaller that that of concave and plan slope.

The safety factor of critical slip surface of convex slope is the smallest one under the same condition while the plan slope's safety factor is the maximum

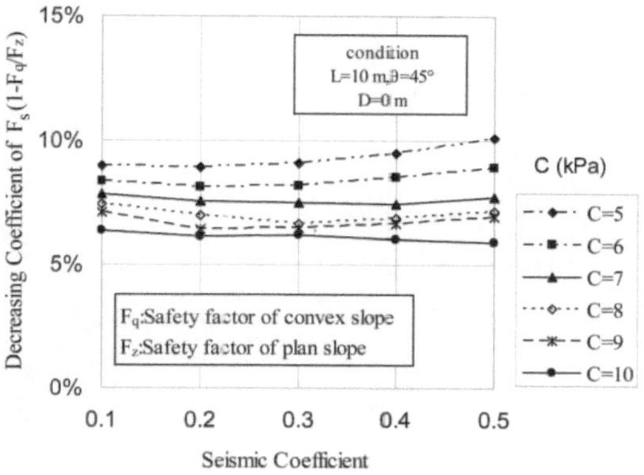

Figure 3. The influence grade of slope shape on factor of safety.

one. As for the decreasing degree by seismic force in each type of slope, the concave slope gets the minimum value. In addition, when slope angle is smaller, the decreasing degree of plan slope's safety factor is bigger, while when slope angle become bigger and/or seismic force become bigger, the decreasing degree of convex slope's safety factor become bigger. The decreasing degree of safety factor under seismic force show a order of convex > plan > concave.

Figure 3 shows the influence grade of slope shape on factor of safety. The horizontal axis is seismic coefficient and the vertical axis is the decreasing degree (1-F_q/F_z) of the ratio of convex F_s and plan F_s. F_q is the safety factor of convex slope and F_z is the safety factor of plan slope. The influence of slope shape on safety factor under different cohesion is 5%–10%. As described before in section 3, the change of slope angle 1.0 degree may convert to 2.5% of decreasing of safety factor. Therefore the influence by slope shape is corresponding to the influence of 2–4 degree of slope angle. The influence by slope shape is smaller than that of slope angle.

5 SOIL PARAMETERS AND LANDSLIDE

The most significant characteristic of earthquake induced landslide is the disseminating of surface slip. There are lots of reports on the concentrated occurrence of surface slip such as Kawabe 1997, Lang & Nakamura 1998, Lin 2008. This paper search the critical slip surface automatically by using of dynamic programming, then get the responding cohesion and internal friction angle theoretically.

Figure 4 shows one example of the investigated slip surface and calculated one in loess. The soil parameters of bedrock are $c = 100$ kPa, $\varphi = 35°$ and for upper loess $c = 1.0$ kPa, $\varphi = 30°$. The calculated slip surface fit well with real slip surface that formed between loess and bedrock. The factor of safety for this slope is 1.167 when seismic coefficient is 0, then it become 0.986 when seismic coefficient is 0.1.

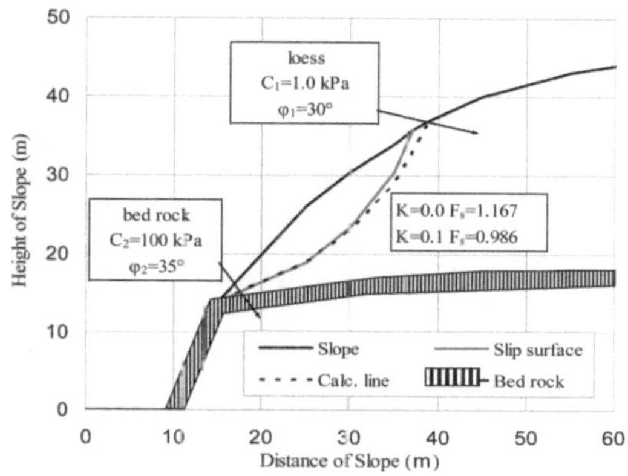

Figure 4. Slip surface and calculated one in loess.

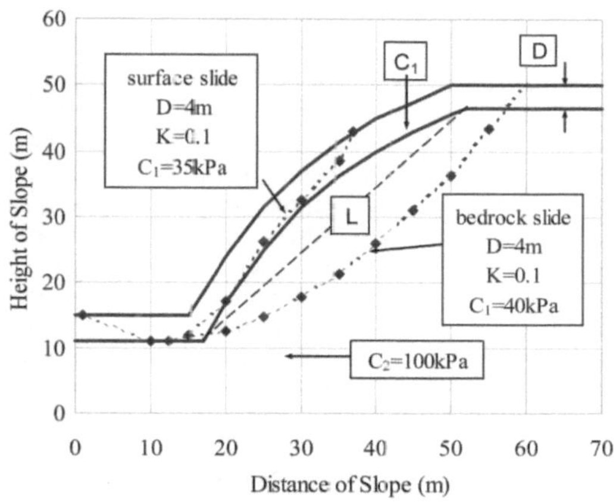

Figure 5. Influence of the depth and cohesion on the shape of slip surface.

The influences of surface depth and cohesion on critical slip surface are calculated by setting three depth of surface soil in model slope as 0 m, 2 m, and 4 m. and cohesion are from 0 kPa to 100 kPa. The soil parameters of bedrock are $c = 100$ kPa, $\varphi = 35°$.

If there are some soft layers in slope they will become slip surface generally. But the slip surface will reach bedrock when cohesion and/or seismic coefficient become big enough whenever the surface layer is deep or not. It means that two kinds of landslide namely surface landslide and bedrock landslide many be formed according to different soil parameters or different seismic coefficient.

Figure 5 shows the influence of the surface depth and cohesion on the shape of slip surface. The calculating conditions are: convex slope, surface depth 4 m, seismic coefficient 0.1, and cohesion of bedrock 100 kPa. The slip surface formed between surface layer and bedrock when surface cohesion c_1 equal to 35 kPa. Whereas the slip surface will occur inside bedrock when surface cohesion c_1 is equal to 40 kPa.

The ratio of shape (depth of slide/length of slope) and the ratio of cohesion (surface cohesion/cohesion

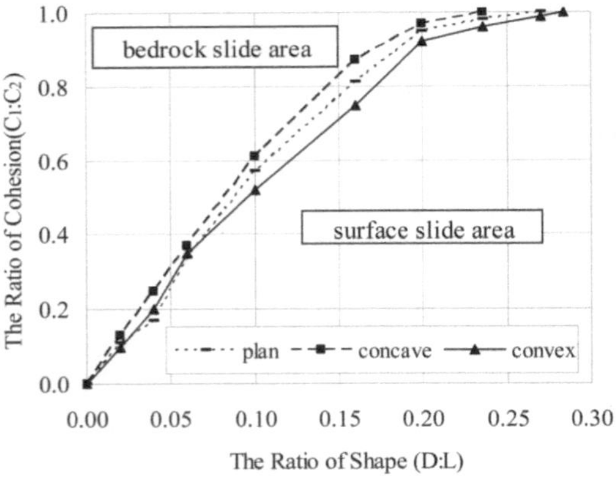

Figure 6. Landslide classification by the relation between the ratio of shape and the ratio of cohesion.

of bedrock) are introduced to investigate the boundary condition of surface slide and bedrock slide. Figure 6 shows the landslide classification by the relation between the ratio of shape and the ratio of cohesion. The horizontal axis is the ratio of shape (D:L) and the vertical axis is the ratio of cohesion (c_1:c_2). The plotted points are the boundary points between bedrock slide and surface slide under that condition. The upper-left area is the area of bedrock slide and the bottom-right area is the area of surface slide. All of the slide will become surface slide when the ratio of shape is bigger than 0.3 in horizontal axes. It means that surface slide will occur when surface depth is deep and cohesion is small even at small seismic coefficient.

6 SEISMIC CRACKS IN LANDSLIDE

There are lots of cracks above the upper area of slip cliff according to field investigation. This may be the result of temporary decreasing of safety factor when earthquake occurred. Figure 7 shows the process of occurrence of seismic cracks in a slope during earthquake. The soil parameters cohesion 6 kPa and friction angle 30° were obtained after calculating repeatedly by different soil parameters until critical slip surface coincident with real slip surface. Then auto-search the critical slip surface continuously on the slope after the occurrence of landslide. The solid line in the figure is the ground surface before sliding and broken line is the slip surface. The dotted line is the critical slip surface after the sliding of lower slope. Letters of a, b, . . . l are the position of real cracks. $F_{sa} - F_{sl}$ are the safety factor of critical slip surface of remained slope after the sliding of lower slope. It can be seen that after the sliding of slope when earthquake occurred, critical slip surface with the minimum safety factor will located to the next b position. Though at the moment of earthquake breaking, the decreasing of safety factor along slope from lower place to higher place may occur then lots of cracks formed but can not reach the

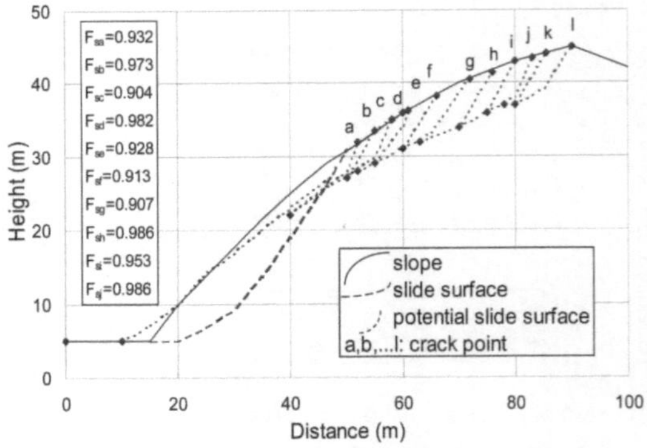

Figure 7. The process of occurrence of seismic cracks.

critical sliding along with the decreasing of seismic force. This is the cracking process that the critical slip surface formed above slip cliff, then became cracks continuously until the place where the safety factor is bigger than 1.0.

7 RELATION BETWEEN EARTHQUAKE FAULT AND LANDSLIDE

Earthquake induced landslide usually concentrated in slopes near the seismic faults but not concentrated in the epicenter. This characteristic may be explained by the distribution of seismic force and their directions. Though there are many reports on the relation between seismic force and landslide, specific answer about this feature is not clearly. Jibson & Keefer 1993 reported that there was a relation between the displacement of earthquake fault and the formation of landslide. The displacement of seismic fault reached a limitation then landslide will occur.

In order to investigate the relation between the displacement of earthquake fault and the formation of landslide authors reviewed the reports of Luhuo M 7.9 Earthquake (Wang et al. 1978) and Yongdeng M5.8 Earthquake (Lang & Nakamura 1997a). The distribution of landslides, the horizontal displacement of earthquake fault, and the vertical displacement of earthquake fault were extracted from these reports and the relations among them were analyzed. It is thought that there are great ground displacements near seismic fault while the displacement may small when the distance between fault and slope are longer.

Because there is a positive correlation between the affecting area of a seismic fault and its horizontal displacement (Newmark 1965, Lang & Nakamura 1997b), this paper divide the displacement of seismic fault to 6 groups and count the landslides inside each group. Figure 8 shows the relation between displacement of seismic fault and number of landslides. The horizontal axis is the number of landslides and the vertical axis is the measured displacement of seismic fault. The black triangle are stand for horizontal displacement and the white quadrangle are stand

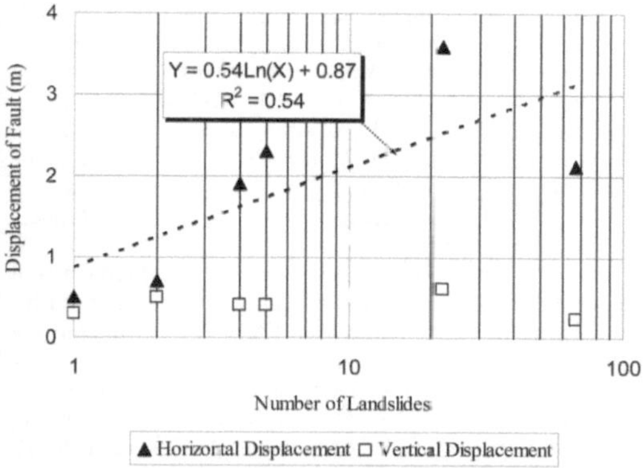

Figure 8. Relation between displacement of seismic fault and number of landslides.

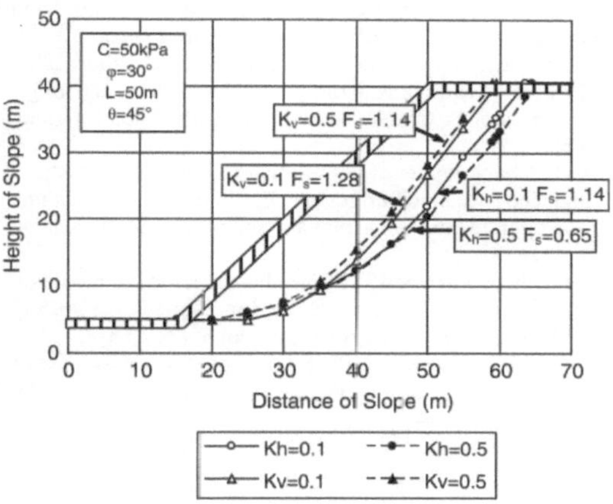

Figure 9. Seismic influence on critical slip surface and the factor of safety.

for vertical displacement. The dashed line shows the correlation between the horizontal displacement of seismic fault and the number of landslides. It means that the bigger the horizontal displacement of seismic fault, the bigger the number of landslides induced.

In order to investigate the influence of vertical seismic force and horizontal seismic force, authors calculated the critical slip surface and safety factor under different combination of seismic coefficient. Figure 9 shows the change of critical slip surface and safety factor under conditions: cohesion 50 kPa, internal friction angle 30°, length 50 m, slope angle 45°, convex slope. The left two curves show the critical slip surface under vertical seismic force only (K_v) and the right two curves show the critical slip surface under horizontal seismic force only (K_h). It is a clear figure that the slip surface become deeper and the safety factor reduce remarkable from 1.14 to 0.65 when horizontal seismic coefficient increases from 0.1 to 0.5. Meanwhile as for the vertical seismic coefficient, the critical slip surface become a little shallower and the safety factor has small change only from 1.14 to 1.28 along with the increase of vertical seismic coefficient from 0.1 to 0.5. It can be seen that the horizontal seismic force has bigger influence not only on the safety factor of slope, but also on the position of critical slip surface.

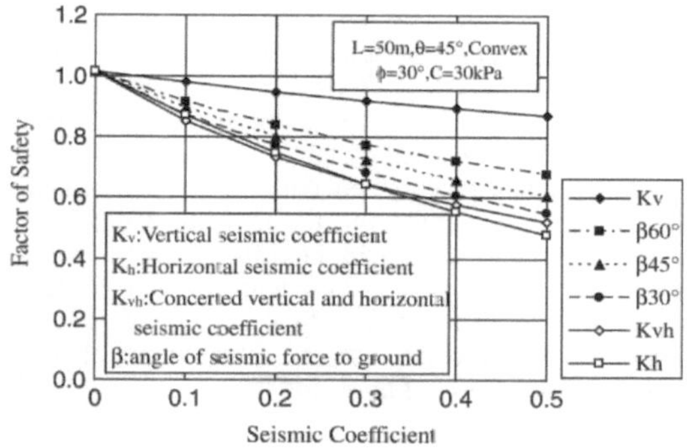

Figure 10. Influence of the direction of seismic force on factor of safety.

Figure 10 shows the influence of the direction of seismic force on factor of safety. The calculation shows the decreasing of safety factor from 1.0 under different seismic combination. Other calculating conditions are: length 50 m, 45° convex slope, internal friction angle 30°, cohesion 30 kPa. K_v is vertical seismic force only, K_h is horizontal seismic force only, and K_{vh} is the combination of the same two directional seismic force. β is the angle between resulting force and horizontal direction. Therefore the smaller the β, the bigger the horizontal part of the seismic force. It is notable from the figure that the vertical seismic force has the minimum influence on safety factor while the horizontal seismic force has the maximum influence.

8 CONCLUSION

This paper summarizes the characteristics of earthquake induced landslide and calculate critical slip surface and safety factor on model slopes to explain the formation of these characteristics theoretically.

As for the feature of concentration of earthquake induced landslide in step slopes, this paper calculated the influence of slope angle and supposed that the increase of 1.0 degree in slope angle may produce 2.5% decrease of safety factor.

As for the feature of concentration of earthquake induced landslide in convex slopes, this paper calculated the influence of slope shape and supposed that the influence of convex shape may equal to the influence of 2°–4° of slope angle and their affecting is smaller than that of slope angle.

As for the feature of small scale surface landslide concentration in earthquake area, this paper makes clear that the influences of depth of surface layer and cohesion of soil are great. And this paper proposed the condition of dividing surface landslide and bedrock landslide quantitatively by introducing of the shape ratio and the cohesion ratio.

As for the cracks of slopes in earthquake area, this paper simulated the process of crack formation along slip cliff and makes the position and extension of cracks clearly by calculation on the shapes of critical slip surface and their safety factors.

As for the feature of concentration of earthquake induced landslide along seismic fault instead of epicenter, this paper calculated the different influence of horizontal seismic force and vertical seismic force and supposed that the horizontal displacement of seismic fault is the main factor which has a positive correlation with the distribution of landslide.

REFERENCES

Jibson R.W., & Keefer D.K. 1993. Analysis of the seismic origin of landslides: examples from the New Madrid seismic zone. Geological Society of America Bulletin Vol. 105. pp.521–536.

Jibson R.W. & Keefer D.K. 1988. Statistical analysis of factors affecting landslide distribution in the New Madrid seismic zone, tennessee and Kentucky. Engineering Geology, Vol.27, pp.509–542.

Kawabe et. al 1997. The landslide distribution in Roko Mountain after Hyogo Earthquake. Journal of Japan Society of Erosion Control Engineering. Vol.49. No.5. pp.12–19.

Keefer D.K. 1984. Rock avalanches caused by earthquakes: source characteristics. Science. Vol.223. pp.1288–1290.

Kobayasi M 1981. Explanation on several earthquake induced landslides. Annual Reports of Disaster Prevention Research Institute. Kyoto University. No.24B-1. pp.401–410.

Lang Yu-hua & Nakamura Hiroyuki 1996. The variation of safety factor during earthquake for re-active landslide and new landslide. Proc. Annual symposium of Japan Society of Erosion Control Engineering. 1996 pp.51–52.

Lang Yu-Hua & Nakamura Hiroyuki 1997a. Influences of the horizontal and vertical seismic force on earthquake induced landslide. Journal of Gansu Sciences China. Vol.9 Sum.No.36. pp91–98.

Lang Yu-hua & Nakamura Hiroyuki 1997b. Characteristics of earthquake induced landslides and the theoretical analysis on model slopes - Take the Yongdeng earthquake as a example. Journal of Japan Landslide Society. Vol.34. No.3. pp.25–33.

Lang Yu-hua & Nakamura Hiroyuki 1998. Characteristics of slip surface of loess landslides and their hazard area prediction. Journal of Japan Landslide Society. Vol.35. No.1. pp.9–18.

Lang Yu-hua, Saeki Takasi, Nakamura Hiroyuki 2002. Evaluation of influence factors in three dimensional slope stability analyses by model study and spot inspection. Journal of Japan Landslide Society. Vol.39. No.2. pp.63–69.

Lang Yu-hua, Luo Xian-qi, Nakamura Hiroyuki 2008. Three dimensional simulation of landslide motion and the determination of geotechnical parameters. Landslides and Engineered Slopes. Chen et al. (eds). Taylor & Francis Group. pp777–781.

Li Tian-chi 1986. The relation between earthquake and landslide and the prediction of landslide induced by earthquake. Selected Papers on Landslide. No.2. China railway Press. Beijing. pp.127–132.

Lin M.L, Wang K.L, Kao T.C. 2008. The effects of earthquake on landslides – A case study of Chi-Chi earthquake, 1999. Landslides and Engineered Slopes. Chen et al. (eds). Taylor & Francis Group. pp193–201.

Nakamura Hiroyuki et al. 1987. Analysis method on critical slip surface using dynamic programming. Documents of Civil Engineering Institute. No.2425. pp.1–69.

Newmark, N.M. 1965. Effects of qarthquakes on dames and embankments. Geotechnique 15 (2). pp.139–160

Towhata Ikuo, Shimomura Tetsuo, Mizuhashi Masanori 2008. Effects of earthquakes on slopes. Landslides and Engineered Slopes. Chen et. al. (eds). Taylor & Francis Group. pp53–65.

WANG Cheng-hua et al. 1978. Investigation on Landslides induced by Luhuo Earthquake. Earthquake and Landslide. Chengdu Institute of Mountain Science, Chengdu, China. pp.23–38.

Yamaguchi & Kawabe 1982. The characteristics of mountain disaster induced by earthquake. Journal of Japan Society of Erosion Control Engineering. Vol.35. No.2 pp.3–15.

Yasue A 1981. Analysis on the features of seismic response of slope. Documents of civil engineering technology. Vol.23. No.4. pp.191–244.

Yeats R.S., Sieh K, Allen C.R. 1997. The geology of earthquakes. Oxford University Press. 568P.

Yoshimatsu 1990. Investigation on the landslides induced by Qiba-oki earthquake. Documents of Civil Engineering Institute. No.2824. pp.1–149.

Yoshimatsu & Kurokawa 1997. A prediction of slip surface depth by using a fuzzy inference method with genetic algorithm. Journal of Japan Landslide Society. Vol.33. No.4. pp.8–13.

Modelling and Computation in Engineering – Zhu (ed)
© 2011 Taylor & Francis Group, London, ISBN 978-0-415-61516-7

Energy and exergy analysis of air cooling systems with consideration of the thermal comfort

F.H. Ge
School of Energy and Power Engineering, Jiangsu University, Zhenjiang, Jiangsu, China

Q.S. Yu & C.Q. Wang
Jilin Architectural and Civil Engineering Institute, Changchun, China

ABSTRACT: Energy and exergy analyses are presented for the air cooling of the dew air supply in buildings, which is applied to a commercial store with a floor area of 2000 m^2 as an example of application in four Chinese cities. A new method is described in this article for selecting indoor temperature and relative humidity to achieve minimum energy and exergy consumption for a required indoor thermal comfort level which is evaluated with indoor effective temperature. The results show that humans are more sensitive to variations in temperature than those of relative humidity, and the influence of indoor relative humidity on exergy consumption is far greater than that of energy consumption, and far greater than that of indoor temperature. Hence energy consumption and exergy consumption in an air-conditioning building may be reduced by choosing larger relative humidity.

1 INTRODUCTION

The energy consumption of buildings in developed countries comprises 20–40% of total energy use (Omer 2007), and a substantial share of this energy is used to maintain room air temperatures and humidity. For an air-conditioned building, a major part of energy consumed is for the heating and cooling system. In some climate regions where cooling takes the major energy consumption in air-conditioned buildings, the cooling load in a central air-conditioning system is usually composed of two parts: sensible heat load and latent heat load, which is influenced by indoor/outdoor thermal and humid environment.

According to PMV-PPD formula(International standard 7730, 1994), the predicted mean vote between the limits of ±0.5 corresponds to the point where 90% of the residents feel satisfied, the indoor thermal environment data (indoor temperature and indoor relative humidity) for each season are determined consequently. This index considers environmental variables and individual factors and the closer to zero the PMV value, the better the occupants' thermal comfort sensation. Another example of thermal comfort index is presented in Ref. (Chu et al. 2005), where a least enthalpy estimator (LEE) is proposed. Such scheme utilizes the concept of thermal comfort and an enthalpy-based theory to provide a suitable setting on the effective temperature line for the HVAC control system to achieve thermal comfort and energy saving purposes.

The outdoor climate is another important factor influencing energy consumption for air conditioning. There are five climatic regions in China, and the dry bulb temperature and wet bulb temperature are in the range of 22.8°C to 36.5°C and 13.5°C to 28.5°C respectively in the design outdoor air conditions in summer season.

Associated to the thermal comfort concept, energy consumption is an important issue related to HVAC systems performance. All macroscopic natural phenomena happening around us involve the dispersion of energy and matter, which in due course change their forms from one to another, but the total amount of energy and matter involved is never consumed but necessarily conserved. This is why we need to use exergy, one of the thermodynamic concepts, to articulate how much exergy is supplied to a system in question, where and how it is consumed, and then how the generated entropy due to exergy consumption is discarded into the environmental space for the system. The exergy of a system in a given environment is the maximum theoretical work that might be extracted from it. Consequently, exergy is a measure of the potential of a given energy flow to be transformed into high quality energy.

The exergy concept has so far been applied to the built environment by relatively few researchers. Saito et al. (2001) reported the results of a pattern of human body exergy consumption related to the exergy balance of various heating and cooling systems and found out which were the low exergy systems. Yildiz et al. (2009) presented energy and exergy analyses for the whole process of space heating in buildings using both the first and second laws of thermodynamics, determining energy and exergy losses in the whole system using exergy analysis method. Shukuya (2009) discussed how a built environmental control system such as space heating and cooling can be described by the concept of "exergy". Wan et al. (2009) and Freire et al.

Table 1. Indoor and outdoor air state.

	Changchun	Beijing	Shanghai	Lhasa
Indoor design parameters (dry-bulb temperature/relative humidity)	25/50%	25/50%	25/50%	25/50%
Outdoor design parameters (dry-bulb temperature/wet-bulb temperature)	30.5/24.2	33.6/26.3	34.6/28.2	24.0/13.5
Indoor cooling load (W/m2)	100	100	100	100
Indoor air enthalpy (kJ/kg)	50.36	50.36	50.36	50.36
Indoor air exergy (kJ/kg)	0.212	1.620	1.825	1.241
Outdoor air enthalpy (kJ/kg)	72.98	81.85	90.62	51.75
Outdoor air exergy (kJ/kg)	0.3625	0.53028	0.42313	1.10482

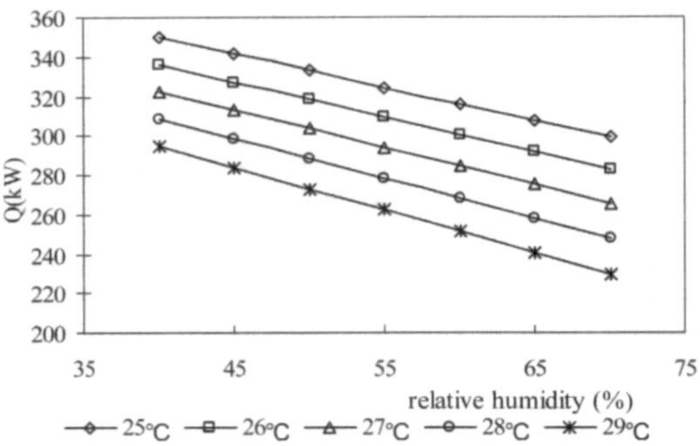

Figure 3. Energy consumption of air handling versus the different temperature and relative humidity.

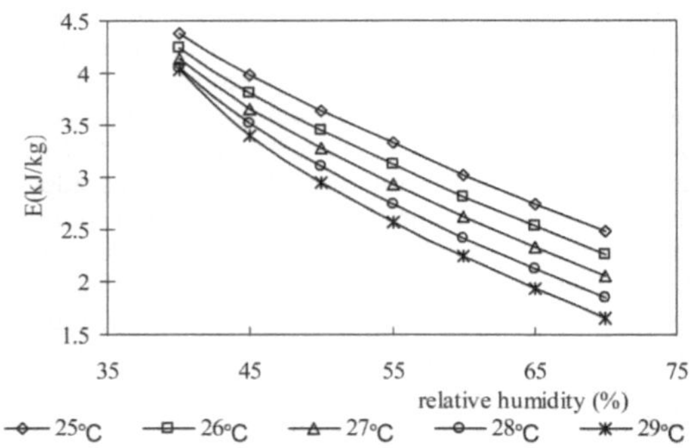

Figure 4. Exergy consumption of air handling versus the different temperature and relative humidity.

It is can be seen from figure 3 and figure 4 that the indoor thermal environment has a direct effect on the building energy and exergy consumption. The energy consumption can be reduced by increasing temperature, the reducing ratio is about 5.55% per 1° under the same relative humidity, and the energy consumption can also be reduced by increasing relative humidity, the reducing ratio is about 12.02%~ 18.25% from 40% humidity ratio to 65% humidity ratio under the same temperature. In the mean time the indoor teperature and relative humidity can produce the similar influence on exergy consumption. The effect of humidity on exergy consumption is far more than the temperature's. The exergy consumption of the relative humidity

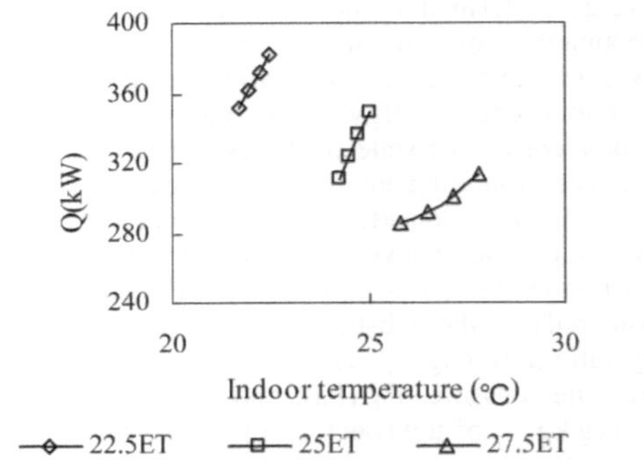

Figure 5. The effect of indoor temperature on system energy consumption under same thermal comfort condition.

40% is almost two times more than that of the relative humidity 65%.

The increase of indoor temperature and relative humidity will however affect the indoor human thermal comfort level. In the air-conditioning system design, the human thermal comfort is related to the concept of effective temperature (ET*) (ASHRAE 1972), which combines temperature and humidity into a single index so that two environments with same ET* evoke the same thermal response even though they have different temperature and humidity. The effective temperature (ET*) indicates the rate at which indoor air temperature can be increased with decreased relative humidity while maintaining the same thermal comfort. Figure 5 and figure 6 show that under the same indoor thermal comfort level (same ET*), although indoor cooling load and fresh-air cooling load decrease with the increase of indoor temperature, the energy and exergy consumption for the chiller increase with the increase of indoor temperature. This is due to the decrease of indoor relative humidity, which is necessary to keep the same ET*. Consequently, the increase of indoor temperature for the same ET* causes an increase of the system cooling load and total energy and exergy consumption. This phenomenon means probably that under the same indoor thermal comfort condition (i.e., ET*), increasing indoor temperature does not save energy and exergy. This conclusion seems to contradict the conventional understanding that increasing indoor temperature can reduce energy consumption. Actually

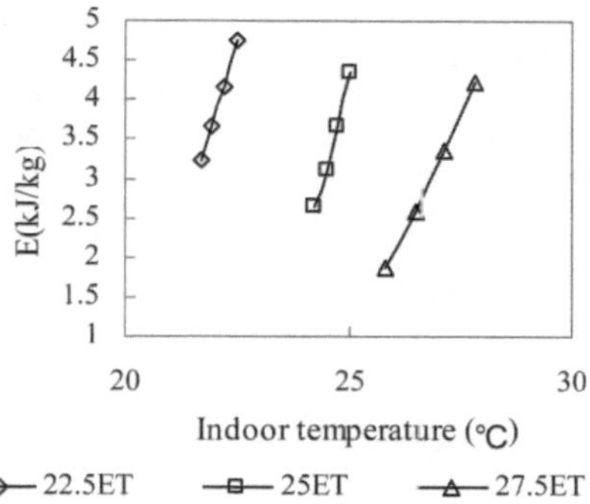

Figure 6. The effect of indoor temperature on system exergy consumption under same thermal comfort condition.

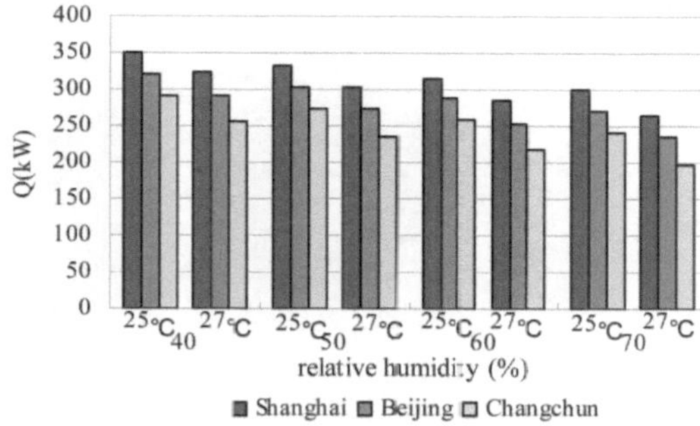

Figure 7. Energy consumption against indoor temperature and relative humidity in some cities.

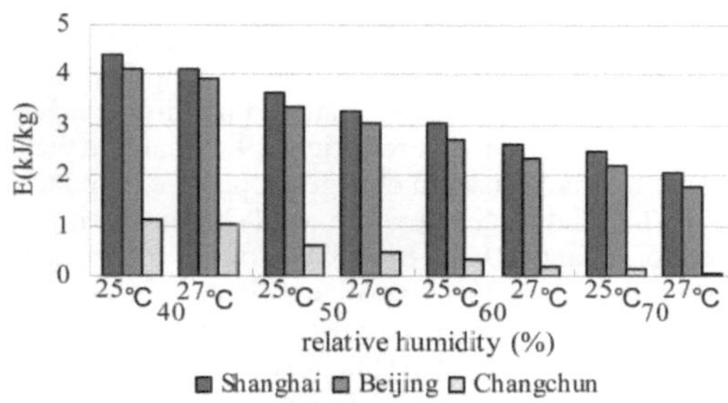

Figure 8. Exergy consumption against indoor temperature and relative humidity in some cities.

under the same ET*, because there are small changes in temperature and major changes in humidity, and the impact of humidity on the energy and exergy consumption is much greater than the impact of temperature's.

At the same time, figure 5 and figure 6 show that the impact of humidity on the exergy consumption is much greater than the impact on the energy consumption under the same indoor thermal comfort level (same ET*).

3.2 *The relationship between outdoor parameters and energy/exergy consumption*

The weather data include temperature (dry-bulb and wet-bulb) have a direct effect on the building energy consumption as well. Figure 7 and figure 8 show the energy consumption and the exergy consumption of chiller respectively versus indoor thermal and moist environment under the different outdoor design air parameters using three cities such as Shanghai, Beijing and Changchun in table 1. The design outdoor air dry-bulb temperature of three cities is 34°C, 33.2°C and 30.5°C respectively, and the design outdoor relative humidity of three cities is 65%, 60% and 57% respectively. It is obtained that although the indoor temperature and humidity are the same, the energy and exergy consumption varies considerably in different outdoor climate. The impact of outdoor climate on exergy consumption is far greater than the impact on energy consumption.

It can be seen from Table 1 that the dry-bulb/wet-bulb temperature under outdoor climate for Lhasa is 24°C/13.5°C, the enthalpy value under the outdoor air design state is much lower than the indoor enthalpy value (air temperature 25°C, relative humidity 50%), and the exergy value under the outdoor air design state is much higher than the indoor exergy value (air temperature 25°C, relative humidity 50%). Conventional air-handling program (figure 1, figure 2) does not apply to this area. Solely from the energy-saving point of view, if the cost of fresh air handling to the air supply state is less than the cost of the return air handling, the running mode of all fresh air should be chosen, on the contrary the return air should be selected to process to the air supply state. As to some specific thermal processes, fresh air can be spontaneously changed into the state of supplying air, so that no additional input power is needed. Similar to the climatic characteristics of Lhasa area, evaporative cooling technology can be used in order to achieve the minimum cost of indoor thermal comfort environment.

4 THE THERMAL COMFORT AND INDOOR TEMPERATURE/RELATIVE HUMIDITY

Conventional methods of determining acceptable indoor thermal conditions have been based mostly on the best-known PMV model (International standard 7730, 1994), which was developed with human subjects exposed to well-controlled environments in climate chambers.

In an ordinary office building for the personnel working state for typing, human metabolic rate values (M) is the 65 W/m^2, mechanical efficiency value is 0, indoor mean radiant temperature ($\bar{t}_r$) can be approximately replaced by the air temperature (t_a), summer clothing thermal resistance (I_{cl}) is 0.5clo, and PMV values can be calculated by substituting these parameters into the thermal comfort equation under various indoor air temperature (t_a) and relative humidity values (φ) (International standard 7730, 1994). The results are calculated and shown in figure 9.

As can be seen from figure 9 that indoor temperature and humidity effects different degrees on thermal

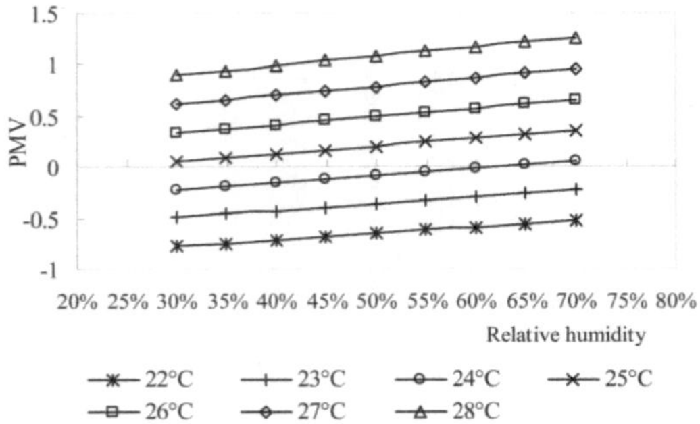

Figure 9. The effect of indoor relative humidity on PMV.

sensation, the effects of air temperature on the thermal sensation is greater than those of humidity, and at a certain comfort range, relative humidity can be varied in a larger range. From figure 9 we can see that human thermal sensation changes about 0.29 for 1°C variation of indoor temperature, while human thermal sensation changes 0.08 for 10% variation of indoor relative humidity. These results indicate that people are more sensitive to the variation of temperature than that of relative humidity. This phenomenon was also observed in the experiment reported by Amai et al. (2007).

In Chinese standard (GB50019-2003), the indoor temperature and indoor humidity were in the range of 22°C ~ 28°C and 40% ~ 65% respectively in summer. Since the effects of humidity on thermal comfort is far smaller than the effects of temperature, and the effects of humidity on energy and exergy is far greater than the effects of temperature, the larger relative humidity values should be made a choice appropriately according to different climate characteristic.

The above discussion indicates that indoor temperature and humidity can be chosen in a certain range. For the hot and humid climate, selecting the upper limit of relative humidity is beneficial to energy-saving under the same temperature. While in the dry regional climate such as Lhasa, indoor temperature and humidity options should be considered based on thermal comfort, because outdoor air can be used as a natural cold-source, energy consumption is not a prominent issue.

What is more, air handling system of dew-point air supply (no reheating) generally makes it too large indoor relative humidity, which is consistent with the ideas of raising indoor relative humidity.

More and more researchers have paid attention to the adaptive model of thermal comfort. de Dear et al. (2001) suggested that occupants should be allowed to be active in modifying their indoor environment as they preferred. They proposed an optional method to determine acceptable indoor thermal conditions, also known as the adaptive model of thermal comfort. The adaptive model has also been included in recent versions of ASHRAE Standard 55 (ASHRAE standard 55 2004, Nicol et al. 2010, Rueylung et al. 2009) for buildings with spaces without mechanical cooling.

There are various types of climates in China due to its vast territory. According to the adaptive model, thermal comfort in all kinds of climate regions may result in relaxed indoor temperature and humidity criteria and may therefore provide a potential means to reduce the consumption of energy used to cool buildings. If adopting uniform standards for the interior design indices in all the territories, which will lead to unimaginable pressure for energy application.

5 CONCLUSIONS

In this paper, principles of energy and exergy analysis in HVAC applications are discussed and are further used to evaluate the performance of cooling system. An approach is described, which is carried out to investigate the effect of indoor temperature and relative humidity on human thermal comfort and energy and exergy consumption in a central air-conditioning system in a few climate regions of China. We can extract some concluding remarks from this study as follows:

The energy and exergy consumption for air-conditioning system can be reduced by increasing indoor temperature and humidity in summer. Under a same indoor thermal comfort level (ET*), a parameter variation study has also been conducted, which suggests that for a slight temperature variation, a combination of high indoor relative humidity and low indoor temperature should be generally taken.

An exergy benchmark for space conditioning in different climates could not only serve as a target for air conditioning processes but also for selection of building design strategies. Given the interaction of humidity and air temperature in the sensation of comfort, a condition that minimizes necessary exergy expenditure could be chosen. Humidity's impact on exergy is far greater than energy consumption, whose impact on energy consumption is far greater than temperature, and whose impact on the thermal comfort is far smaller than temperature.

Humans are more sensitive to variations in temperature than those of relative humidity. Hence energy consumption in an air-conditioning building may be reduced by choosing larger relative humidity. Furthermore, indoor temperature may be determined in priority and indoor relative humidity may be determined according to outdoor climate change in different regions. The closer indoor and outdoor climate parameters are, the less energy consumption and exergy expenditure are.

In the comfort air conditioning system, the dew air supply which is an energy-saving model because of no using reheat type usually causes the larger indoor humidity which is consistent with the ideas of raising indoor relative humidity.

REFERENCES

Angelotti, A. & P. Caputo (2007). The exergy approach for the evaluation of heating and cooling technologies;

first results comparing steady-state and dynamic simulations, in: *Proceedings of the 2nd PALENC and 28th AIVC Conference, vol. I*, Crete Island, Greece, pp. 59–64.

ASHRAE. *ASHRAE fundamentals handbook*. ASHRAE Inc.; 1972.

Amai, H., S-i. Tanabe, T. Akimoto, & T. Genma (2007). Thermal sensation and comfort with different task conditioning systems. *Building and Environment. 42*,3955–3964.

ASHRAE. *ASHRAE standard 55*: thermal environmental conditions for human occupancy. American Society of Heating, Refrigerating and Air-conditioning Engineers; 2004.

Chu, C.M., T.L. Jong, & Y.W. Huang (2005). A study of thermal comfort control using least enthalpy estimator on HVAC systems, in:*Proceedings of the IEEE American Control Conference* (ACC'05), Portland, OR, USA, pp. 3665–3670.

Chengquin, R., L. Nianping, & T. Guanfa (2002). Principles of exergy analysis in HVAC and evaluation of evaporative cooling schemes, *Building and Environment. 37*, 1045–1055.

de Dear, R.J., & G.S. Brager (2001).The adaptive model of thermal comfort and energy conservation in the built environment. *International Journal of Biometeorology. 45*,100–108.

Freire, R.Z., G.H.C. Oliveira, & N. Mendes (2008). Predictive controllers for thermal comfort optimization and energy savings. *Energy and Buildings. 40*, 1353–1365.

Guadalupe-Alpuche, M., C. Heard, R. Best, & J. Rojas (2005). Exergy analysis of air cooling systems in buildings in hot humid climates, *Applied Thermal Engineering. 25*, 507–517

ISO. International standard 7730: moderate thermal environments determination of the PMV and PPD indices and specification of the conditions for thermal comfort. *International Standards Organization*; 1994.

Ministry of Construction of the People's Republic of China. *Code for design of heating ventilation and air conditioning (GB50019-2003)*. Beijing: China Planning Press, 2003.

Nicol, F. & M. Humphreys (2010). Derivation of the adaptive equations for thermal comfort in free-running buildings in European standard EN15251. *Building and Environment. 45*, 11–17.

Omer, A.M. (2007). Energy, environment and sustainable development. *Renewable and Sustainable Energy Reviews*, Available online 2 July.

Rulung, H., C. Mingjen, L. Tzuping, & H. Mingchin (2009). Thermal perceptions, general adaptation methods and occupant's idea about the trade-off between thermal comfort and energy saving in hot–humid regions. *Building and Environment. 44*, 1128–1134

Saito, M. & M. Shukuya (2001). The human body consumes exergy for thermal comfort. *LowEx News 2*, 6–7.

Shukuya, M. (2009). Exergy concept and its application to the built environment. *Building and Environment. 44*, 1545–1550.

Torio, H. & D. Schmidt (2008). Exergetic assessment and contribution of solar energy systems to the energy performance of buildings, in: *Proceedings of the Nordic Symposium of Building Physics 2008, vol. 2*, Copenhaguen, Denmark, pp. 637–644.

Wan, J.W., K.L. Yang, W.J. Zhang, & J.L. Zhang (2009). A new method of determination of indoor temperature and relative humidity with consideration of human thermal comfort. *Building and Environment*. 44, 411–417.

Yildiz, A. & A. Gungor (2009). Energy and exergy analyses of space heating in buildings. *Applied Energy. 86*, 1939–1948.

Modelling and Computation in Engineering – Zhu (ed)
© 2011 Taylor & Francis Group, London, ISBN 978-0-415-61516-7

Application of soil-water characteristic curve taking stress influence into considerations in unsaturated seepage analysis

Sh.M. Wang & H.B. Qin
Three Gorges University, the Three Gorges Reservoir Region Geologic Hazard Critical Laboratory of Ministry of Education, Yichang City, Hubei Province, China

G. Wang
Liaoning Institute of Engineering Investigation, Jinzhou City, Liaoning Province, China

ABSTRACT: Based on the soil-water characteristic curve functions which considering the stress state, combined with the experience functions between the permeability functions of unsaturated soil and the soil-water characteristic curve proposed by Brooks & Corey, authors derived the unsaturated permeability functions available in considering the stress influence. This paper applied it in unsaturated seepage analysis and proposed a calculating method which can couple stress field with seepage field together. Taking a simple slope as an checking example, authors get following conclusion: under the same precipitation conditions, compared with the uncoupled method, the coupled method of stress field and seepage field calculated that groundwater level becomes low while the metric suction increases, which is much closer to the actual situations of slope seepage. This research provide theory and method evidence in analyzing the seepage field of unsaturated soil more accurately.

1 INTRODUCTION

Hydraulic conductivity in unsaturated soil is the functions of saturation and volumetric water content. With the purpose of emphasizing this kind of functions relation, it is called permeability functions in unsaturated soil mechanics. It is generally believed that there is a strong relation between soil-water characteristic curve and permeability functions. In contrast to unsaturated permeability functions, the actual measurement for soil-water characteristic curve is relatively much easier to operate. Therefore, it is general to derive the permeability functions via soil-water characteristic curve in unsaturated seepage analysis.

Christensen (1943) brought forward the permeability functions of exponential functions type firstly. Averjanon (1954) put forward the "taking air entry value as the boundary point to conduct segmented description towards unsaturated permeability coefficient and saturated infiltration coefficient, at the same time considering the pore size effects on unsaturated permeability coefficient". Gardner (1956) put forward the functions relationship of power functions type between permeability coefficient and volumetric water content, based on which Campbell (1973) made the amendment. The experience formula of power functions type proposed by Brooks & Corey (1964) took the effect of air entry value and the affection of pore size towards permeability functions into consideration, on the basis of which Philip (1986) made some modifications to the formula. Davidson et al. (1969) proposed a exponential type functions fix the related coefficients by experiments.

It can be seen that, because of not considering the stress influence in these soil-water characteristic curves, the unsaturated soil permeability functions derived from soil-water characteristic curves also fail in considering the stress, which falls short of the actual situation, because in actual engineering the soil is in a stress state totally. Consequently, to establish the considering-the-stress-influence permeability function of unsaturated soil becomes a quite necessary basic work.

2 THE DERIVATION OF THE CONSIDERING-THE-STRESS-INFLUENCE UNSATURATED SOIL PERMEABILITY FUNCTION

Brooks & Corey (1964) put forward the following experience formula of power function type which not only takes the effecting of air entry value, but also the affection of pore size towards permeability functions into consideration, consequently, this formula is extensively applied in rock and soil field.

$$\begin{cases} K_w = k_s, u_a - u_w < (u_a - u_w)_b \\ K_w = k_s \left[\dfrac{(u_a - u_w)_b}{u_a - u_w} \right]^{\eta}, u_a - u_w > (u_a - u_w)_b \end{cases} \tag{1}$$

Table 1. Water content under different consolidation stress and suction.

Water content % Consolidation stress /kPa	Suction /kPa	S=0	S=50	S=100	S=150	S=250	S=300
100		21.23	17.94	16.88	15.99	14.69	14.42
200		18.77	16.79	15.96	15.37	14.49	14.37
300		18.45	15.84	15.21	14.89	14.15	13.86
400		17.59	15.31	14.74	14.15	14.13	13.85

Where $k_s =$ permeability coefficient of saturated soil; $\eta =$ pore size coefficient; and $(u_a - u_w)_b =$ air entry value.

Now we use the soil-water characteristic curve function which can reflect stress influence and the permeability function of unsaturated soil proposed by Brooks & Corey to derive the considering-the-stress-influence permeability function of unsaturated soil.

Wang Shi-mei (2007) used experimental data (Table 1) of soil-water characteristic curve measured in different consolidation stress by direct shear instruments of unsaturated soil, got the following two kinds of soil-water characteristic curve functions after fitting the data to formula (3) and formula (3) respectively:

$$\theta = B_0 + B_1 e^{-u/t} \quad (2)$$

$$\theta = f(u) = \frac{A_1 - A_2}{1 + (u/u_0)^P} + A_2 \quad (3)$$

Where $\theta =$ gravity water content; $u =$ ground mass suction; $B_0, B_1, t, A_1, A_2, u_0$ and P are fitting parameters.

Authors established the permeability function of unsaturated soil considering the consolidation stress by studying the relation between consolidation stress P and the fitting parameters of formula (2) and formula (3),

$$\theta = 14.1 + (-0.0113P + 7.94)e^{-u/(0.144P+113.7)} \quad (4)$$

$$\theta = \frac{1.48\times10^{-5}P^2 - 0.0194P + 9.3}{1 + \left[u/(-0.146P + 101.5)\right]^{(1.15\times10^{-6}P^2 + 1.79\times10^{-4}P + 1.23)}} + 13.5 \quad (5)$$

Formula (4) and formula (5) are respectively the exponential function type and sigmoidal half logarithmic function type of soil-water characteristic curve functions considered the stress influence. The fitting precision of the two functions can be above 97%. Formula (4) is simple but it cannot fix the air entry value u_b while formula (5) is complicated but from it we can fix the air entry value u_b. Consequently, we combine the two formulas, namely substituting exponential function formula (4) and the air entry value u_b fixed by sigmoidal half logarithmic function into the unsaturated soil permeability function (1), consequently deriving the permeability function of unsaturated soil available in taking stress into consideration, as the following formula (6):

$$k_w = k_s \left[\frac{(0.144p - 113.7)(15.6 - 0.023p)}{\ln\left((7.94 - 0.0113p) \big/ (\omega - 14.1) \right)} \right]^{\eta}, u > u_b \quad (6)$$

Where $k_s =$ permeability coefficient of saturated soil; $\eta =$ empirical constant. And it has the following relation with core size distribution index λ:

$$\eta = \frac{2 + 3\lambda}{\lambda} \quad (7)$$

Therefore, only if we fix λ, we can calculate η. Corey (1977) proposed the relation between effective saturation S_e and ground mass suction as formula (8):

$$S_e = \left\{ \frac{(u_a - u_w)_b}{(u_a - u_w)} \right\}^{\lambda}, (u_a - u_w) > (u_a - u_w)_b \quad (8)$$

If $(u_a - u_w)_b = u_b, u_a - u_w = u$, when $u > u_b$, formula (8) can be expressed as:

$$S_e = \left\{ \frac{u_b}{u} \right\}^{\lambda}, u > u_b \quad (9)$$

Owning to the existence of the residual water content in unsaturated soil, effective saturation S_e can be also defined as:

$$S_e = \frac{\theta - \theta_r}{\theta_s - \theta_r} \quad (10)$$

where $\theta_r =$ residual water content; $\theta_s =$ saturated water content.

Substitute (10) into (9), we get:

$$S_e = \frac{\theta - \theta_r}{\theta_s - \theta_r} = \left\{ \frac{u_b}{u} \right\}^{\lambda}, u > u_b \quad (11)$$

So λ is the slope of half logarithm curve S_e.

After fitting the data in Table 1 using formula (3), we can get 4 parameters under different stress showed in Table 2.

From Table 2 we can know: $\theta_r = A_2$; $\theta_s = A_1$; in which A_2 is not related to stress, while A_1 and u_b

Table 2. The fitting parameters of function.

Pressure (kpa) Parameters	100	200	300	400
A_1(%)	20.96	19.57	18.24	17.43
A_2(%)	13.11	13.41	13.62	13.74
$u_0(kPa)$	86.38	73.25	57	43.02
$P(kPa)$	1.26	1.31	1.39	1.48
$u_b(kPa)$	13.2	7.2	6.1	3.1
R^2	0.992	0.999	0.998	0.994

Table 3. Index λ with different consolidation stress and suction.

P(kPa)	100	200	300	400
λ	0.7476	0.7326	1.1512	1.5120

will decrease in pace with stress. Substitute the data in Table 1 into formula (11) respectively, we get:

$$\begin{cases} \dfrac{\theta-0.131}{0.296-0.131}=\left\{\dfrac{13.2}{u}\right\}^{\lambda},(u>13.2, p=100kp_a) \\ \dfrac{\theta-0.134}{0.196-0.134}=\left\{\dfrac{11.2}{u}\right\}^{\lambda},(u>11.2, p=200kp_a) \\ \dfrac{\theta-0.136}{0.182-0.136}=\left\{\dfrac{8.7}{u}\right\}^{\lambda},(u>8.7, p=300kp_a) \\ \dfrac{\theta-0.137}{0.174-0.137}=\left\{\dfrac{6.5}{u}\right\}^{\lambda},(u>6.5, p=400kp_a) \end{cases} \quad (12)$$

Substitute the experimental data under different stress in Table 1 into formula (12), we can get the λ values under different stress showed in Table 3.

For easy analyzing, we choose the average value $\lambda = 1.04$ under different stress. Substitute $\lambda = 1.04$ into formula (7) it can get $\eta = 4.9$. Then substitute $\eta = 4.9$ into formula (4) we get the considering-the-stress-influence unsaturated permeability function expression (13) as follows:

$$k_W = k_S\left[\frac{(0.144p-113.7)(15.6-0.023p)}{\ln\left((7.94-0.0113p)\Big/(\omega-14.1)\right)}\right]^{4.9}, u>9.9 \quad (13)$$

Hence the relative permeability coefficient k_r is

$$k_r = \frac{k_w}{k_s} = \left[\frac{(0.144p-113.7)(15.6-0.023p)}{\ln\left((7.94-0.0113p)\Big/(\omega-14.1)\right)}\right]^{4.9} \quad (14)$$

The relative permeability coefficient acquired is not only the function of water content, but also the function of stress p. Since the unsaturated permeability function is related to the stress state, it brings a problem to the calculation of unsaturated seepage, that is, the problem of coupling stress field with seepage field. It is necessary to adjust the unsaturated permeability coefficients for each-time seepage calculation according to the change of stress field.

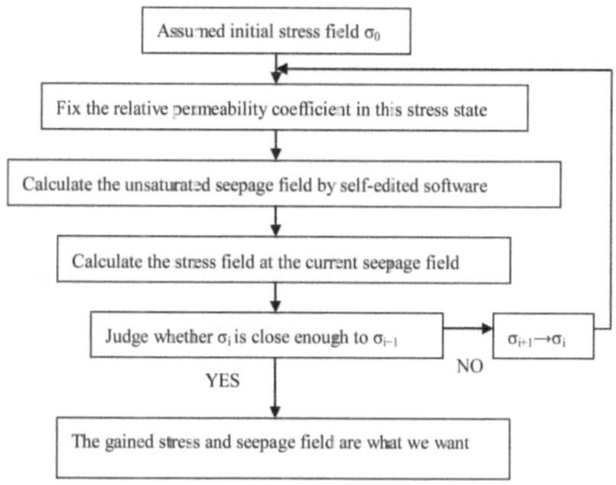

Figure 1. Calculating flow diagram.

3 THE COUPLED ANALYSIS METHOD OF STRESS FIELD AND UNSATURATED SEEPAGE FIELD

At present, the general uncoupled analysis method (calculating the stress field and seepage field separately) is: first calculate the seepage field, then add it into the analysis of stress field as a seepage load or a factor that can change medium mechanic parameters. But in the analysis of unsaturated seepage, the permeability coefficients of materials will change in pace with water content. So the permeability coefficient is a function of water content. When fixing the permeability function without considering of stress field, it is an uncoupled issue. If the unsaturated permeability function taking stress into consideration is adopted, the calculation will be done according to the present stress field, which is then related to the coupled problem of stress field and seepage field. Adopting iterative method and based on the calculation model for unsaturated and non-steady seepage analysis as well as the unsaturated permeability functions taking stress into consideration, we compiled a calculating program which can couple stress field and seepage field in seepage analysis of unsaturated soil. The coupled calculating flow is as Figure 1.

4 EXAMPLE COMPUTATION

By using the self-edited calculating program for coupled analyzing on stress field and seepage field, taking the slope of ACADS in slope stability analysis standard exam question 1(a) as an example, this paper calculate

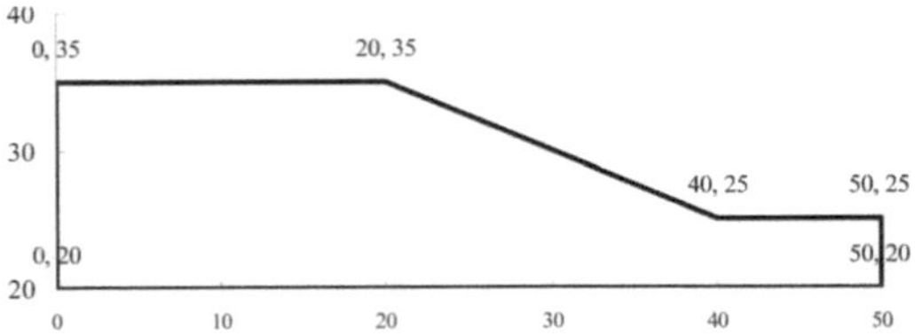

Figure 2. Slope used in Australian standard exam question 1(a).

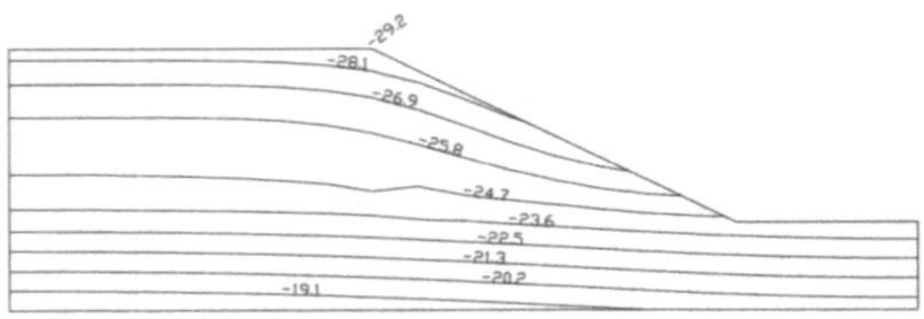

Figure 3. Hydraulic head isoline of the seepage field in slope in initial time.

the change of seepage field in the slope under specified rainfall intensity, contrast the results with that of uncoupled method and analyze the difference as well as the causes.

4.1 *Calculation model and calculation condition*

Figure 2 shows the uniform slope of calculation model and its length unit is meter.

The rainfall intensity is 0.0165 m/h. Its left, right and bottom side are impervious boundaries. The calculation takes upside as flow (rainfall) boundary,

Suppose the rainfall period is 100 h, it calculate the changing situation of seepage field in slope within 50 hours after the rainfall penetration. Choose natural water content as 15%, and adopt formula (14) for relative permeability coefficient which is the function of water content and the stress state in slope.

4.2 *Calculating results and analysis*

(1) The effects of rainfall on slope seepage field
In order to gain the law of rainfall effect on seepage field in slope, we select the hydraulic head isoclines in initial time to make the analysis (Figure 3). From the initial time in the figure we can see that there is no underground water while there is large distribution of ground mass suction in slope.
(2) The results contrast of coupled and uncoupled method of stress field and seepage field
Draw respectively the hydraulic head isoline calculated by the two kinds of methods at the 50th hour during the rainfall on the same figure as shown in Figure 4. From the figure we can see that when the rainfall lasts 50 hours, the calculated groundwater tables in slope by two methods both ascend to the medium part of the slope while the former groundwater table more lower than the latter. Having considered the stress, which diminishes the permeability coefficient of unsaturated soil and slows down the rainwater's permeating rate, accordingly the groundwater level becomes slow when rising. Evidently, this is much closer to the actual situations of seepage in slope.

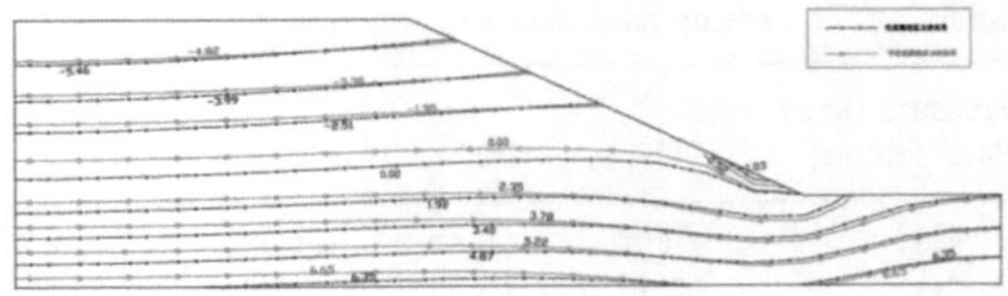

Figure 4. Hydraulic head isoline of the seepage field in slope when time is at the 50th hour (contrast of the results of being with and without considering consolidation pressure).

5 CONCLUSION

(1) Based on the permeability functions of unsaturated soil and experience functions of soil-water characteristic curve proposed by Brooks & Corey, authors derived the unsaturated permeability functions available in considering stress influence by using the soil-water characteristic curve functions available in considering stress influence.
(2) Adopting iterative method and based on the calculation model for unsaturated and non-steady seepage analysis as well as the unsaturated permeability functions taking stress into consideration, we self-edited the calculating program for coupled analyzing on stress field and seepage field in seepage analysis of unsaturated soil.
(3) From the seepage calculation on model slope, it gained the following conclusion: under the same precipitation conditions, compared with the uncoupled method, the coupled method of stress field and seepage field calculated lower groundwater level and higher ground mass suction. This is much closer to the actual situation of seepage in slope.

ACKNOWLEDGEMENT

Thanks to the National Natural Science Foundation of China in supporting the project (50839004) "research on disaster-causing mechanism, risk assessment and disaster reduction of landslide induced by rainstorm" and project (50879004) "Research on the features of unsaturated rheology of landslide debris".

REFERENCES

Averjanov, S. F. 1950. About permeability of subsurface soils in case of complete saturation. *English Collection*, 19–21.
Brooks, R. H. and Corey, A. T. 1964. Hydraulic properties of porous media. *Colorado State University, Hydrology Paper*, 3(3).

Campbell, J. D. 1973. *Pore pressures and volume changes in unsaturated soils*. Ph.D. Thesis, University of Illinois at Urbana-Champaign, IL.

Chen Z.Y. 2003. *Soil slope stability analysis—principle method procedure*. Beijing: China Waterpower Press. (In Chinese)

Davidson, J. M., Stone, L. R., Nielson, D. R. and Larue, M. E. 1969. Field measurement and use of soil-water properties. *Water Resources Research*, 5, 1312–1321.

Gardner, W. R. 1956. Calculation of capillary conductivity from pressure plate outflow data. *Soil Sci.Soc.Am.,Proc.*, 20, 317–320.

Philip, J. R. 1986. Linearized unsteady multidimensional infiltration. *Water Resources Res.*, 22, 1717–1727.

Wang SH.M. 2007. *Landslide stability analysis and engineering application based on unsaturated soil mechanics*. Wuhan: Wuhan University. (In Chinese)

Modelling and Computation in Engineering – Zhu (ed)
© 2011 Taylor & Francis Group, London, ISBN 978-0-415-61516-7

Upper bound analysis and calculation comparison for rock slope stability with Hoek–Brown failure criterion based on strength reduction technique

L.H. Zhao, D.P. Deng, F. Huang & Y.L. Lin
College of Civil and Architectural Engineering, Central South University, Changsha, China

ABSTRACT: The evaluation index based on the limit analysis theorem for rock slope stability, which differs from the evaluation index based on other common methods such as the limit equilibrium method and the numerical analysis method, is inconvenient for application during the engineering practice. Based on the upper bound limit analysis theorem and the shear strength reduction technique, the equation for expressing the critical limit-equilibrium state is employed to define the safety factor and its corresponding critical failure mechanism for a given rock slope by means of the kinematical approach of limit analysis theory. The nonlinear Hoek–Brown shear strength parameters (internal friction angle φ_t and cohesive force c_t) are treated as variable parameters and a knematically admissible failure mechanism is considered for the calculation schemes. The traditional definition of factor of safety is adopted in the present method so that the results from the limit analysis technique can be directly compared with the normal method (such as the limit equilibrium method and the numerical method) and could be applied in engineering practices. The case study and comparative analysis have shown that the solutions presented in this paper agree well with available finite element limit analysis methods, and the effectiveness and validity of the analysis method in this paper could be illuminated.

1 INTRODUCTION

Hoek-Brown failure criterion has been widely used in slope stability analysis since this criterion is proposed by Hoek & Brown (1980), and the validity of this application has been proved (Collins et al. 1988, Hoek, 1983, Hoek et al. 2002). The present stability analysis methods of rock slope subjected to Hoek-Brown failure criterion including limit equilibrium method, numerical analysis methods (finite element method and finite difference method) and limit analysis method (upper bound theorem, lower bound theorem). Among these methods, Collins et al. (1988) employed tangential technique to achieve the upper bound analysis of rock slope subjected to Hoek-Brown failure criterion for the first time. Based on the tangential technique, YANG et al. (2004a,b, 2006, 2007, 2009) used the instantaneous strength parameters c_t and φ_t to calculate the upper bound values of stability ratio of rock slope. At present, the new development is reported which applys limit analysis finite element method to analyze the stability of rock slope. By developing the failure surface approximation method proposed by Abbo & Sloan (1995), Merifield et al. (2006) and Li et al. (2008, 2009) adopt this method to compute the upper and lower bound values of static and dynamic stability ratio for rock slope.

Though the validity of applying limit analysis method to analyze the stability of rock slope subjected to Hoek-Brown failure criterion has been proved, the critical height (H_{cr}) and stability ratio (N_n) is mainly used to represent stability evaluation index of rock slope, which is not in accordance with conventional slope stability evaluation index (slope safety of factor F_s). Thus, it is difficult to apply this evaluation index directly in actual projects. Furthermore, though the slope safety of factor (F_s) equals 1.0 when the slope reaches critical height (H_{cr}), the study of convertion between slope safety of factor and slope critical height under noncritical condition is rare. Therefore, it is inconvenient to analyze the stability of rock slope by using limit analysis method.

Due to the application of Hoek-Brown failure criterion is constrained in isotropic rock mass, the analysis objects in this paper is an isotropic material which subjected to certain application conditions [4].

According to the former research, the slope safety of factor (F_s), in accordance with existing rock slope evaluation index, is obtained by using strength reduction upper bound theorem in conjunction with Hoek-Brown failure criterion and non-associated flow rule in this paper.

2 ASSUMPTIONS OF LIMIT ANALYSIS WITH HOEK-BROWN FAILURE CRITERION

2.1 *Tangential technique*

The tangential technique proposed by Collins et al. (1988) is illustrated in Fig. 1, and its main ideas can be expressed as: the values of c and φ are determined by drawing a tangential line to the Hoek-Brown failure curve at a random point.

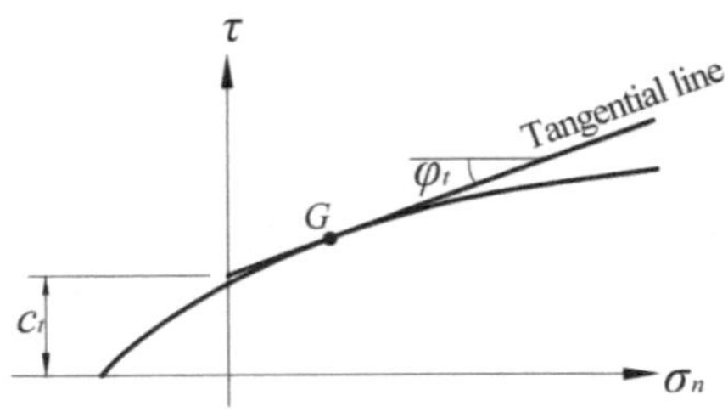

Figure 1. The tangential line for Hoek–Brown failure criterion.

The intercept between the tangential line and τ axial is c, and the inverse tangent value of the slope for the tangential line is φ. The tangential technique is employed to establish the relationship between c and φ, but not to determine the strength index of rock mass. Since the values of c and φ should be derived by the minimum energy consumption theorem, the c and φ which are expressed as c_t and φ_t are instantaneous values rather than constants. The envelope of shear failure can be written as

$$\frac{\tau}{\sigma_c}=\frac{\cos\varphi_t}{2}\left[\frac{mn(1-\sin\varphi_t)}{2\sin\varphi_t}\right]^{\frac{n}{1-n}} \tag{1}$$

$$\frac{\sigma_n}{\sigma_c}=(\frac{1}{m}+\frac{\sin\varphi_t}{mn})\left[\frac{mn(1-\sin\varphi_t)}{2\sin\varphi_t}\right]^{\frac{1}{1-n}}+\frac{s}{m} \tag{2}$$

where τ and σ_n are the normal and shear stresses on Hoek-Brown failure envelope respectively, φ_t is the angle between tangential line and horizontal line at a random point. The tangential line equation can be expressed as follows:

$$\tau=c_t+\sigma_n\tan\varphi_t \tag{3}$$

By iterative computation, the relationship between c_t and φ_t can be written as

$$\frac{c_t}{\sigma_c}=\frac{\cos\varphi_t}{2}\left[\frac{mn(1-\sin\varphi_t)}{2\sin\varphi_t}\right]^{\frac{n}{1-n}}+\frac{s}{m}\tan\varphi_t -\frac{\tan\varphi_t}{m}(1+\frac{\sin\varphi_t}{n})\left[\frac{mn(1-\sin\varphi_t)}{2\sin\varphi_t}\right]^{\frac{1}{1-n}} \tag{4}$$

The nonlinear shear strength index φ_t in equations (1) to (4) is an unknown parameter, which represented the variability of tangential line to the Hoek-Brown failure curve. Moreover, the nonlinear shear strength index c_t is determined by equation (4), and is also an unknown parameter.

2.2 *Strength reduction method subjected to Hoek-Brown failure criterion*

The strength reduction method is originally proposed by Bishop (1955), and he defined the soil slope safety factor as follows, the slope would reach critical limit equilibrium state by reducing the original shear strength parameter, and the reduction coefficient is the factor of safety (F_s). According to the tangential line equation of Hoek-Brown failure criterion, the equation of shear strength index using strength reduction method is obtained

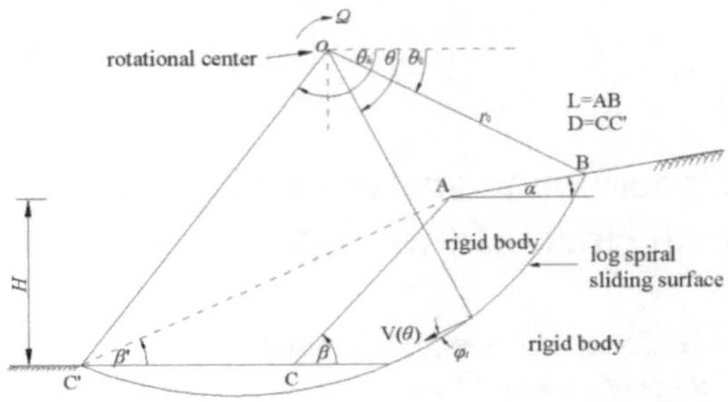

Figure 2. Rotational failure mechanism for slope stability analysis.

$$c_f=c_t/F_s, \varphi_f=\arctan\left(\tan\varphi_t/F_s\right) \tag{5}$$

where nonlinear shear strength index φ_t is an unknown parameter in upper bound calculation process, and c_t can be derived by equation (8).

Combining the upper bound theorem of limit analysis and strength reduction method, we obtain the upper solutions of safety factor (Fs), which is in accordance with existing rock slope stability evaluation index.

3 UPPER BOUND FAILURE MECHANISM AND SLOPE STABILITY ANALYSIS

3.1 *Failure mechanism and energy calculation*

As the failure surface of rock slope satisfied isotropic assumption is similar to log-spiral surface, the rotational failure mechanism is employed in slope upper bound analysis, which is shown in Fig.2.

Based on the upper bound theorem, the external rate of work equals the internal energy rate of dissipation, so the slope safety factor F_s (θ_h, θ_0, α, β, β', φ_t) is obtained

$$F_S=\frac{c_t}{\gamma H}\frac{e^{2\left[(\theta_h-\theta_0)\tan\varphi_f\right]}-1}{2\tan\varphi_f\left(f_1-f_2-f_3-f_4\right)}\left(\frac{H}{r_0}\right) \tag{6}$$

where f_1, f_2, f_3 and f_4 are functions related to θ_h, θ_0, α, β, β', φ_t, and the detail expressions can be found in documents (Chen 1975, Yang et al. 2004b, Yang 2007). Being different from these documents, the friction angle in this paper is φ_t, which is introduced by tangential technique subjected to Hoek-Brown failure criterion. Furthermore, the $\varphi_f=$ arctan (tan φ_t/F_s), and the friction angle in f_1, f_2, f_3, f_4, H/r_0 are all substituted by $\varphi_f=$ arctan (tan φ_t/F_s).

3.2 *Analysis process*

Based on upper bound theorem, the upper solution of slope safety factor (*Fs*) is given by equation (6).

$$\begin{cases} \frac{\partial F_S}{\partial \theta_h}=0, \quad \frac{\partial F_S}{\partial \theta_0}=0, \quad \frac{\partial F_S}{\partial \beta'}=0, \quad \frac{\partial F_S}{\partial(\tan\varphi_t/F_s)}=0 \\ \text{and } H_{cr}=H_{actual}, \quad \theta_0<\theta_h, \quad \beta'\le\beta, \quad \tan\varphi_t\ge 0 \end{cases} \tag{7}$$

When the condition expressed in equation (7) is satisfied, the extreme value of function F_s (θ_h, θ_0, α, β, β', ϕ_t) is obtained, and the upper solution of slope safety factor (*Fs*) is derived. By using sequential quadratic programming algorithm in optimization theory, equation (6) is optimized in this paper[0].

4 COMPARING CALCULATION

4.1 *Comparison with lower bound finite element method*

Li et al. (2008, 2009) develops the limit analysis finite element method proposed by Abbo & Sloan (1995) and Merifield et al. (2006). He used this method to analyze the static and dynamic stability of rock slope and compare the results with the solutions calculated by limit equilibrium method. The compared results shows that the exact solution of stability ratio of rock slope can be confined in the variation range of ±9% by using limit analysis finite element method, and the solution difference between lower finite element and limit equilibrium method is confined in the variation range of ±4.6%.

When other parameters of rock slope are constants, rock slope reachs the critical stability ratio (N), and the safety factor (F_s) of rock slope is 1.0. Using the same parameters of generalized H-B failure criterion, the compared results of this paper and former documents are illustrated in Table 1.

It is can be found from table 1 that the safety factor *Fs* calculated by upper bound theorem are all slightly larger than the solutions computed by lower finite element method, and the maximum absolute difference is less than 7% with average difference being 4.5%. Li et al. (2008, 2009) state that the upper and lower bound theorem of limit analysis finite element can make the exact solution of stability ratio of rock slope confining within ±9% variation range of the upper and lower solutions. Therefore, the good agreement shows that the slope stability analysis based on Hoek-Brown failure criterion in this paper is valid.

Considering typical slope angle ($\beta = 15°, 30°, 45°, 60°, 75°$), the critical sliding surfaces presented in this paper and by upper bound finite element method are compared in Fig. 3. It is found that the critical sliding surfaces calculated by these two methods are roughly the same, but the critical sliding surfaces presented in this paper is slightly shallow.

The calculation of slope safety factor and the search of critical sliding surface are the core contents for slope stability analysis. According to the compared results between this paper and upper bound finite element method, it can be concluded that the method used in this paper is effective and valid. Furthermore, the method in this paper is more simple and no complex calculation model is required. Thus, large quantity of optimization calculation can be avoided and the computing time is shortened.

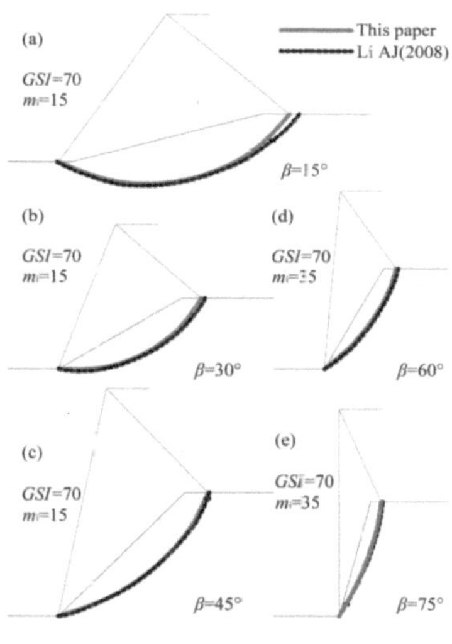

Figure 3. Comparison of critical sliding surfaces based on Hoek–Brown failure criterion.

Table 1. Comparison of safety factor with finite element limit analysis methods based on Hoek–Brown failure criterion.

			Calculation methods							
			Lower finite element limit analysis		Limit equilibrium: SLIDE-equivalent M-C criterion				This paper	
Basic parameters			Nonlinear H-B		Nonlinear H-B	Linear M-C			Nonlinear H-B	
β/°	*GSI*	m_i	$\sigma_c/\gamma H$	F_{S1}	F_{S2}	F_{S3}	F_{S4}	F_{S5}	F_S	error/%
75	70	35	0.717	1.0	1.016	1.394	1.156	–	1.03208	3.208
60	70	35	0.200	1.0	1.005	1.183	1.047	–	1.03779	3.779
45	70	15	0.176	1.0	1.012	1.080	1.002	1.098	1.04635	4.635
30	70	15	0.075	1.0	1.023	0.966	–	1.028	1.06002	6.002
15	70	15	0.026	1.0	1.010	1.087	–	0.987	1.05599	5.599

Table 2. Comparison of safety factor with normal methods based on Hoek–Brown failure criterion.

Problem no.	γ kN/m^3	H /m	σ_c /MPa	β/°	GSI	m_i	D	m	s	n	EMC parameters φ/°	EMC parameters c/kPa	Source	Calculation methods	Failure criterion	F_s
1	25	10	30	35.5	5	2	0.0	–	–	–	20.89	20	Benz et al. (2008)	Bishop	EMC	1.37
														FEM	EMC	1.37
														FEM	GHB	1.51
														FEM	HBMN	1.72
													This paper	UBLA	EMC	1.3744
														UBLA	GHB	1.4244
2	26	32	40	75	45	10	0.9	–	–	–	38	180	Benz et al. (2008)	Bishop	EMC	1.55
														FEM	EMC	1.54
														FEM	GHB	1.00
														FEM	HBMN	1.02
													This paper	UBLA	EMC	1.7856
														UBLA	GHB	1.4823
3	25	10	30	45	5	2	0.0	0.067	2.5×10^{-5}	0.619	20.89	20	Hammah et al. (2005)	FEM	GHB	1.150
														FEM	EMC	1.150
														Bishop	EMC	1.153
														Spencer	EMC	1.152
													This paper	UBLA	EMC	1.1591
														UBLA	GHB	1.1798
														UBLA	GHB	1.1731
4	25	300	10.64	40	–	–	–	0.422	2.404×10^{-4}	0.5	22	317	Zheng et al. (2009)	FEM	GHB	1.06
														FEM	EMC	1.10
														Spencer	EMC	1.08
													This paper	UBLA	EMC	0.9991
														UBLA	GHB	1.0008

4.2 *Comparison with conventional methods*

In order to evaluate the validity of the method used in this paper, several existing project examples are employed to compare the solutions calculated by the method in this paper. Based on generalized Hoek–Brown failure criterion and equivalent parameters of Mohr-Coulomb criterion, the compared results between limit analysis and existing conventional method are illustrated in Table 1.

The abbreviation are illustrated as follows: EMC, equivalent Mohr-Coulomb criterion; GHB, generalized Hoek–Brown failure criterion; HBMH, developed Hoek–Brown failure criterion; Bishop, simplified Bishop method; Janbu, Janhu method; Spencer, Spencer method; FEM, Finite Element method Strength Reduction; UBLA, Upper Bound Limit Analysis.

Table 2 presents that when using Hoek–Brown failure criterion, the safety factors F_s calculated by upper bound theorem with Hoek–Brown shear strength parameter introduced by tangential technique in examples 2 and 3 are larger than existing achievements, and the maximum difference excesses 48%. However, the safety factors Fs in this paper of examples 1 and 4 are slightly smaller than existing results, and the maximum difference being less than 12.7%. When the failure criterion using equivalent Mohr-Coulomb criterion, the difference among each method is large, and the maximum difference is greater than 22%. Furthermore, the maximum difference between the upper solutions in this paper and the existing results is also greater than 19%. Due to the upper bound theorem can provide rigorous variation range of upper solution, the comparison presents that the results of conventional methods are instable.

5 CONCLUSIONS

The actual failure of rock material follows a nonlinear Hoek–Brown failure criterion, and the values of strength parameters have significant influence on estimating the safety of rock mass. According to upper bound theorem, the existing tangential technique of nonlinear failure criterion with limit analysis theory is developed in this paper. Combining with strength reduction method, we analyzed the stability of rock slope subjected to Hoek–Brown failure criterion. Based on the above comparing analysis, following conclusions can be drawn:

(1) By using tangential technique, the safety factor and sliding surface calculated in this paper in accordance with the results obtained by limit analysis finite element method on the same conditions, and the maximum difference of safety factor less

than 7.5% and average difference is 4.4625%. The good agreement shows that the method used in this paper is effective and valid.

(2) Numerous rock slope examples illustrates that, the variation range of safety factor calculated by conventional method is too wide, which presents that the results of conventional methods are instable.

(3) Comparing with existing method, the method in this paper is more simple and no complex calculation model is required. Moreover, because large quantity of optimization calculation is avoided, the computing time is shortened. Therefore, the method in this paper can be widely used in rock slope stability analysis.

ACKNOWLEDGMENT

The author is grateful to T. Zhang for her assistance.

REFERENCES

Abbo, A.J. & Sloan, S.W. 1995. A smooth hyperbolic approximation to the Mohr–Coulomb yield criterion. *Comput Struct* 54(3):427–41.

Benz, T., Schwab, R., Kauther, R. A. & Vermeer, P. A. 2008. A Hoek–Brown criterion with intrinsic material strength factorization [J]. *International Journal of Rock Mechanics & Mining Sciences* 45(2):210–222.

Bishop, A.W. 1955. The use of the slip circle in the stability analysis of slopes. *Geotechique* 5(1):7–17.

Chen, W. F. 1975. *Limit analysis and soil plasticity*. Amsterdam: Elsevier Scientific Company.

Collins, I.F., Gunn, C.I., Pender, M.J. & Yan, W. 1988. Slope stability analyses for materials with nonlinear failure envelope. *Int J Numer Anal Methods Geomech* 12(6):533–550.

Hammah, R.E., Curran, J.H., Yacoub, T.E. & Corkum, B. 2004. Stability analysis of rock slopes using the finite element method. //In: Schubert, editors. *Eurock 2004 and 53rd Geomechanics Colloquium*, Salzburg.

Hammah, R.E., Yacoub, T.E., Corkum, B.C. & Curran, J.H. 2005. The Shear Strength Reduction Method for the Generalized Hoek-Brown Criterion. // Alaska Rocks. *The 40th U.S. Symposium on Rock Mechanics (USRMS): Rock Mechanics for Energy, Mineral and Infrastructure Development in the Northern Regions*. Alaska: ARMA:1–6.

Hoek, E. & Brown, E.T. 1980. Empirical strength criterion for rock masses. *ASCE J Geotech Eng* 106(GT9):1013–1036.

Hoek, E. 1983. Strength of jointed rock masses. *Geotechnique* 33: 187–223.

Hoek, E., Carranze-Torres, C. & Corkum, B. 2002. Hoek-Brown failure criterion-2002 edition. In: *Proceedings of the North American Rock Mechanics Society Meeting*, pp 267–273.

Li, A.J., Merifield, R. S. & Lyamin, A.V. 2008. Stability charts for rock slopes based on the Hoek–Brown failure criterion. *International Journal of Rock Mechanics & Mining Sciences* 45(5): 689–700.

Li, A.J., Lyamin, A.V. & Merifield, R S. 2009. Seismic rock slope stability charts based on limit analysis methods. *Computer and Geotechnic* 38(1):135–148.

Merifield, R.S., Lyamin, A.V. & Sloan, S.W. 2006. Limit analysis solutions for the bearing capacity of rock masses using the generalised Hoek–Brown criterion. *International Journal of Rock Mechanics and Mining Sciences* 43(2): 920–937.

Yang, X.L., Li, L. & Yin, J.H. 2004a. Stability analysis of rock slopes with a modified Hoek–Brown failure criterion [J]. *International Journal for Numerical and Analytical Methods in Geomechanics* 28(2): 181–190.

Yang, X.L., Li, L. & Yin, J.H. 2004b. Seismic and static stability analysis of rock slopes by a kinematical approach. *Geotechnique* 54(8): 543–549.

Yang, X.L. & Zou, J.F. 2006. Stability factors for rock slopes subjected to pore water pressure based on the Hoek–Brown failure criterion. *International Journal of Rock Mechanics and Mining Sciences* 43(7): 1146–1152.

Yang, X.L. 2007. Seismic displacement of rock slopes with nonlinear Hoek–Brown failure criterion[J]. *International Journal of Rock Mechanics and Mining Sciences* 44(6): 948–953.

Yang, X.L. & Yin, J.H. 2009. Slope Equivalent Mohr–Coulomb Strength Parameters for Rock Masses Satisfying the Hoek–Brown Criterion. *Rock Mech Rock Eng*. DOI 10.1007/s00603-009-0044-2.

Zhao, L.H. 2009. *Energy analysis study on slope stability and reinforcing design*. Changsha: Central south university.

Zheng, H., Sun, G.H. & Liu, D.F. 2009. A practical procedure for searching critical slip surfaces of slopes based on the strength reduction technique. *Computers and Geotechnics* 36: 1–5.

Modelling and Computation in Engineering – Zhu (ed)
© 2011 Taylor & Francis Group, London, ISBN 978-0-415-61516-7

Study on contribution rate of equipment manufacturing industry to the Northeast China's economic increase

Wenwu Shao
School of Economy and Management, Shenyang Aerospace University
Business School, Jilin University

Liyu Chen & Weiwei Zhou
School of Economy and Management, Shenyang Aerospace University

ABSTRACT: The contribution rate of equipment manufacturing industry to northeast China's economic increase was researched. The time series data of equipment manufacturing industry production value and GDP in the Northeast China was used. The induction force coefficient and influence coefficient of the equipment manufacture in the Northeast China were measured with the input-output theory. And the comprehensive contribution rate of equipment manufacturing industry to the Northeast China economy increase was measured and analyzed. So some results were deduced that the equipment manufacturing industry in the Northeast China was the base industry, and the role in boosting economic growth has been increasing year by year.

1 INTRODUCTION

Equipment manufacturing industries was the base industry in the national economy. And it could provide technical equipment to other industry. The Northeast China is the main equipment manufacturing base. In the equipment manufacturing industry there was abundant industrial base in the Northeast China. A large number of key equipment and supporting enterprises were centralized in this area. And some enterprises owned leading product research and manufacturing capacity in the domestic level, heavy equipment products in domestic had irreplaceable position. In 2003, revitalization plan of the Northeast China old industrial bases was brought forward. Along with the implementation of the revitalization policy unceasingly, the revitalization of effect appeared gradually. In recent years, the economic growth in northeast China appears good situation in the Northeast China. The equipment manufacturing industry occupies a large scale in the region economy. So the revitalization policies to the Northeast China old industry base come into being a large stimulation use. Opportunities of national policy were sized tightly. And the equipment manufacturing industry development planning was worked out in the Northeast China. The Northeast China would be completed the advanced equipment manufacturing base.

While the equipment manufacture in developed countries and the present domestic coastal areas would be transferred to developing region. And the Northeast China had the capability to receive these transferred industry because there were better equipment manufacturing industry base and good policy support in this region. The equipment manufacturing industry in Northeast China was studied in order to analyze the developing trend of this industry with quantitative way. And on the other hand the status of equipment manufacturing industry in Northeast China economy could be evaluated objectively. The relative status change of the equipment manufacturing industry in the Northeast China industrial structure in recent years embodied it's the importance to the Northeast China economy growth.

2 ANALYZING THE STATUS OF EQUIPMENT MANUFACTURING INDUSTRY IN NORTHEAST CHINA

The Northeast China is one of important equipment manufacturing bases. The competition capability of industries in the Northeast China was obvious difference. And industries competitive level of three provinces in the Northeast China was obvious difference too.

Firstly, in field of industry, the comprehensive capability of come industries was very strong such as transportation equipment manufacturing industry, manufacture of general purpose machinery, manufacture of communication equipment and computers and so on. These industries were leading industries of the equipment manufacturing industry in Northeast China. Some industries developed rapidly such as manufacturing of automobile, ships, floating device, aviation aircraft industry and so on. These industries formed industrial cluster model initially.

Secondly, in the field of layout, equipment manufacture in Liaoning province was better than that of other provinces in the Northeast China in following aspects, competitive advantage, development potential and development mode etc. The equipment manufacturing industry in Liaoning province outstanding started in late 19th and early 20th century. In that time Liaoning was the equipment manufacturing industry scientific research and production base of all over the country. Equipment manufacturing industry in Liaoning included eight big industry countries delimited by the government. its outstanding advantage of equipment manufacture in Liaoning province existed in there were a array of leading enterprises in the domestic and well-known in international, such as light passenger vehicle, diesel engine, numerical control machine, ocean shipping equipment and railway locomotive and bearings industry and so on. The production value of machine tool in Liaoning was 26% of all country in 2008. Output of some production was the largest in all country such as diesel locomotive, refrigeration equipment, pneumatic tools and so on. The output of petroleum equipment was in second place in the country. Although the overall strength of the equipment manufacturing industry in Jilin province was not strong, the transportation equipment manufacturing was the leading industries of the Northeast China. The advantage industries of Jilin province were cars and railway transport equipment manufacturing, lorry manufacturing. Among these industries Jilin auto industry sales revenue accounted for about 13% of the country. So Jilin province was the core of the Northeast China transportation industry. Some equipment manufacture industries of Heilongjiang were situated in the top five. They were truck manufacturing industry, airplane manufacturing, metallurgy industry and special equipment manufacturing, generator manufacturing, steam turbine manufacturing, boiler manufacturing, farming machine manufacturing, mini-car manufacturing, cutting tool manufacturing etc. A large scale and complete kind equipment manufacture were formed in these three provinces of the Northeast China. Each province had their dominant production. And these dominant productions in every province were complementary.

After the reform and opening, structure adjustment mechanism management system and the market economic aspects in Northeast China didn't adapt to development. The station of equipment manufacture industry in the Northeast China was challenged by other region. Adjacent to Hong Kong and Macao and the Pearl River delta region in Southeast Asia, with open policy support, communication equipment and computer equipment manufacturing industry cluster was emerged. The value of these industries was more than 30% of all country. Since the 1990s, the Yangtze River delta region equipment manufacturing developed rapidly. A new highland of equipment manufacture that was typical of automobile and automobile parts manufacture was formed. So the equipment manufacturing center was moved to south China gradually (Wu et al. 2007).

Now the main problems existing in equipment manufacturing industry of the Northeast China are as follows. The firstly, there was the enterprise unreasonable structure. System of organization reform lagged behind. The proportion of state-owned economy in Northeast China was too large. And the proportion of private enterprises and joint venture enterprises was too low. Under the influence of traditional planned economy system, some problems exist in following fields. The idea was obsolete. Market consciousness was scarce. The management system was rigid. The manufacturing industry mode was single. The organization structure was unreasonable. The innovation mechanism wasn't flexible enough. Because there were different interests subject in each region respectively these subject couldn't cooperate mutually. And production could not adapt to the demand of the market. Secondly, the equipment manufacture industry cluster level was low. The system of dividing the work was not reasonable. The Northeast China equipment manufacturing lacks developed system of industrial division. Support industry development lagged behind. Specialization of cooperation degree was low. The industry chain wasn't formed. All these problems restricted the equipment manufacturing industry development of the Northeast China in a certain extent. Thirdly, independent innovation ability was not strong. In equipment manufacture enterprises above designated size by industrial sector investment accounts for sales income less than 1%. But in main developed countries the index was 10% to 15%. Most equipment manufacturing enterprises in Northeast China was built in early time. The problem of old equipment and old technology in enterprises was very serious. High value-added products were short. The cooperation ability was very poor. All of these seriously affected low enterprise's production and innovation ability. The equipment the equipment manufacturing industry enterprises nearly 5% only reached the international advanced level.

3 MEASURING THE EQUIPMENT MANUFACTURING INDUSTRY CONTRIBUTION RATE TO THE NORTHEAST CHINA ECONOMY INCREASE

3.1 *The measuring model analyzing*

The method of input-output was used to measure the equipment manufacture contribution rate to the Northeast China economic growth. This way was used constantly in industrial economy research. In a region, the change of industrial structure was slow. So in the paper the 2007 input-output table being used was reasonable.

(1) Input-output model

Input-output method was brought forward by Vassili. Leontief. The Input-output table was divided to three parts. These were left top part, right top part and left bottom part. These three parts was named respectively I, II and III quadrant. These three quadrants represented respectively the intermediate input production part, finally used production part and increase value parts such as Table 1.

Table 1. Input–output value table.

		intermediate use part (j)				end use	total output
		1	2	…	n		
intermediate inputs (i)	1	x_{11}	x_{12}	…	x_{1n}	Y_1	X_1
	2	x_{21}	x_{22}	…	x_{2n}	Y_2	X_2
	…	…	…	…	…	…	…
	n	x_{n1}	x_{n2}	…	x_{nn}	Y_n	X_n
value-added		N_1	N_2	…	N_n		
total inputs		X_1	X_2	…	X_n		

Table 2. The input–output value table combined from n region.

		intermediate use part (j)				end use	total output
		1	2	…	n		
intermediate inputs (i)	1	X_{11}	X_{12}	…	X_{1n}	Y_1^L	X_1^L
	2	X_{21}	X_{22}	…	X_{2n}	Y_2^L	X_2^L
	…	…	…	…	…	…	…
	n	X_{n1}	X_{n2}	…	X_{nn}	Y_n^L	X_n^L
value-added		N_1^L	N_2^L	…	N_n^L		
total inputs		X_1^L	X_2^L	…	X_n^L		

In the table, x_{ij} represented that the product from$i$ department into the j department, or the production in department j used i department production quantity. Y_i represented eventually application amount of i department. N_j represented the increasing output number of j department. The core ideas of input–output table were as follows.

$$\sum_{j=1}^{n} x_{ij} + Y_i = X_i$$

$$\sum_{i=1}^{n} x_{ij} + N_j = X_j$$

It was said that in the region i industry intermediate input production number added on end use production number equal to the total value of output. And j department intermediate use production number added on increase number equal to the total input production number in this industry (SU, 2005). If input–output tables of n regions were combined the relationship was constant. The input–output value table being combined was as Table 2.

In the table 2, X_{ij}—the total number that L administrative regions the i department input production into the j department, or the j department's using part of the i department's product quantity.

$$X_{ij} = \sum_{l=1}^{L} x_{ij}^{l}$$

Y_i^L—the i department production number that was used finally in L administrative regions.

$$Y_i^L = \sum_{l=1}^{L} Y_i^{l}$$

X_i^L—the i department total production value of L administrative regions.

$$X_i^L = \sum_{l=1}^{L} X_i^{l}$$

N_j^L—the j department increase production value.

$$N_j^L = \sum_{l=1}^{L} N_j^{l}$$

X_j^L—the j department total input production value of L administrative regions (Wang et al. 2006).

$$X_j^L = \sum_{l=1}^{L} X_j^{l}$$

Since each part of the combined input-output table was formed by the same part added on of each input-output of n administration regions (Hang et al. 2005). So it kept to relationships as follows.

$$\sum_{j=1}^{n}\sum_{l=1}^{L} x_{ij}^{l} + \sum_{l=1}^{L} Y_i^{l} = \sum_{l=1}^{L} X_i^{l}$$

$$\sum_{i=1}^{n}\sum_{l=1}^{L} x_{ij}^{l} + \sum_{l=1}^{L} N_j^{l} = \sum_{l=1}^{L} X_j^{l}$$

(2) The direct consumption coefficient matrix

The direct consumption coefficient matrix in the input–output table was used to reflect technology and economy relation between different departments in relevant industry.

The formula was used to calculate the direct consumption coefficient matrix as A.

a_{ij} was named input-output coefficients.

$$a_{ij} = \frac{\sum_{l=1}^{L} x_{ij}^{l}}{\sum_{l-1}^{L} X_{j}^{l}}$$

$$A = \begin{bmatrix} a_{11} & a_{12} & \cdots & a_{1n} \\ a_{21} & a_{22} & \cdots & a_{2n} \\ \cdots & \cdots & \cdots & \cdots \\ a_{n1} & a_{n2} & \cdots & a_{nn} \end{bmatrix}$$

Further the matrix A was used to calculate the Leontief matrix as $(I - A)$ and the inverse matrix $B_{ij} = (I - A)^1$ was deduced.

(3) The model to estimate the influence force and induction force

Influence force coefficient and induction force coefficient were important parameters to judge correlative degree. The influence coefficient could reflect that the finally production value in a certain industry increased one unit to affect other industry degree. If the influence coefficient of certain industry more than 1 shows the industry pulling economy growth capability was more that the average level of all industries. The more the value the more it pulling economy growth capability was. Induction force coefficient reflects that each industry increased one unit production value it need a certain industry supply number. The value of this industry more than 1 shows that all industries demand for the certain industry is more than the average level of other industries. The greater its value the larger the certain industry accelerate to the economy.

The formula to calculate influence force coefficient is as follow.

$$T_j = \frac{\frac{1}{n}\sum_{i=1}^{n} B_{ij}}{\frac{1}{n^2}\sum_{i=1}^{n}\sum_{j=1}^{n} B_{ij}}$$

In the formula T_j—the influence force coefficient of the j industry; B_{ij}—the i line and the j row coefficient in the Leontief inverse matrix.

The formula to calculate induction force is as follow.

$$S_j = \frac{\frac{1}{n}\sum_{j=1}^{n} B_{ij}}{\frac{1}{n^2}\sum_{i=1}^{n}\sum_{j=1}^{n} A_{ij}}$$

In the formula, S_j—the j industry induction force coefficient; and the meaning of B_{ij} is the same with above.

(4) The model to estimate a certain industry contribution rate to the regional economic growth

The industry contribution to region economy not only embodies for direct contributions, but also embodies for indirect influence that effect would be realized though the certain industry affecting other industry in the region, namely the spillover effect. The inverse matrix coefficient table$(I\text{-}A)^{-1}$was used to calculate when the certain industry production value changed one unite it deduced output change sum of all industry direct and indirect output. The output change sum of all industry direct and indirect output was stand for C. According to the certain production value change $\Delta\gamma$ the formula $\Delta G' = \Delta\gamma \bullet C$ was used to calculate under the effect of $\Delta\gamma$ the region GDP change $\Delta G'$. Further according to the region GDP total change number the certain industry contribution rate to the region economic growth was calculated, namely E (Shao et al. 2006).

$$E = \frac{\Delta G'}{\Delta G}$$

3.2 *Estimating the equipment manufacturing industry contribution rat to the Northeast China economic growth*

In calculating equipment manufacturing industry contribution rate to the Northeast China some data was used such as the input-output table of 2007, the Northeast China GDP from 2004 to 2008, the Northeast China equipment manufacture production total value from 2004 to 2008.

(1) Measuring the induction force coefficient and influence coefficient of equipment manufacturing industry.

The combined input-output table was used. According to formula the direct consume coefficient table A of three provinces in the Northeast China was deduced. Base on the matrix A software of MATLAB and SPSS was used to calculate the Leontief inverse matrix A^{-1}. Then the influence force coefficient model was used to calculate the Northeast China equipment manufacture influence force coefficient from 2004 to 2008. The result was $T_j = 1.3217$. The induction force coefficient model was used to calculate the Northeast China equipment manufacture induction force coefficient from 2004 to 2008. The result was $S_j = 1.1183$.

From the result the influence coefficient T_j and induction coefficient S_i were more than 1. It was said that in all industries the equipment manufacture industry influence force and induction force were above average in the Northeast China. Through equipment manufacturing induction force coefficient and influence of numerical analysis the conclusion could be deduced as follows. The equipment manufacturing industry in the Northeast China was situated in base industrial group. It s use was safeguard to the Northeast

Table 3. The equipment manufacturing industry pulling to the Northeast China economic growth.

	2004	2005	2006	2007	2008
The Northeast China GDP (Billion)	14735.30	17140.80	19715.20	23373.20	28195.66
Equipment manufacture industry production value (Billion)	4312.84	5148.60	6712.40	8788.23	10994.43
Equipment manufacturing industry output value pulling GDP increase value (Billion)	9547.76	11397.97	14859.91	20340.9	24339.48
Equipment manufacturing industry contribution rate to GDP (%)	64.79	66.49	75.37	83.23	86.32

Data sources: Liaoning province statistics yearbook (2005–2009), Jilin province statistics yearbook (2005–2009), Heilongjiang province statistics yearbook (2005–2009).

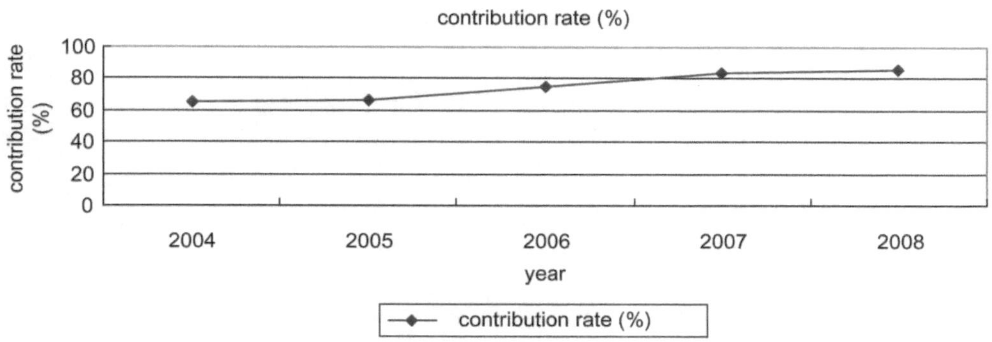

Figure 1. The equipment manufacturing industry contribution rate to the Northeast China economy growth.

China economy growth. It had strong influence to the whole economy and it was influenced by other industry seriously. If the Northeast China economy growth rate is high the equipment manufacture industry develops more quickly. The equipment manufacturing industry in Northeast China economy can have Main leading role. The production value of equipment manufacture in the Northeast China average annual increase was 26.45% from 2004 to 2008. It pulled the Northeast China economy rapid growth.

(2) Measuring equipment manufacturing industry contribution rate to the Northeast China economy growth

The input-output table of 2007 and the related data were based on, and the inverse matrix of $(I - A)^1$ was used to calculate if the equipment manufacturing industry production value changed one unite all industries were brought direct and indirect output value change summation, namely C. C being calculated was 2.2138. The result being calculated that equipment manufacturing industry contribution value and contribution rate was showed in Table 3

From Figure 1 it shows that the equipment manufacturing industry contribution rate to GDP of the Northeast China was raising per annual, which indicates that the equipment manufacturing industry was situated more and more important role to the whole industries. In recent years, with the beneficial national policies and good opportunities of at home and abroad, the equipment manufacturing industry in Northeast China showed fast growth momentum.

4 CONCLUSIONS

The equipment manufacturing industry in Northeast China had good industrial base. From the influence force coefficient and the induction force coefficient the equipment manufacturing industry was the fundamental industry in the Northeast China. It has greater influence to the Northeast China's economic growth in recent years. with vitalizing the northeast old industrial base policy implementation, and accepting home and abroad equipment manufacturing industry transfer ability constantly strengthen the equipment manufacturing industry increase speed of the Northeast China was 26.45% annually. Meanwhile the contribution rate to economic growth was rising year by year. So the equipment manufacturing industry structure in the Northeast China occupies the core status. The revitalization of northeast old industrial base depend on the revitalization of the equipment manufacturing industry in a certain extent.

REFERENCES

Hang Wei, Zhang Eling, Zhang Xiaohua. Analyzing and comparing spillage effect of industry GDP increase between different regions in China. Economy Science, 2005(3): 15–28. (in Chinese)

SHAO Wenwu, XIA Enjun, REN Peimin. Measuring the contribution rate of mechanical and electrical equipment manufacturing to Liaoning economy. Transactions of Beijing Institute of Technology, 2006,06. (in Chinese)

SU Dongshui. Industry economy. Peking: Higher education publishing company, 2005,12. (in Chinese)

Wang Yongjun, Peng Xuejun, Shao Wenwu. Research the estimation method of enterprise group contribution rate to regional economy. Market Modernization, 2006,10. (in Chinese)

Wu Zhengzheng, Wu Dianting, Feng Xiaojie. Research on the status and change of the equipment manufacturing in northeast china. Human Geography, 2007,01. (in Chinese)

Modelling and Computation in Engineering – Zhu (ed)
© 2011 Taylor & Francis Group, London, ISBN 978-0-415-61516-7

Application of fiber wall element model in nonlinear analysis of steel high performance concrete shear walls

Liang Bai
School of Civil Engineering, Chang'an University, Xi'an, Shaanxi Province, P.R. China

Xingwen Liang
Key Laboratory for Structural Engineering and Earthquake Resistance of China Education Ministry, Xi'an University of Architecture and Technology, Xi'an, Shaanxi Province, P.R. China

ABSTRACT: Study on seismic behavior of steel high performance concrete (SHPC) shear walls with different parameters was performed in this paper. Firstly, the fiber wall element model in CANNY program was used to analyze the non-linear performance of SHPC shear walls. The results of numerical simulation were agreement with experimental results. Also, some parameters on seismic behavior of SHPC shear walls were analyzed. The simulated analysis showed that the bearing capacity improved with the increase of the axial load ratio. The deformation capability decreased with the increase of the axial load ratio. Moreover, when the value of axial load ratio reached to 0.5, the displacement ductility ratio is less than 3.0. The deformation capability improved with the increase of the length of restricted edge members. The bearing capability improved with the increase of the steel ratios, however, the deformation capability improved not so significant.

1 INTRODUCTION

High performance concrete has the advantages of high strength, better fluidity, high durability and impermeability which meet the need of modern engineering structure for large span, high rise, heave load and bad environmental condition. High performance concrete leads an important developing trend in concrete technology (Zhaoyuan 1997; Lü 2004). Because of fragility of high performance concrete and poor deformability of shear wall, its seismic performance, especially deformability, needs to be improved. Steel high performance concrete shear walls was a combination of steel and high performance concrete (SHPC). They can give full play to the advantages of steel and high performance concrete and have better dynamic behaviour, such as ductility and ability of dissipating energy.

The numerical simulation was used as an analysis tool in civil engineering. An analytical structural engineering model for simulating the nonlinear hysteretic behaviour was presented and all characteristics of the hysteretic behaviour of RC walls are explicitly modelled (Youssef & Lestuzzi 2007). Nonlinear finite element (FE) analysis of steel reinforced concrete (SRC) shear wall was performed using software MSC.Marc. Based on the proposed FE model, parametric analysis wag carried out to study the effect of the axial compression ratio and steel ratio on seismic behaviour of SRC shear walls (Zuozhou, Zhiyuan & Jiaru 2009). Two types of short-leg shear walls were introduced through static tests on the same type of four four-story 1:3 scaled short-leg shear wall specimens respectively. According to damage energy release rate-based elasto-plastic damage constitutive model for concrete, nonlinear behaviour of short-leg shear walls were investigated (Jie & Kuiming 2009).

Above on, the study on the seismic behavior of SHPC shear walls is deficient and the experiment that can be used for model analysis is also not adequate. Therefore, the fiber wall element model considering stiffness degradation, strength deterioration and pinching behavior in CANNY program was used to analyze the non-linear performance of SHPC shear walls. The simulated results were verified to be in close agreement with the experimental ones, then a series of shear walls with different parameters were analyzed and compared by using this method so as to search for the parameters on seismic behavior of SHPC shear walls, such as axial load ratio, the length of restricted edge members and ratio of steel.

2 EXPERIMENTAL SURVEY

Three cantilever SHPC shear walls were constructed and the cross section is 1000 mm × 100 mm. The specimens are numbered as SHPCW-01, SHPCW-02 and SHPCW-03. The axial load ratio of SHPCW-01 and HPCW-03 is 0.17 and that of SHPCW-02 is 0.25. The shear span ratio of SHPCW01 and SHPCW01 is 2.1 and that of SHPCW03 is 1.5. The value of compressive strength of concrete is 88.5 MPa, 83.6 MPa and 87.7 MPa respectively. The strength of steel and

Table 1. The measured strength of materials.

Steel	Steel type	Yield strength/ MPa	Ultimate strength/MPa
channel	2[5	411.0	512.1
Distribution Bars	$\phi 6.5$	428.8	557.1
	$\phi 8$	440.1	530.0
	$\phi 10$	443.2	519.7
hoops	d5	—	713.0

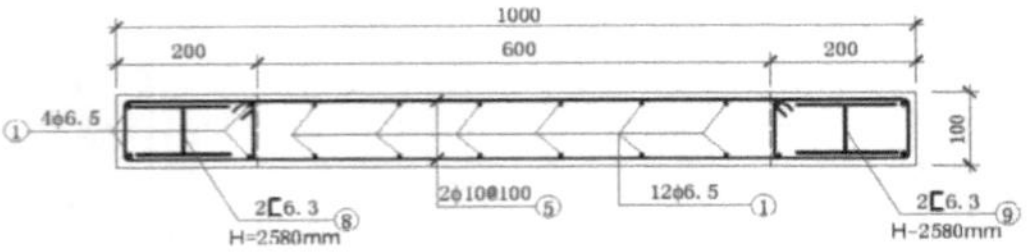

Figure 1. SHPCW01.

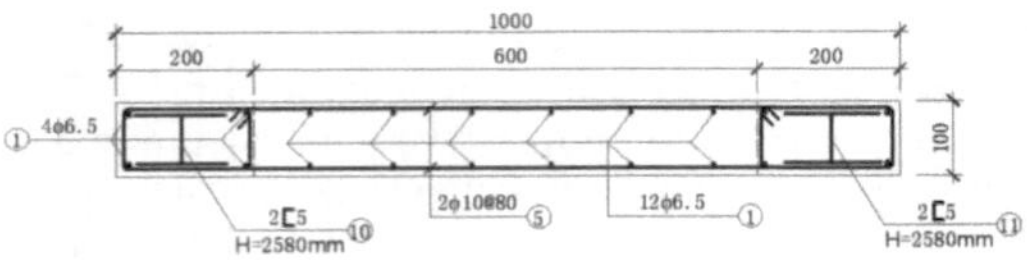

Figure 2. SHPCW02.

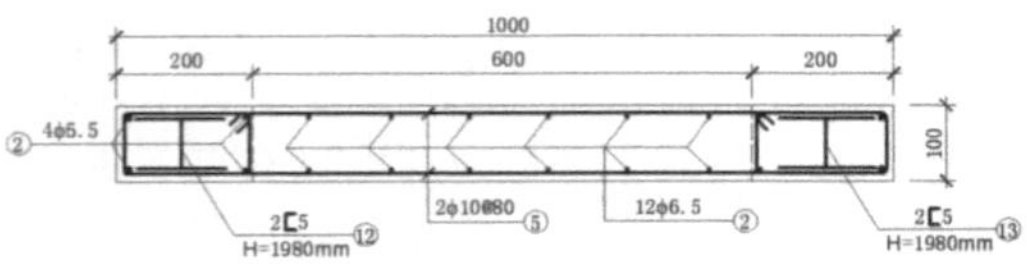

Figure 3. SHPCW03.

reinforcements were determined before the experiment, shown in Table 1. The concrete loading beam was arranged on the top of the shear wall and steel plates were embedded in both sides of the loading beam. The dimensions of specimens and arrangement of reinforcements was shown Figure 1 to Figure 3.

3 FIBER WALL ELEMENT

The above wall element is idealized by the fiber model that base on the material stress-strain relationships. There are two fiber slices considered at the base and top critical sections, as shown in Figure 5. Linear distribution of the element flexibility between the two fiber slices is assumed. The specific distribution curve of axial flexibility formed integral wall unit stiffness matrix, in which the wall for plane shear deformation with shear spring to simulate. The fiber model considered fourteen independent displacement components shown in Figure 4 and Figure 5, which were four shear vertical displacement d_{z1}, d_{z2}, d_{z3}, d_{z4}; horizontal displacement for top and bottom d_t, d_x, out-wall-plane horizontal displacement of four nodes d_{y1}, d_{y2}, d_{y3},

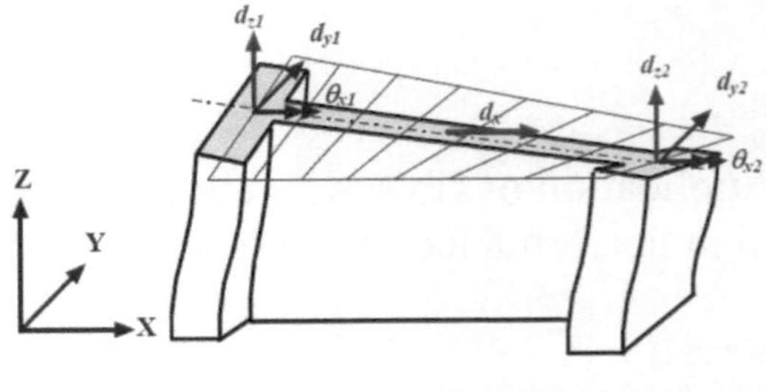

Figure 4. Displacement parameters for fiber element.

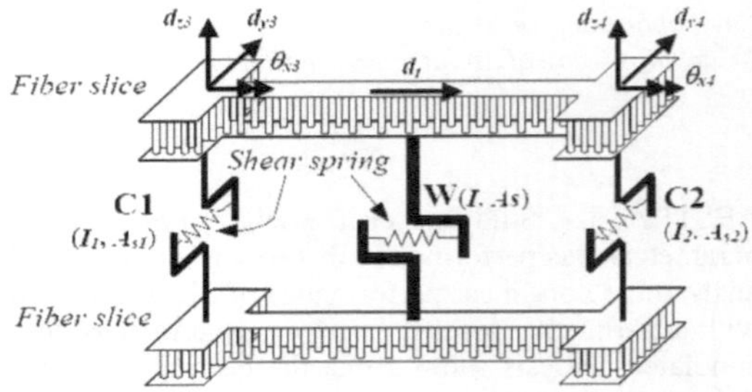

Figure 5. Fiber wall element model.

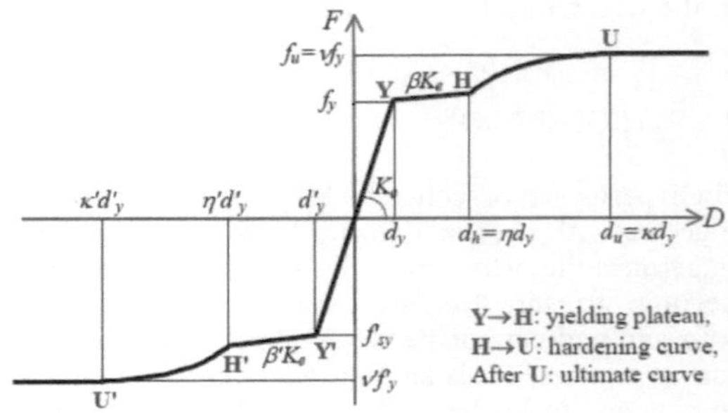

Figure 6. Skeleton curve of steel.

d_{y4} and angular displacement θ_{x1}, θ_{x2}, θ_{x3}, θ_{x4} (Kang Ning & Kubo T 1998).

4 CONSTITUTIVE RELATIONS OF MATERIALS

As shown in Figure 6, skeleton curve of steel material is specified by the yielding strength (f_{sy}, f'_{sy}), the yielding plateau stiffness parameter (β, β'), the factor (η, η') for the displacement at the beginning point of hardening curve, and the factor (κ, κ') and (ν, ν') for the ultimate point (U, U').

The skeleton curve of concrete can be shown in Figure 7. The concrete strength decay in the descending branch C-U and in the post-crack tension branch T-T′ is approximated using step-down rule that takes zero stiffness in computation and corrects the unbalance to meet the real force-displacement relation.

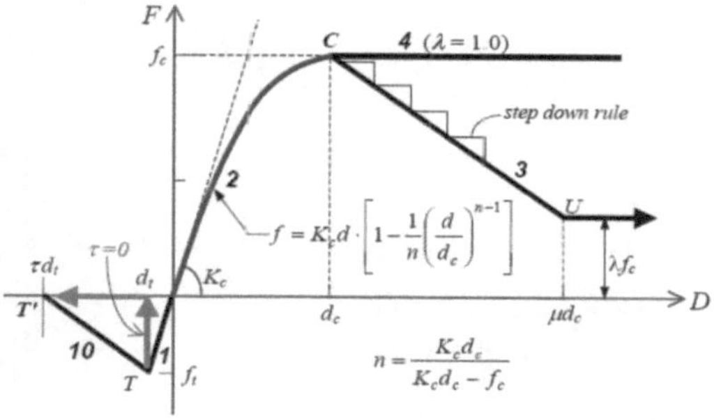

Figure 7. Skeleton curve of concrete.

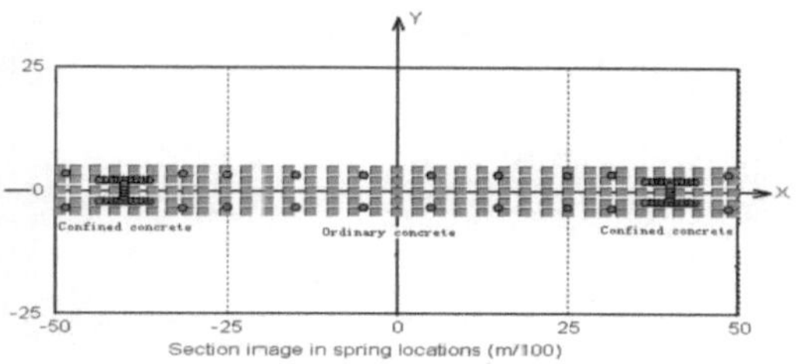

Figure 8. Fiber element partition of wall model.

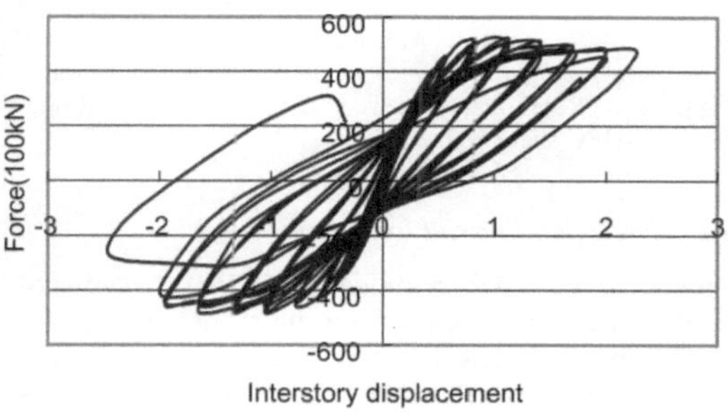

Figure 9. Experimental hysteretic curve of SHPCW-01.

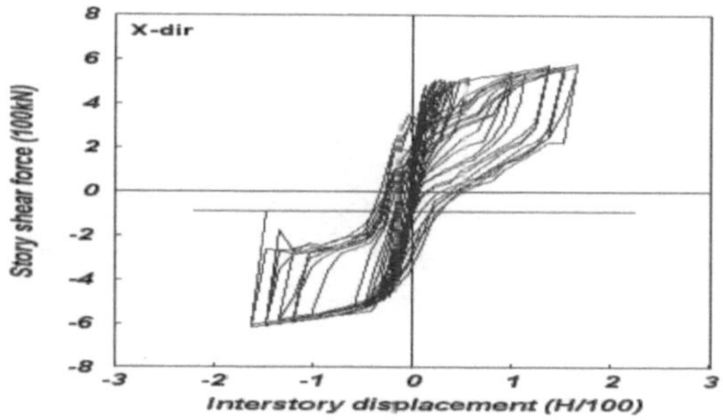

Figure 10. Simulation hysteretic curve of SHPCW-01.

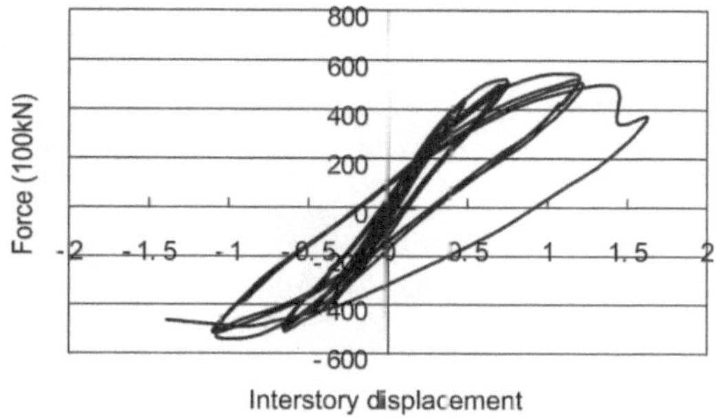

Figure 11. Experimental hysteretic curve of SHPCW-02.

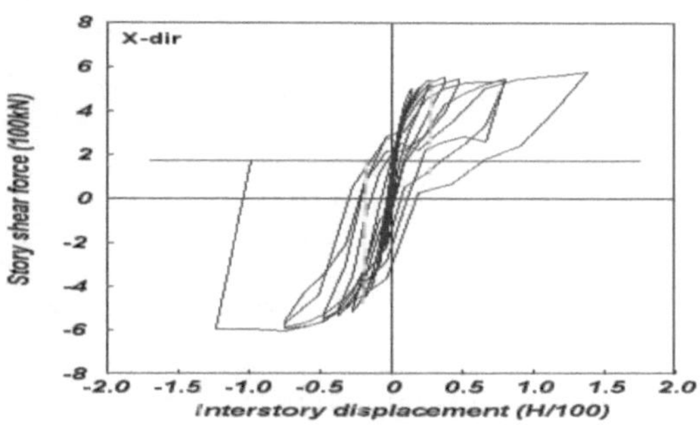

Figure 12. Simulation hysteretic curve of SHPCW-02.

5 NUMERICAL RESULTS AND EXPERIMENTAL RESULTS

According to constitutive relation of materials, the numerical stimulation of SHPC shear walls was performed by fiber wall element model provided by CANNY program. Fibers division of the specimen cross-section can be shown in Figure 8. The vertical loads were applied firstly on the specimen model then the horizontal loads were applied. Finally, the hysteretic curve of numerical simulation and experiment can be shown in Figure 9 to Figure 14. The key points of numerical simulation compared with the experimental values were shown in Table 2.

By comparison, the simulation hysteretic curve of specimen SHPCW-01 is more consistent with the experimental results. The stiffness and ultimate load from simulation analysis results are slightly greater than experimental results, because the base of the specimen slip resulting in calculated bias. The simulation hysteretic curve of specimen SHPCW-02 can be well in line with the experimental results with obtained pinching and sliding effects. The simulation hysteretic curve of specimen SHPCW-03 in the waist is fuller. Numerical calculation of the yield strength, ultimate strength, yield displacement and ultimate displacement are similar to the experimental results. The comparison showed that the fiber wall element model is reasonable and efficient in the nonlinear analysis of SHPC shear walls.

6 STUDY ON THE SEISMIC BEHAVIOR OF SHPC SHEAR WALLS WITH DIFFERENT PARAMETERS

Because of restrictions on the number of test specimens, some parameters influenced on the seismic

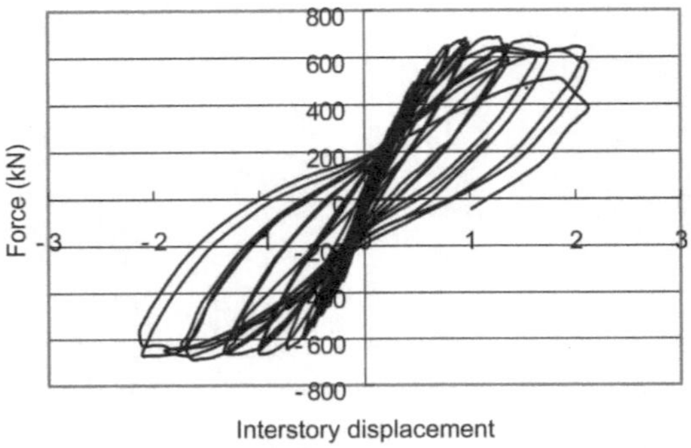

Figure 13. Experimental hysteretic curve of SHPCW-03.

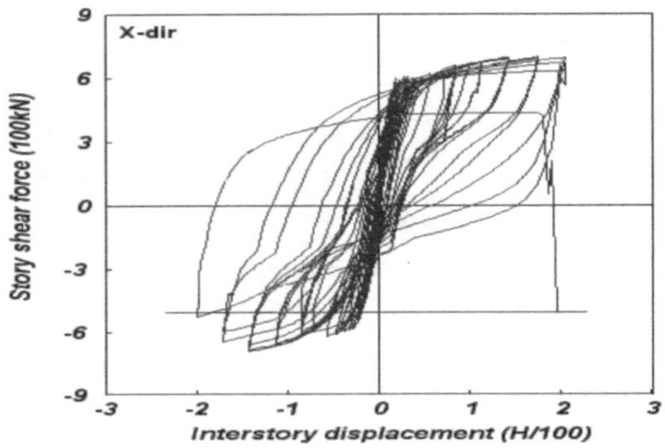

Figure 14. Simulation hysteretic curve of SHPCW-03.

Table 2. Comparison of key values.

Specimen		Ultimate load V/kN	Ultimate displacement/ mm
SHPCW01	Experimental Values	519.8	47.7
	Simulation Values	566.5	39.8
	Error	8.9%	16%
SHPCW02	Experimental Values	546.5	29.7
	Simulation Values	583.8	27.6
	Error	6.8%	7.0%
SHPCW03	Experimental Values	679.2	31.7
	Simulation Values	685.6	30.3
	Error	1.0%	4.6%

behavior of SHPC shear walls can not be a detailed analysis. On the basis of fiber wall element model, the numerical simulation was performed to study the seismic behavior of SHPC shear walls with different parameters, such as axial load ratio, the length of restricted edge members and ratio of steel.

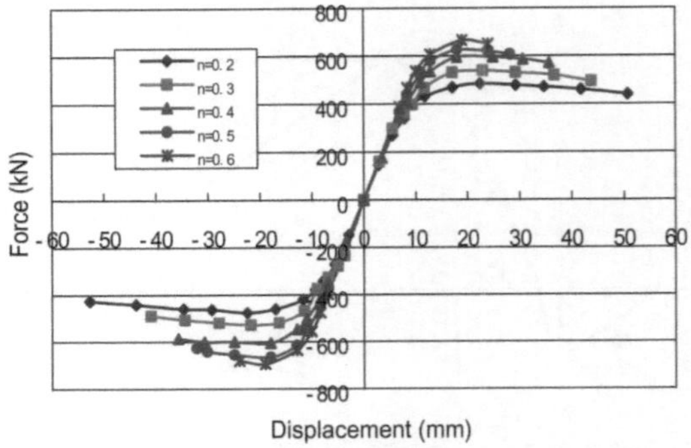

* where n is design axial load ratio.

Figure 15. Load-displacement skeleton curves under different axial load ratio.

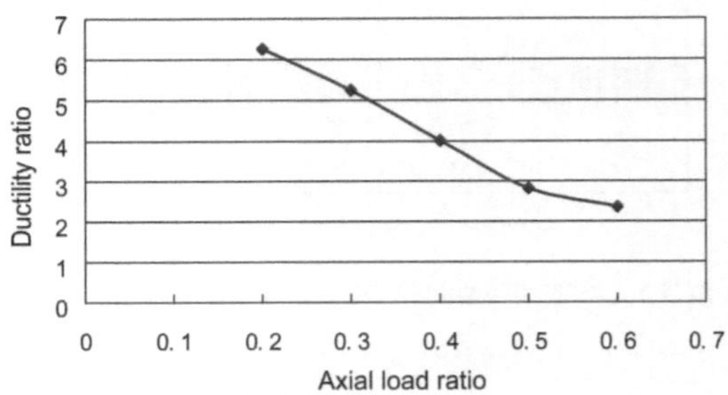

Figure 16. Relationship between axial load ratio and ductility of SHPC shear wall.

6.1 *The effect of axial load ratio*

On the basis of analytic model SHPCW01, a series of vertical loads were applied on the model. The Load-displacement skeleton curves under calculated axial load ratio for 0.11, 0.17, 0.22, 0.28 and 0.33, corresponding to the design of axial load ratio for 0.2, 0.3, 0.4, 0.5 and 0.6 was obtained shown in Figure 15

The results showed that the bearing capacity of SHPC shear walls increased with the axial load ratio increased. As axial load ratio increased, the ultimate displacement decreased and deformation capacity became poorer. The relationships between axial load ratio and displacement ductility ratio were shown in Figure 16. When design axial load ratio reached to 0.5, the displacement ductility is less than 3. Therefore, the limit value of design axial load ratio is 0.5.

6.2 *The effect of different length of restricted edge members*

The numerical simulation of SHPC shear walls with different length of restricted edge members was performed, whose axial load ratio is 0.5. The Load-displacement skeleton curves under different length of restricted edge members was shown in Figure 17. It is indicted that the deformation capability of SHPC shear walls under high axial load ratio improved with the increase of the length of restricted edge

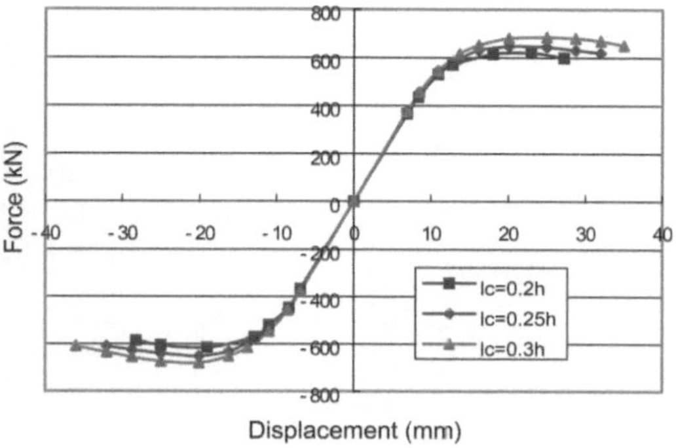

Figure 17. Load-displacement skeleton curves under different length of restricted edge members.

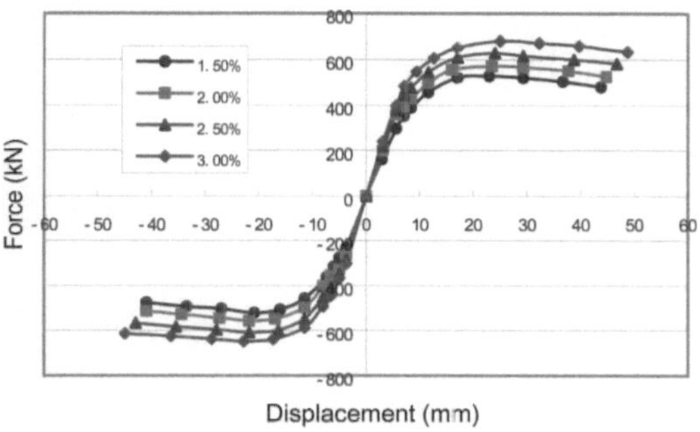

Figure 18. Load-displacement skeleton curves under different ratios of steel.

members. For example, when $l_c = 0.20\,h$, the ultimate displacement is 28.1 mm; when $l_c = 0.30\,h$, the ultimate displacement reached to 34.3 mm.

6.3 *The effect of steel ratio*

The Load-displacement skeleton curve under different ratios of steel was shown in Figure 18. It is indicted that the bearing capability of SHPC shear walls improved with the increase of steel ratio. For instance, when steel ratio increased from 1.5% to 3.0%, bearing capacity increased by 29%, however, ultimate displacement only increased by 8%. By contrast, as steel ratio increase, the deformation capability of the SHPC shear walls improved not so significant.

7 CONCLUSION

The fiber wall element model provided by CANNY program was used to analyze the non-linear performance of SHPC shear walls. The results of numerical simulation were agreement with experimental results and it proved that the fiber wall element model was reasonable in the nonlinear analysis of SHPC shear walls. The analytical results showed that the bearing capacity of SHPC shear walls improved and deformation capacity decreased with axial load ratio increase. It is recommended that the limit value of axial load ratio is 0.5. The deformation capacity of SHPC shear walls under high axial load ratio improved with the length of restricted edge members. The bearing capability and deformation capability of the SHPC shear walls improved with the increase of the steel ratio. However, the deformation capability of the SHPC shear walls improved not so significant.

ACKNOWLEDGEMENTS

This research was financially supported by Special Foundation for Basic Research Program of Chang' and University (Grant No. CHD2009 JC154) and National Natural Science Foundation of China (Grant No.50678146) The financial support is gracefully acknowledged.

REFERENCES

Jie Li & Kuiming Li. 2009. Nonlinear analysis of RC short-leg shear walls.*Journal of Building Structure*.30(1): 23–30.

Kang Ning Li & Kubo T. 1998. Reviewing the mufti-spring model and fiber model. *Proceedings of the 10th Japan Earthquake Engineering Symposium*. (2):2369–2374.

Lü Zhou. 2004. General review on the development of high-performance concrete(HPC). *Building Structure*. 34(6): 65–72.

Youssef Belmouden & Pierino Lestuzzi. 2007. Analytical model for predicting nonlinear reversed cyclic behaviour of reinforced concrete structural walls. *Engineering Structures*.29:1263–1276.

Zuozhou Zhao, Zhiyuan Liang & Jiaru Qian. 2009. Study on seismic behavior of SRC shear walls under high axial compression ratio. *Building Structure*. 39(1): 41–44.

Zhaoyuan Chen. 1997. The development and application of high-strength and high-performance concrete. *China Civil Engineering Journal*. 30(5): 3–11.

Modelling and Computation in Engineering – Zhu (ed)
© 2011 Taylor & Francis Group, London, ISBN 978-0-415-61516-7

Upper bound multi-rigid-body limit analysis on positive soil pressure based on the slip-line field theory

L.H. Zhao, F. Yang & L. Li
College of Civil and Architectural Engineering, Central South University, Changsha, Hunan, China

J. Zhou
College of Civil Engineering and Mechanics, Xiangtan University, Xiangtan, Hunan, China

ABSTRACT: The positive soil pressure was studied by means of the kinematical approach of limit analysis theory in this paper. A slip failure mechanism with mesh-like rigid blocks system has been constructed, in which the velocities changes on radial and tangent directions in transition area are allows. The failure modes are agree with the actual morphology and soil pressure fit the actual better, since the velocity compatibility of the failure mechanism can be satisfied by the corresponding velocity field. The objective functions of the positive soil pressure are obtained and transformed as an upper-bound mathematic optimization model for soil pressure as low as possible. The upper bound solutions for the objective functions have been obtained by use of the nonlinear sequential quadratic programming. From the numerical results and comparative analysis, it can be seen that the solutions presented in this paper compare reasonably well with the results presented in the existing literatures and the proposed method gives better upper bound calculation results. Finally, the design charts are presented.

1 INTRODUCTION

Calculation of soil pressure for earth retaining structures is a studying hot point in the academic and engineer circles, and a thorough understanding of it according to the actual engineering practice is required.

A number of studies, dealing with the positive soil pressure have been reported in literature in recently years (Sokolovskii 1965, Soubra & Regenass 2000, Soubra & Macuh 2002, Zhao et al. (2009). There are many kinds of research methods to study the passive earth pressure of retaining wall, and one of the most widely used method is the upper bound limit analysis.

Among the known solutions mentioned above for positive soil pressure factor K_p, Sokolovskii (1965) have employed characteristics method to obtain K_p. Davis (1968) and Chen (1975) obtained the upper bound solution of K_p by using simple failure mechanism. While Soubra & Regenass (2000), Soubra & Macuh (2002), Yang & Yin (2006) and et al. (2009) obtained the upper bound solution of K_p by introducing the one-dimensional velocity field.

Although all the upper bound solution was comparatively accurate results, however, due to the artificially simplified feature as well as the non-integrity feature of the failure mechanism, some analysis precision gaps still exist between the upper bound solution and Sokolovskii's characteristic solution. Therefore, to calculate the better positive soil pressure solutions with a new two-dimensional velocity field is the main purpose of this paper.

2 RIGID BLOCK FAILURE MECHANISM

On the basis of Yang, Yang & Zhao (2008, 2010), a new reformative slip kinematics admissibility failure mechanism with mesh-like rigid blocks system has been constructed to calculate the better positive soil pressure solutions, in which the velocities changes on radial and tangent directions in transition area are allows, and the brief configuration of the failure mechanism and velocity compatibility is presented here.

2.1 *Configuration of failure mechanism*

The velocity discontinuities configuration of failure mechanism presented here for positive retaining wall is shown in Figure.1.

As can be seen, when the plastic flow occurs, the soil behind retaining wall has the some horizontal velocity of footing while its vertical velocities are non-zero. And the other rigid blocks are all moving with different velocities. In both radial and tangent directions, the velocity discontinuities become curved lines when the number of rigid blocks increased, which directly result in the variation of velocity vectors. From the top of wall and along the radial directions, the velocity values decreased, well along the tangent directions from the ball of the wall, the velocity values increased. These patterns of failure mechanism could generate a smaller failure zone which are more suited for retaining wall with horizontal velocity and obtain accurate upper bound values, as can be seen in the following section.

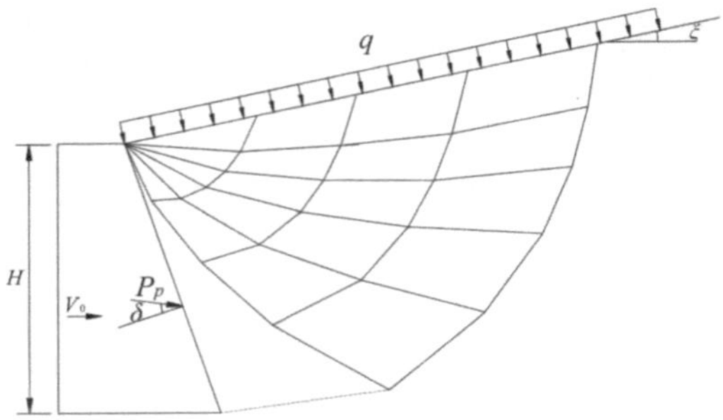

Figure 1. Velocity discontinuities configuration of failure mechanism for positive retaining wall.

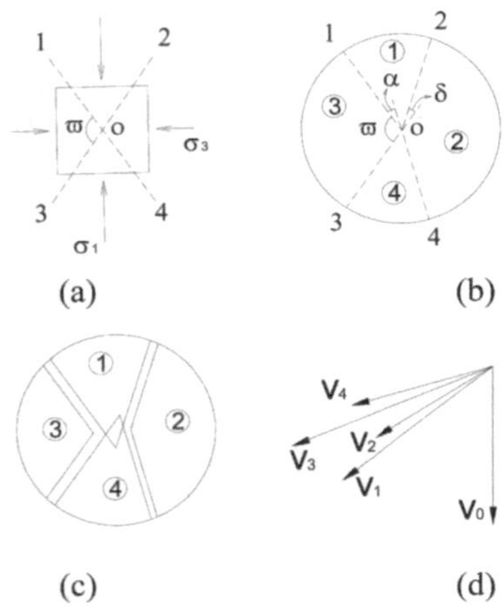

Figure 2. Geometry of the adjacent rigid blocks.

The geometry of the adjacent rigid blocks of failure mechanism is shown in Figures 2. Considering a small element (Fig. 1), for frictional material satisfying the Mohr-Coulomb criterion, the major and minor principal stresses are σ_1 and σ_3, and the maximum shear planes are on lines 1–4 and 2–3. The angle between 1-o and 3-o is ω, which equals to $\pi/2+\varphi$. If the element yield, the lines 1–4 and 2–3 are the slip lines, on which the four different parts moves separately, under the constraints of flow rule and velocity compatibility requirements.

Since the whole plastic zone is composed of rigid blocks, the above requirements can not be satisfied anywhere perfectly. However, a revised geometry of the adjacent rigid blocks of a magnified part of Figure 2(a) is shown in Figure 2(b). It can be see form Figure 2(b) that the lines 1-o and 3-o are remain the same as Figure 2(a), with the angle between them is $\omega=\pi/2+\varphi$. However, the line 4-o rotated in clockwise with an angle of α, well the line 2-o rotated in counter-clockwise with an angle of δ, compare to Figure 2(a). On this assumption, the velocity value would increase from right to left and from lower to upper. This regulation can be seen form the Figure 2(c), which is a magnified part of Figure 2(b). And

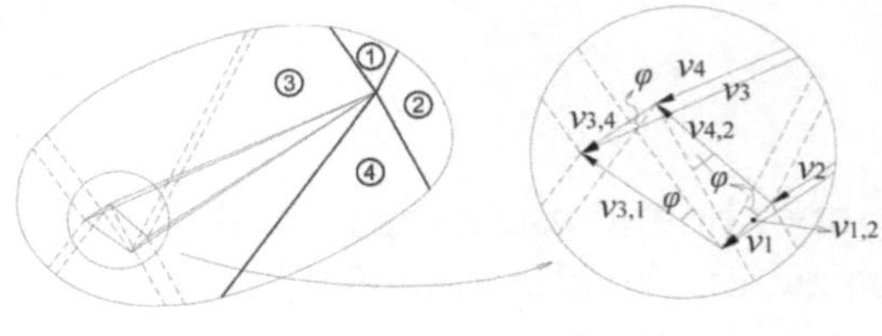

Figure 3. Magnified part of failure mechanism.

Figure 2(d) shows the obsolete velocities of the adjacent rigid blocks and footing. The relative velocity on velocity discontinuity has two directions, and either of them can meet the associated flow rule. However, for considering the requirement that the velocity vector should close, the relative velocity direction can be fixed on the above assumption. It should be note that the angle δ is the function of α, as the whole rigid blocks are considered. Thus, the failure mechanism of positive retaining wall shown in Figure 1 can be formed. It is obvious that with the increase of rigid blocks, the geometry of adjacent rigid blocks in the whole plastic zone meet the requirements of Mohr-Coulomb criterion approximately.

2.2 *Velocity compatibility of the rigid blocks*

The proposed failure mechanism and corresponding velocity field should satisfy the velocity compatibility requirements, i.e. rigid blocks do not have gaps or overlaps. Because velocity field can be deduced directly from the failure mechanism, thus compatibility requirements can be met if all the velocity vectors of rigid blocks are closed. To demonstrate this requirement, a magnified part of the failure mechanism (see Fig. 1) is shown in Figures 3.

It can be seen that the four adjacent rigid blocks are marked as ①, ②, ③ and ④, with absolute velocities v_1, v_2, v_3, and v_4, respectively. The dashed lines show the movement of the four rigid blocks when plastic flow occurs. The relative velocities $v_{3,4}$, $v_{3,1}$, $v_{1,2}$, and $v_{4,2}$ are all make an angle of φ on the corresponding velocity discontinuities, so as to meet the associate flow rule requirement. The absolute and relative velocity vectors of the four rigid blocks are all closed which means that the velocity compatibility of the rigid blocks can be met. For all rigid blocks in failure mechanism move as some regulations of the velocity vectors shown Figures 3, it is expected that velocity compatibility requirements can be satisfied perfectly.

3 DESCRIPTION OF FAILURE MECHANISM

In this section, the geometry of failure mechanism and corresponding velocity field are described for preparation of rigid blocks upper bound solution of positive retaining wall.

For positive retaining wall, the failure mechanism and its corresponding velocity field with variables are shown in Figures 4.

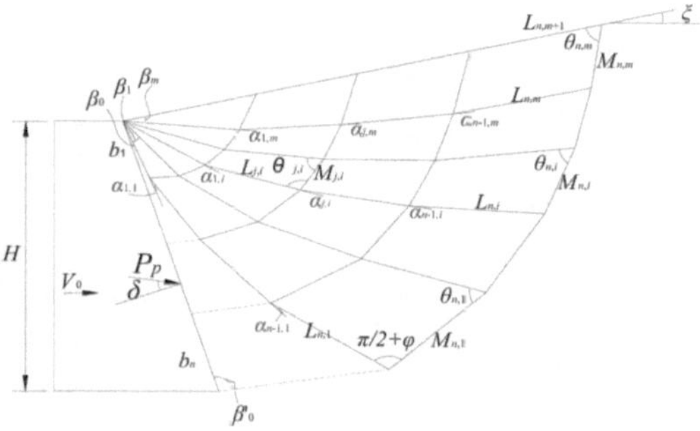

(a) multi-wedge failure mechanism

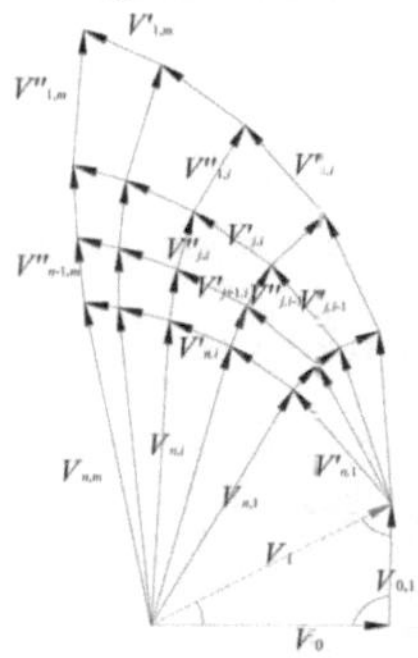

(b) Velocity field

Figure 4. Two-dimensional rigid block failure mechanism for passive earth pressure (for relative rough wall).

As shown in Figures 4, the non-plastic filling soil wedge behind retaining wall is composed of a triangular rigid block and $(n-1)$ quadrangular rigid blocks translating as an entire rigid body with the same initial horizontal velocity as retaining wall. It should be noted that b_i $(j=1,\ldots,n)$ are not velocity discontinuities. From the top of the retaining wall, and across the failure zone, a radial shear zone is formed and composed of m triangular rigid blocks near the wall top and the $m\times(n-1)$ quadrangular rigid blocks in the outer part as shown in Figure 4(a). The velocity discontinuities on the radial directions are $L_{j,i}$ $(j=1,\ldots,n;$ $i=1,\ldots,m+1)$ and on the tangential directions are $M_{j,i}$ $(j=1,\ldots,n;$ $i=1,\ldots,m)$. The angles between each pair of $L_{j,i}$ and $M_{j,i}$ in counter-clockwise are assumed equal to $\pi/2+\varphi$, as mentioned above.

The failure mechanism can be determined by a series of angles of β_0, β_i $(i=1,\ldots,m)$, $\alpha_{j,i}$ $(j=1,\ldots,n-1;$ $i=1,\ldots,m)$ and the length of b_j $(j=1,\ldots,n-1)$, as shown in Figures 5. Once these variables are known, the values of $N_j (j=1,\ldots,n)$, $L_{j,i}$ $(j=1,\ldots,n;$ $i=1,\ldots,m+1)$, $M_{j,i}$ $(j=1,\ldots,n; i=1,\ldots,m)$ as well as the areas of rigid blocks, S_i $(j=1,\ldots,n)$ and $S_{j,i}$ $(j=1,\ldots,n;$ $i=1,\ldots,m)$, can be calculated.

When the initial failure of plastic flow occurs, the entire triangular and quadrangular rigid blocks move as the velocity field requires (see Figs. 4). The velocity of retaining wall is v_0. The velocity of non-plastic wedge is v_1. The velocities $v_{j,i}$ $(j=1,\ldots,n; i=1,\ldots,m)$ are absolute velocities corresponding to the rigid blocks, and velocities $v'_{j,i}$ $(j=1,\ldots,n; i=1,\ldots,m)$ and $v''_{j,i}$ $(j=1,\ldots,n-1; i=1,\ldots,m)$ are relative velocities on the discontinuities $L_{j,i}$ $(j=1,\ldots,n; i=1,\ldots,m)$ and $M_{j,i}$ $(j=1,\ldots,n-1; i=1,\ldots,m)$, respectively. The geometric relations of velocity vectors are expressed as the functions of β_0, β_i $(i=1,\ldots,m)$, $\alpha_{j,i}$ $(j=1,\ldots,n-1;$ $i=1,\ldots,m)$.

4 CALCULATION PROCEDURE OF UPPER BOUND LIMIT ANALYSIS

The Energy dissipation calculation including two aspects. It can be seen from Figures 4 that the external rate of work are power of positive soil pressure E_{Pp}, and work rate of failure block produced by weight E_γ respectively.

$$E_{P_p}=P_P\cdot v_0\cdot\cos(\delta-\eta) \tag{1}$$

$$E_\gamma=\gamma\left(v^y{}_1\sum_{i=1}^{n}s_j+\sum_{j=1}^{n}\sum_{i=1}^{m}s_{j,i}v^y_{j,i}\right) \tag{2}$$

where v_0 is the velocity of retaining wall, P_p is the positive soil pressure, δ is the friction angle at the soil-wall interface, η is the dorsal wall angle, $\gamma=$ soil unit weight, s_j, $s_{j,i}=$ areas of the rigid blocks, and $v^y_1, v^y_{rmj,i}=$ vertical compound of absolute velocities.

Since the backfill is regarded as perfectly rigid and no general plastic deformation is permitted to occur, the contribution to energy dissipation P_c is provided along the discontinuous line N_j $(j=1,\ldots,n)$, $L_{j,i}$ $(j=1,\ldots,n;$ $i=1,\ldots,m)$, $M_{j,i}$ $(j=1,\ldots,n;$ $i=1,\ldots,m)$ and the soil-wall interface.

$$P_c=c\cos\varphi\left(\begin{aligned}&N_nv_1+\sum_{j=1}^{n}\sum_{i=1}^{m}L_{j,i}v'_{j,i}\\&+\sum_{j=1}^{n-1}\sum_{i=1}^{m}M_{j,i}v''_{j,i}+\sum_{i=1}^{m}M_{n,i}v_{n,i}\end{aligned}\right) \tag{3}$$

For soil pressure of retaining wall, the roughness feature on the back of wall must be take into account in the recurrence formula. δ is the friction angle at the soil-wall interface, and for the relative clean soil-wall surface $\delta<\varphi$, and for the relative rough soil-wall surface $\delta\geq\varphi$. Then the power of soil-wall friction E_{Pf} is,

for the relative clean soil-wall surface,

$$E_{P_f}=E_{AB}=P_P\cdot\sin\delta\cdot V_{0,1} \tag{4}$$

for the relative rough soil-wall surface,

$$E_{P_f}=E_{AB}=c\cdot(H/\cos\eta)\cdot V_{0,1}\cdot\cos\varphi \tag{5}$$

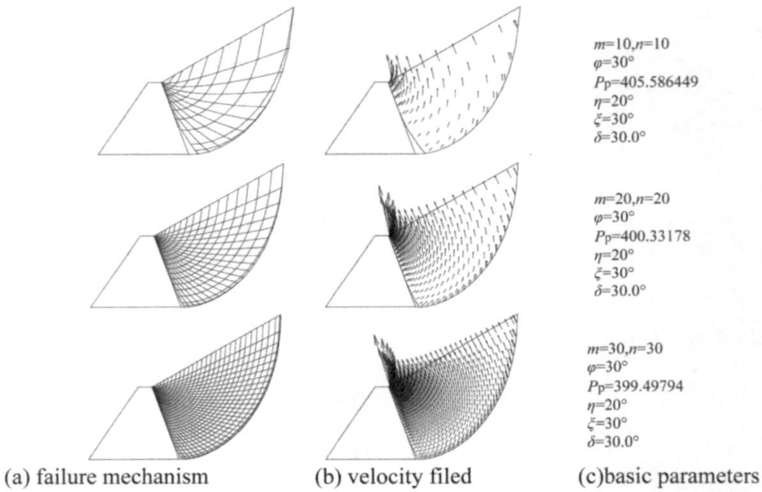

Figure 5. Mesh density (m× n) vs. the failure mechanism and the corresponding velocity filed.

Considering external load on boundary and using the virtual power theory, the expression of positive soil pressure of retaining wall P_p subjected to horizontal load is deduced by equating the external work rates to the total internal dissipation rates which expressed as follows:

for the relative clean soil-wall surface,

$$P_P = \left(E_c - E_\gamma - E_q\right) / \left[v_0 \cdot \cos(\delta - \eta) - \sin\delta \cdot V_{0,1}\right] \tag{6}$$

for the relative rough soil-wall surface,

$$P_P = \left(E_c + E_{Pf} - E_\gamma - E_q\right) / \left[v_0 \cdot \cos(\delta - \eta)\right] \tag{7}$$

5 CONSTRAINT CONDITIONS OF FAILURE MODE AND OPTIMIZATION PROCEDURE

In Equations (6) and (7), $H, \theta, \gamma, c, \varphi$ are regarded as invariables, $\beta_0, \beta_0', \beta_i\ (i = 1, \ldots, m), \alpha_{j,i}\ (j = 1, \ldots, n-1; i = 1, \ldots, m)$ and $b_j\ (j = 1, \ldots, n)$ are regarded as independent variables. As shown in Figs.4, the independent variables of $\beta_0, \beta_0', \beta_i\ (i = 1, \ldots, m)$, $\alpha_{j,i}\ (j = 1, \ldots, n-1; i = 1, \ldots, m)$ and $b_j\ (j = 1, \ldots, n)$ should assure the geometry conditions of failure mechanism (the constraint conditions of geometric) and velocity hodograph (the constraint conditions of velocity vector).

It is well-known that the litter the upper-bound solutions the more close to the truth value in the process of the limit analysis. Therefore, the solving of the upper-bound of the passive earth pressure is converted to a constrained extreme value of the multivariate nonlinear function, and an optimization procedure with sequential quadratic programming algorithm (*SQP*) had been compiled to get the minimum P_p in the present paper.

In engineering practice, passive earth pressure P_p was often expressed in the passive earth pressure coefficient K_p:

$$K_p = 2P_p / \left(\gamma H^2\right) \tag{8}$$

6 EXAMPLES AND DISCUSSIONS

An sandy soil example has been cited to illustrate the validity of this method. The parameters in this example are as follows: retaining wall height $H = 2$ m, slope top gradient $\alpha = 0°$, soil bulk density $\gamma = 18$ kN/m^3, internal friction angle φ varying from 10° to 40°, the dorsal wall angle θ equal to 110° and the friction angle at the soil-wall interface δ varying from 5° to 40° . Comparisons were made by different methods (Sokolovskii's haracteristics method, Chen's upper bound Method and multi-wedge failure mechanism solutions) on passive earth pressure coefficient K_p, which is outlined in Table 1.

As can be seen from Table 1, the values of K_p calculated by means of the present study are almost all remain lower as compared to the solutions of other upper bound method, and the present solutions are in agreement well with the existing Sokolovskii's haracteristics solutions as well. In term of the upper bound limit analysis method, the smaller passive earth pressure coefficient K_p is, the better upper bound solutions will be, and characteristic solution usually has the nature of lower bound answer. Therefore, the comparison shows that the results obtained here give better upper bound values of passive earth pressure and the proposed method is an effective method for evaluating the passive earth pressure.

The passive earth pressure P_p, the passive earth pressure coefficient K_p, the failure mechanism and corresponding velocity filed of failure mechanism for a relative rough retaining wall is illustrated in

Table 1. Comparison calculation of passive earth pressure coefficient K_p.

$\theta=110°, \beta=0°$					
				Present method	
φ (°)	δ(°)	Chen (1975)	Sokolovskii (1965)	multi-wedge failure mechanism ($n=100$)	Two-dimensional rigid block failure mechanism ($m=30$,n=30)
10	5	1.96	1.95	1.981	1.93782
	10	2.16	2.10	2.153	2.08273
20	10	3.91	3.80	3.857	3.80444
	20	5.04	4.62	4.865	4.61101
30	15	4.93	4.45	4.731	4.48691
	30	14.4	12.3	13.357	12.4178
40	20	25.5	22.4	24.501	23.4519
	40	56.6	44.7	50.627	46.1064

Table 2. Mesh density ($m \times n$) vs. passive earth pressure P_p and passive earth pressure coefficient K_p.

$\theta=70°, \beta=30°, \varphi=30°, \delta=30°, \gamma=18\,\mathrm{kN/m^3}, H=2\,\mathrm{m}$						
m	5	10	15	20	25	30
n	5	10	15	20	25	30
P_p	424.4665	405.5864	401.6993	400.3318	399.7275	399.4979
K_p	11.79074	11.26629	11.15831	11.12033	11.10354	11.09716

Table 2 and Figures 5 with an increase the mesh density varying from $m \times n=5 \times 5$ to $m \times n=30 \times 30$. And owing to the particularity of the upper bound limit analysis method, the magnitudes and the directions of the velocities were plotted at all centers of rigid blocks.

It can be expected from Table 2 that the proposed rigid block failure mechanism with upper bound solution is actually an effective way to approach the exact solution of characteristics with higher values of m and n. But for calculation of passive earth pressure, the division of mesh density on the passive earth pressure is not significantly affected P_p and K_p after the mesh density reaches a certain number, usually 15×15 grid density can achieve sufficient accuracy.

It was noted from Figures 5 that the magnitudes of the velocities along the ground surface near the top of wall were found to be significantly greater than the velocity (v_0) of the toe of wall, and the velocity vectors near the top of wall are changed rapidly and much higher than far away from the side of the wall as well, this two-dimensional transformation failure mode can match engineering practice more closely.

7 CONCLUSIONS

(1) The values of K_p calculated by means of the present study are almost all remain lower as compared to the solutions of other upper bound method, and the present solutions are in agreement well with the existing Sokolovskii's haracteristics solutions as well. The comparison shows that the results obtained here give better upper bound values of passive earth pressure and the proposed method is an effective method for evaluating the passive earth pressure.

(2) For calculation of passive earth pressure, the division of mesh density on the passive earth pressure is not significantly affected P_p and K_p after the mesh density reaches a certain number, usually 15×15 grid density can achieve sufficient accuracy.

(3) The assumed failure mechanism beforehand has apparent influence on precision of the results obtained by upper bound solution, and the presented rigid block failure mechanism is more "rigorous" and can be used effectively to analyze the stability of geotechnical structures.

ACKNOWLEDGMENT

The author is grateful to T. Zhang for her assistance.

REFERENCES

Chen, W.F. 1975. *Limit analysis and soil plasticity*. Amsterdam: Elsevier Science.

Davis, E.H. 1968. *Theories of Plasticity and the Failure of Soil Masses*. Soil Mechanics-Selected Topics, Lee I K, ed. New York: American Elsevier:341–380.

Sokolovskii, V.V. 1965. *Statics of Soil Media*. Translator Jones R, Schofield A. London: Butterworths Science.

Soubra, A.H. & Regenass, P. 2000. Three-dimensional passive earth pressure by kinematical approach. *Journal of Geotechnical and Geoenvironmental Engineering. ASCE* 26(11): 969–978.

Soubra, A.H. & Macuh, B. 2002. Active and passive earth pressure coefficients by a kinematical approach.

Proceeding of the Institution of Civil Engineers, Geotechnical Engineering 55(2): 119–131.
Yang, F. & Yang, J.S. 2008. A revised failure mechanism of strip footings for upper bound solution[J]. *Electronic journal of geotechnical engineering, Bundle F* 13: 1–17.
Yang, F., Yang, J.S. & Zhao, L.H. 2010. Collapse mechanism and support pressure for shallow tunnel face. *Chinese Journal of Geotechnical Engineering* 32(2): 279–284.
Yang, Xi. L. & Yin, J.H. 2006. Estimation of seismic passive earth pressures with nonlinear failure criterion. *Engineering Structures* 28: 342–348.
Zhao, L.H., Luo, Q. & Li, L. et al. 2009. The Upper Bound Calculation of Passive Earth Pressure based on Shear Strength Theory of Unsaturated Soil //*Geo-Hunan International Conference*. American: ASCE, 151–157.

Modelling and Computation in Engineering – Zhu (ed)
© 2011 Taylor & Francis Group, London, ISBN 978-0-415-61516-7

Fabrication and testing of frequency selective surface based on fabrics

Chuanyou Li, Qun Wang, Zhanghong Tang & Jingyu Han
College of Material Science and Engineering, Beijing University of Technology, Beijing, China

Meiwu Shi & Maohui Li
The Quartermaster Equipment Research Institute of the General Logistics Department of the PLA, Beijing, China

ABSTRACT: In this paper, a new method to use the selective electroless plating on fabric to prepare frequency selective surface (FSS) is presented. Fabric was selected as the substrates. Different from other substrates which are compact layers and can form close coatings, the coating on the fabric will form holes and gaps between fibers. The effect of holes and gaps to the transmission of electromagnetic waves was analyzed and results show that the holes and gaps of the fabric do not affect the performance of FSS. By changing the viscosity of the passivation in the process of the fabrication, the edge errors of elements can be greatly reduced which is satisfying the performance of FSS and a sample of FSS was successfully prepared. Test result shows that the selective frequencies of FSS were agree with that designed.

1 INTRODUCTION

Frequency selective surfaces (FSS) are well known for their filtering characteristics at microwave and millimeter wave frequencies. FSS are being investigated here for use as infrared (IR) and-reject and band-pass filters with application to low-cost, high-performance devices including beam-splitters, filters, radomes, and polarizers (Munk 2000, Wu 1995).

There are many methods to prepare FSS such as ink jet printing, micro pen dispensing, printed circuit board, plastering sheet metal so on. Ink jet and micro pen dispensing are perhaps the oldest, most established DW techniques. For both techniques, a conductive material (e.g., silver particles in a polymer binder) or dielectric material is dispensed onto the substrate in liquid form. A piezoelectric transducer is used for ink jet dispensing, and a motorized pump is used for micro pen work. In both cases, the material is dispensed onto the substrate through a narrow orifice at the desired dispensing resolution. The printed materials require drying/curing with heat or UV to achieve high conductivities approaching those of copper foil and bulk silver. As needed, printed ink patterns can be laser-trimmed down to the appropriate size or shape (Marhefka et al. 2007). The method has better accurate, but it also has the problem of high cost and complex preparation, which limited its application. Plastering sheet metal is usual plasting copper or other metals patch on dielectric substrate, which is applied to the fabrication of simple FSS (Bozzi et al. 2006). The printed circuit board is the most common method which has low cost, however, it has poor flexibility, which is hard to applied to the fabricate curved surface.

In this paper, a new method to prepare FSS on flexible materials such as fabric is presented. This method bases on the selective electroless plating. The results indicated that the fabrication process is feasible and verifies the modeling method used. This method has lower cost, and is simple and convenient. Furthermore, because the fabric has good flexibility, it can be used in curved FSS.

2 EXPERMENTRAL PROCEDURE

Figure 1 illustrates schematically the experimental procedures of selective electroless copper plating. Polyethylene terephthalate (PET) fabrics (35×45 count/cm^2, taffeta fabric) were used as substrates. In order to boost the binding force, the clean PET should be immerged into the NaOH solution at 80°C for 10 min. After coarsening, the fabric was activated in $PdCl_2$ solution which induced the copper plating. The principle of selective electroless plating is to print some passivation chemicals selectively after the activation of palladium, the chemicals reacted with palladium, which makes the substrates loss the ability for electroless deposition. Basing on different designs of frequency selective surfaces, the chemicals was printed selectively by screen printing. The accuracy of the elements' edges was controlled by changing viscosity of the chemicals and size of model elements (Bhansali & SoodS 1996). Figure 2(a) is the part picture of the fabricated FSS, Figure 2(b) is the designed element, whose outer and inner diameter are 16 mm, and 12 mm respectively. The arrangement of the element is square and the spaces between two elements

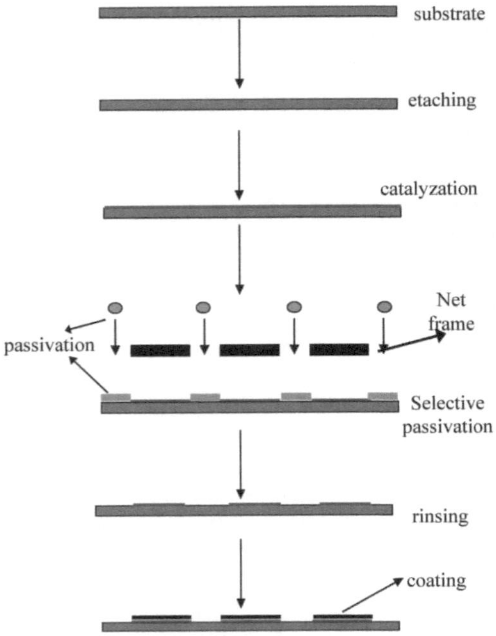

Figure 1. Schematic diagram of a new area selective electroless plating for fabricating Cu patterns using screen printing prepared on fabrics.

are dx = dy = 10 mm. After passivation, the copper can deposit on the PET selectively. The method of selective electroless copper plating can be applied to fabricate the FSS with different elements with complex shapes.

In Figure 2, the exact edges of elements are necessary for the performance of FSS, any edge error can lead to the draft of the resonant peak (Lü et al. 2005). Figure 3(a) is the edge of selective electroless copper plating fabrics, the left part with bright color of the fabric is Cu coating, while the right part with gray color is the passivation part without coating. From the SEM photo it can be found that the edge is exact with the error no more than a branch of fibers, which is about 200 μm. The errors caused by capiliary diffusion effect can not be eliminated completely. However, by improving the viscosity of the passivation liquid and adding some chemicals which can inhibit effect of capillary proliferation, the precision can be improved, which is shown in the Figure 3(b) the edge is in the middle of one fiber, the error is only several microns.

3 RESULT AND DISCUSSION

Usually the FSS has the periodic structure with close coating of every structure. However, the coating of fabric based FSS has many holes and gaps because of fibers. Figure 4 shows SEM image of the fabrics structure. It can be seen that the size of holes is about 100 μm and the size of gaps between fibers is

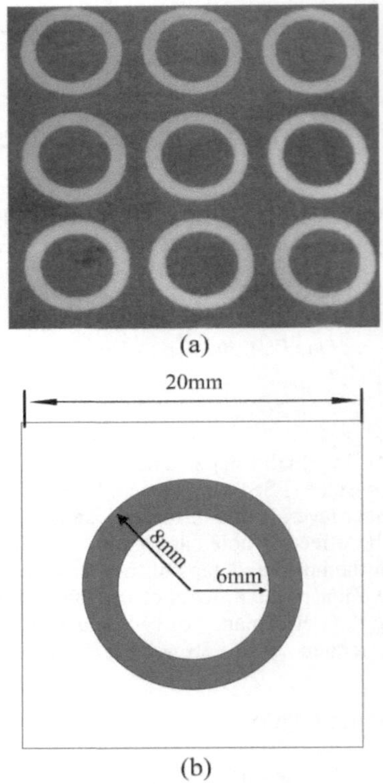

Figure 2. Picture of the fabricated FSS based on PET (a) fabricated FSS (b) cirque unit cell.

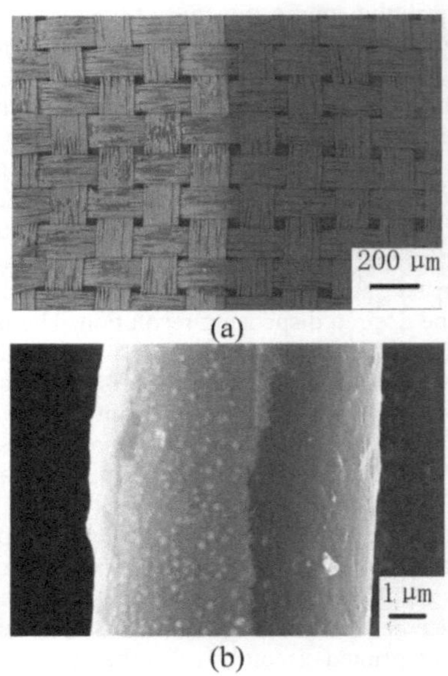

Figure 3. SEM of the edge of the unit cell on PET (a) before improvement (b) after improvement.

Figure 4. SEM of electroless copper plating fabric with gaps.

about 1 μm. These holes and gaps could change the electromagnetic characteristics of FSS.

To analyze the influence of holes and gaps, a periodic element model was established to simulate the transmission characteristics of FSS. By applying the periodic boundary conditions (Han et al. 2009).

$$E_{\alpha n}(x+d_1, y) = E_{\gamma n}(x, y)e^{-jkd_1\cos\theta}$$
$$H_{\beta n}(x, y+d_2) = H_{\delta n}(x, y) \quad (1)$$

As shown in Figure 5(a), it is simulation of one hole on the fabric which have many similar periodic holes, so α, β, γ and δ are periodic boundary conditions, n is the normal direction and τ is the incident wave. The curve of reflection coefficient/frequency of FSS is calculated and the result is shown in Figure 5(b).

The reflection coefficient curve of Fig.5(b) shows that the holes and gaps of fabric almost have no influence to the electromagnetic waves transmitting on the FSS because in the wide frequency range 50 GHz ~ 500 GHz, the reflection coefficient is close to 1 (the largest reflection is 6×10^{-3}dB, which is close to 0 dB).

On the other hand, basing on the electromagnetic wave theory proposed by Munk 2000, the resonant frequency of the periodic structure with element size 0.3 mm is near 100 GHz, but there is no any resonant frequency near 100 GHz from Figure 5(b), this is because that the separation of the horizontal bars is large and the equivalent capacitance C is too small to produce a 'real' resonance. So the holes and the gaps can be ignored when the FSS designed on the fabric is analyzed.

The reflection characteristics of FSS were tested by the arch test system with incident angle 15°, Figure 6 is comparison of testing results and the design. In design, there are two resonant frequencies, which are 9.1 GHz and 14.9 GHz respectively. Measured results show that the resonant frequencies of real FSS sampling are 8.82 GHz and 15.3 GHz respectively, which have good agreement with that of designed. However, because of bad performance of antennas (used in the arch test system) at high frequencies, the amplitude of two curves isn't agree with well around the second resonance frequency.

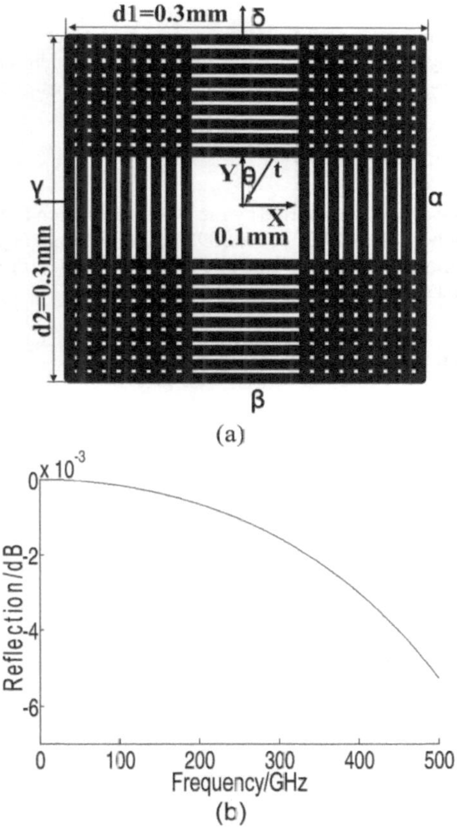

Figure 5. Simulation the influence of holes and gaps to the FSS, (a) periodic element of fabric and the incident wave, (b) reflection coefficient curve.

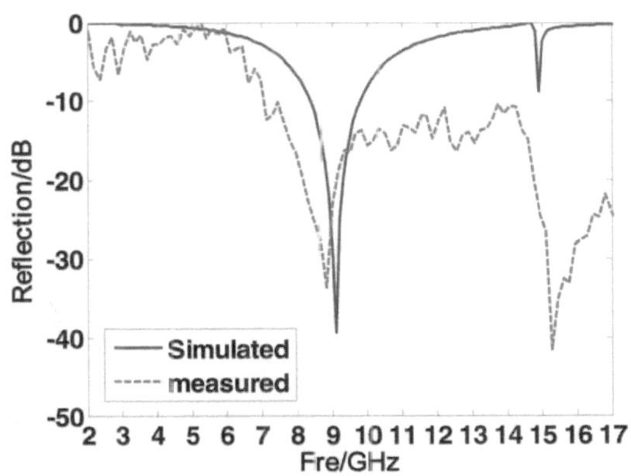

Figure 6. Comparision of simulation to testing results.

4 CONCLUSION

In this paper a new method to fabricate the FSS was presented. Unlike the traditional FSS, the fabric was selected as the material of FSS, the simulation suggest that the holes and gaps between fibers have not affect the performance of FSS used, by controlling the tenichics,the errors of the edges can be decreased greatly. Test results show that the resonant frequency of the prepared FSS sampling is agree with that designed.

ACKNOWLEDGMENT

This work was supported by the Doctor Research Fund of School (No. X0009011200905) and National High Technology Research and Development Program of China (No. 2007AA06Z320).

REFERENCES

Bhansali, S. & SoodS, D.K. 1995. A novel technique for fabrication of metallic structures on polyimide by selective copper plating using ion implantation. *Thin Solid Films* 270(2):489–492.

Bozzi, S., Biber M., Günther, O., Perregrini, L. & and L.-P. 2006. Schmidt, IEEE Trans. *Antennas Propag.* 54, 2638.

Han Jingyu, Wang Qun, Guo Hongxia & Tang Zhanghong, 2009. A Method to Simulate the Electromagnetic Parameters of Materials with Quasi-periodic Microstructures and its Applications. *Proceedings CEEM'2009, 9:16–20, Xi'an, China.*

Lü Mingyun, Zhu Ming, Wang Huanqing, et al. 2005. Digital machining system of complex curved-surface FSS. *Acta Aeronautica et As-tronautica Sinica*, 26(4):524–527.

Marhefka, Ronald J., Young, Jonathan D. & Towle, Jonathan P. 2007. Design, Fabrication and Measurement of an FSS Antenna Ground Plane. *Antennas and Propagation Society International Symposium, 2007 IEEE*: 3972–3975.

Munk, B.A. 2000. *Frequency Selective Surfaces: Theory and Design*. NewYork: Wiley.

Wu, T.K. 1995. *Frequency Selective Surface and Grid Array*. NewYork: Wiley.

Modelling and Computation in Engineering – Zhu (ed)
© 2011 Taylor & Francis Group, London, ISBN 978-0-415-61516-7

Apply grey relational analysis to microstructure and mechanical property of weld metal

L. Zhao, L.T. Yang & K.J. Dai
School of Power and Mechanical Engineering of Wuhan University, Wuhan, Hubei, China

ABSTRACT: In order to study the influence of chemical composition and welding parameter on microstructure and mechanical properties of weld metal, this paper addresses grey relational analysis to analyze the experimental data. Results indicate acicular ferrite has good influence on mechanical properties of weld metal. Mn and Ni are the most important factors to the formation of acicular ferrite.

1 INTRODUCTION

The mechanical properties of weld metal are strongly dependent on their microstructures. (Du et al. 1999) proposed that there were several different microstructures during solidification and cooling of the weld metal of low-carbon steel, including grain boundary ferrite, side plate ferrite, acicular ferrite, pearlite, granular bainite, martensite. The microstructures of weld metal are directly determined by the weld chemical composition of weld metal and the welding parameters.

(Huang et al. 2003) analyzed grain boundary ferrite was generally undesirable in weld metal, since it distributed in the austenite grain boundary and formed in veins which provide for a relatively continuous, easy path through which a crack could expand. (Ichikawa et al. 1997, Jones et al. 1998) thinked side plate ferrite declined the toughness of the weld metal because of connecting with grain boundary ferrite. (Fang et al. 1981) proposed there were not only blocky or needle ferrite in granular bainite, but also high carbon austenite like isolated island surrounded by these ferrites. The island organization enhance yield strength and tensile strength of the weld metal. (Lee et al. 2000) thinked acicular ferrite had good strength and toughness. So it was desirable to obtain enough acicular ferrite in the weld metal.

Grey system theory was pioneered by Deng JuLong in 1982. (Deng 2005) proposed that it mainly depended on the developing trend of the inner factors and measured the relevancy between different factors. Grey system theory studied problems of non-deterministic systems with incomplete or unknown information. Grey system theory had no special requirement and restriction on raw data. (Liu et al. 2006) thinked the strong applicability and practicability of grey system theory were necessary for analyzing and judging various objects, programming macroscopic system, prediction.

Table 1. Chemical composition of base metal and filler wire.

Material	C	Mn	Si	S	P	O
Base metal	0.223	0.770	0.250	0.032	0.020	350
Filler wire	0.102	0.556	0.054	0.023	0.012	380

The unit of O is ppm and the units of other chemical composition are wt%

However few attempts have been made to correlate the chemical composition to the microstructure of the weld metal by using the grey relational analysis. (Kanjilal et al. 2006) hoped to find the relationship between the chemical compositions and the microstructures of weld metal by second-order regression model in terms of mixture related variables (MRV). In this work, authors hoped to find the relation of each chemical composition to the microstructure of the weld metal and the relations of the microstructures to the mechanical properties of the weld metal. The data was quoted from (Kanjilal et al. 2006). The establishment and testing of the grey model have been done using 16 experimental data. The results of grey model are compared with the experimental data. The method was compared with the second-order regression model.

2 EXPERIMENTAL DATA

The chemical composition of the base metal and filler wire is shown in Table 1 and the obtained data from the experiment is shown in Table 2 and Table 3. In Table 3, these symbols are as follows: YS, yield strength; UTS, ultimate tensile strength; %EL, percent elongation; IT, impact toughness; Hv, Vickers hardness; GBF, grain boundary ferrite; PF, polygonal ferrite; SPF, side plate ferrite; AF, acicular ferrite; FAS, ferrite with aligned second phase (upper bainite).

Table 2. Welding parameters and Chemical compositions of weld metal samples.

	Welding Parameters*			Chemical compositions of weld metal						
No.	I	U	v	C	Mn	Si	S	P	Ni	O
1	350	28	25	0.080	0.71	0.13	0.026	0.028	0.75	330
2	350	28	50	0.057	0.40	0.11	0.029	0.018	0.11	500
3	350	32	25	0.054	0.35	0.11	0.034	0.019	0.09	550
4	350	32	50	0.120	0.74	0.16	0.028	0.024	0.22	340
5	450	32	50	0.065	0.42	0.12	0.026	0.019	0.12	480
6	350	32	50	0.100	0.82	0.15	0.021	0.020	0.41	310
7	350	32	25	0.059	0.46	0.13	0.035	0.022	0.11	480
8	450	32	25	0.060	0.48	0.08	0.025	0.026	0.54	390
9	350	28	50	0.054	0.40	0.12	0.035	0.018	0.12	520
10	450	28	50	0.090	0.59	0.18	0.023	0.020	0.27	420
12	350	28	25	0.090	0.54	0.24	0.032	0.021	0.37	350
13	450	28	25	0.050	0.39	0.11	0.041	0.025	0.13	500
14	450	28	50	0.098	0.54	0.19	0.024	0.018	0.44	380
15	450	32	25	0.075	0.46	0.17	0.033	0.021	0.40	450
16	450	32	50	0.052	0.35	0.15	0.033	0.019	0.10	580

*I = current, (A); U = voltage, (V); v = speed, (cm/min); The units of chemical compositions are same as table 1.

Table 3. Microstructure content and Mechanical properties.

	Microstructure content (%) Mechanical properties									
No.	GBF	PF	SPF	AF	FAS	YS (MPa)	UTS (MPa)	El (%)	IT (J)	Hv*
1	27.5	20.6	14.4	30.2	7.3	386.4	489.4	26.01	24.67	146
2	30.2	33.4	17.7	13.5	5.2	334.1	453.4	14.90	12.67	163
3	33.6	35.8	19.7	7.9	3.0	254.5	358.0	23.90	10.14	144
4	25.2	24.8	17.3	27.3	5.4	381.8	493.2	23.64	20.67	174
5	36.7	29.6	21.4	9.6	2.7	302.3	440.7	19.85	13.67	146
6	21.8	23.1	15.8	31.2	8.1	391.7	511.6	22.85	25.34	188
7	32.3	30.5	19.4	14.6	3.2	314.0	403.0	13.85	14.10	144
8	30.5	31.5	15.4	18.5	4.1	314.0	441.3	23.37	21.34	144
9	32.6	33.8	21.6	8.5	3.5	350.0	440.7	16.67	9.44	169
10	24.8	29.2	17.7	23.7	4.6	344.4	451.0	17.83	23.34	164
11	31.3	28.7	15.3	20.2	4.5	254.6	382.6	21.10	20.01	134
12	27.5	31.7	15.8	18.5	6.5	323.8	443.7	16.65	19.34	156
13	29.6	34.5	18.6	14.2	3.1	286.4	421.6	19.25	14.17	135
14	28.6	34.4	15.6	17.5	3.9	339.2	490.6	24.10	14.67	175
15	36.4	32.3	18.8	9.2	3.3	242.3	393.7	27.28	9.34	145
16	36.4	29.5	20.2	11.1	2.8	310.2	426.3	15.40	12.67	152

* The unit of Hv is VHN.

3 GREY THEORETICAL CALCULATION

In this study, the experimental data are processed by grey relational analysis. From the analysis, the most important factor could be found.

The grey relational analysis is used to explore the qualitative and quantitative relationships among abstract and complex sequences and to capture their dynamic characteristics during the development process. It can make use of relatively small data sets and does not demand strict compliance to certain statistical laws, simple or linear relationships among the observable variables.

In the grey relational analysis, the first performed is a data preprocessing to normalize the raw data for analysis. In this study, a linear normalization of the experimental results was processed and the range is from 0 to 1, which is also called "grey relational generation" (GRG). The original sequence is normalized as follows:

$$\Delta_i(k) = \left| \frac{x_0(k)}{x_0(1)} - \frac{x_i(k)}{x_i(1)} \right| \quad (i=1\ldots7, k=1\ldots16) \tag{1}$$

In the formula, x_0 and x_i are reference and compare sequence, respectively. In reference sequence, number

k is $x_0(k)$. And in the *ith* compare sequence, number k is $x_i(k)$.

Next, the grey relational grade is calculated to express the relationship between the reference sequence and the compare sequence. There are two steps to calculate. The grey relational coefficient $\xi_i(k)$, the first step, is defined as follows:

$$\xi_i(k) = \frac{\min_{j \in I}\min_{k}|\chi_0(k) - \chi_j(k)| + \zeta \max_{j \in I}\max_{k}|\chi_0(k) - \chi_j(k)|}{|\chi_0(k) - \chi_i(k)| + \zeta \max_{j \in I}\max_{k}|\chi_0(k) - \chi_j(k)|}$$

$$(i=1\ldots7, k=1\ldots16, \chi_0(k) = \frac{x_0(k)}{x_0(1)}, \chi_i(k) = \frac{x_i(k)}{x_i(1)}) \quad (2)$$

ζ is the identification coefficient which is defined in the range $0 \le \zeta \le 1$. In this paper, we defined it as 0.5. The grey relational grade $\gamma(x_0, x_i)$ is calculated as follows:

$$\gamma(x_0, x_i) = \frac{1}{n}\sum_{k=1}^{n}\xi_i(k) \quad (3)$$

4 DATA ANALYSIS

4.1 *The influence of chemical composition on microstructure of weld metal*

From Table 2, seven kinds of chemical composition are chosen as a compare sequence, from left to right, marked as y_C, y_{Mn}, y_{Si}, y_S, y_P, y_{Ni}, and y_O. In Table 3, considering one of the five microstructures of weld metal as a reference sequence, there are five reference sequences in column Microstructure content, from left to right denoted as x_{GBF}, x_{PF}, x_{SPF}, x_{AF} and x_{FAS}. A grey relational grade matrix is structured as the following:

$$\gamma = \begin{array}{c} \begin{matrix} y_C & y_{Mn} & y_{Si} & y_S & y_P & y_{Ni} & y_O \end{matrix} \\ \begin{bmatrix} 0.627 & 0.610 & 0.689 & 0.831 & 0.671 & 0.500 & 0.761 \\ 0.603 & 0.561 & 0.706 & 0.743 & 0.556 & 0.459 & 0.826 \\ 0.666 & 0.604 & 0.724 & 0.818 & 0.610 & 0.481 & 0.838 \\ 0.699 & 0.835 & 0.620 & 0.584 & 0.773 & 0.778 & 0.541 \\ 0.727 & 0.833 & 0.632 & 0.579 & 0.769 & 0.750 & 0.536 \end{bmatrix} \end{array} \begin{matrix} x_{GBF} \\ x_{PF} \\ x_{SPF} \\ x_{AF} \\ x_{FAS} \end{matrix}$$

where data of every row means a grey relational grade between chemical composition and one of microstructures of weld metal. Such as, $\gamma(y_C, x_{GBF}) = 0.627$ indicates the grey relational grade of *C* and *GBF*. The rest can be done in the same manner.

From the results of above, carbon, manganese, sulfur and oxygen have important implications on microstructures of weld metal.

4.2 *The influence of welding parameter on microstructure of weld metal*

From Table 2, current, voltage and speed are regarded as a compare sequence, marked as z_I, z_U and z_v, respectively. The reference sequences are still the five microstructures of weld metal, x_{GBF}, x_{PF}, x_{SPF}, x_{AF}, and x_{FAS}. A grey relational grade matrix is structured as the following:

$$\gamma = \begin{array}{c} \begin{matrix} z_I & z_U & z_v \end{matrix} \\ \begin{bmatrix} 0.825 & 0.848 & 0.620 \\ 0.623 & 0.573 & 0.485 \\ 0.749 & 0.746 & 0.567 \\ 0.633 & 0.659 & 0.532 \\ 0.633 & 0.667 & 0.536 \end{bmatrix} \end{array} \begin{matrix} x_{GBF} \\ x_{PF} \\ x_{SPF} \\ x_{AF} \\ x_{FAS} \end{matrix}$$

4.3 *The influence of microstructure on mechanical properties of weld metal*

In this calculation, microstructures of weld metal is shown as a compare sequence, marked as m_{GBF}, m_{PF}, m_{SPF}, m_{AF} and m_{FAS}, correspondingly. Mechanical properties make up five reference sequences, marked as n_{YS}, n_{UTS}, n_{EL}, n_{IT} and n_{Hv}, from left to right in column Mechanical properties. A grey relational grade matrix is structured as the following:

$$\gamma = \begin{array}{c} \begin{matrix} m_{GBF} & m_{PF} & m_{SPF} & m_{AF} & m_{FAS} \end{matrix} \\ \begin{bmatrix} 0.674 & 0.501 & 0.606 & 0.708 & 0.717 \\ 0.705 & 0.512 & 0.639 & 0.645 & 0.659 \\ 0.658 & 0.472 & 0.581 & 0.729 & 0.702 \\ 0.636 & 0.503 & 0.591 & 0.864 & 0.843 \\ 0.721 & 0.550 & 0.712 & 0.507 & 0.524 \end{bmatrix} \end{array} \begin{matrix} n_{YS} \\ n_{UTS} \\ n_{EL} \\ n_{IT} \\ n_{Hv} \end{matrix}$$

5 CONCLUSIONS

(1) The influence of Oxygen element on the formation of GBF, PF and SPF is significant.
(2) In the formation of AF, Mn and Ni have the most obvious effect.
(3) In welding parameters, voltage has significant on the formation of microstructure of weld metal.
(4) AF has good influence on mechanical properties of weld metal.

REFERENCES

Deng, J.L. 2005. *The Primary Methods of Grey System*. Wuhan: HUST press.

Du, Z.Y. et al. 1999. Microstructure of weld metal of low-carbon and low-alloy steel and influencing factors. *Iron and Steel* 34(12): 67–71.

Fang, H.S. et al. 1981. The granular bainite of low carbon Fe-Mn-B steel and its strength and toughness. *Materials for Mechanical Engineering* 5(2): 5.

Huang, A.G. et al. 2003. Prediction of PF volume fraction in low-alloy steel welding seam. *Electric Welding Machine* 33(10):9–14.

Ichikawa, K. et al. 1997. Modeling of allotriomorphic ferrite in steel welds. *Mathematical Modeling of Weld Phenomena* (3): 181–197.

Jones, S.J. et al. 1998. Kinetics of the widmanstatten ferrite transformation in steel. *Proceedings of the 1996 International Conference on Displacive Phase Transformations and their Applications in Materials Engineering; Urban IL USA, 8–9 May.* Warrendale:The Minerals, Metals & Materials Soc.

Kanjilal, P. et al. 2006. Combined effect of flux and welding parameters on chemical composition and mechanical properties of submerged arc weld metal. *Journal of Materials Processing Technology* (171): 223–231.

Lee, T.K. et al. 2000. Effect of inclusion size on the nucleation of acicular ferrite in welds. *ISIJ International* 40(12):1260.

Liu, S.F. et al. 2006. *Grey Information: Theory and Practical Applications*. London: Springer-Verlag.

Modelling and Computation in Engineering – Zhu (ed)
© 2011 Taylor & Francis Group, London, ISBN 978-0-415-61516-7

Optimum study on wind-induced vibration control by IGA

D.Y. Wang & Y. Zhou
School of Civil Engineering, GuangZhou University, Guangdong, P.R. China

ABSTRACT: Optimum analysis of wind-induced vibration control of high-rise buildings is an important yet unsolved problem in structural wind engineering. IGA (Improved Genetic Algorithms) was introduced and mainly studied in this paper to optimize structural responses with damping control. Six target functions were presented in IGA based on spectrum analysis. Structural optimum analysis programs were developed based on Matlab software to calculate wind-induced responses. A high-rise steel building with 20-storey was adopted and 10 cases were employed to perform comparison analysis to validate the feasibility and validity of the optimum method considered. The results show that IGA could distribute damping coefficients along structural height rationally and make full use of energy dissipation capability of dampers. Structural responses are reduced greatly by IGA optimum cases, especially for case "IGA-*TF5*". Influence of higher modes is suggested to be considered in structural analysis to obtain reasonable and reliable results.

1 INTRODUCTION

In recent years, many experts and scholars at home and abroad have performed a lot of theoretical and experimental studies on placement and parametric optimization of structures with viscous dampers and highly effective results were obtained, which established favorable foundation for engineering applications of dampers in all kinds of structures. Takewaki (1997) & Moreschi (1998) used a gradient-based approach to search for the optimal solution that would minimize a desired system transfer function. Tsuji & Nakamura (1996) proposed an algorithm to find both the optimal story stiffness distribution and the optimal damper distribution for a shear building model subjected to a set of spectrum-compatible earthquakes. Silva (1981) presented a gradient algorithm for the optimal design of discrete passive dampers in the vibration control of a class of flexible systems. The algorithm are aimed at making the modal damping and natural frequencies of the system reach the preassigned values. Ou (1998) studied parameter influence of passive energy dissipators, viscous, viscoelastic, metallic and frictional dampers in series with bracing members on suppressive effectiveness of structural vibration under earthquake and suggested the rational range of parameters for the design of passive energy dissipation systems. Zhou & Xu (1998) proposed five different optimum design methods for installation of dampers in structure and presented an example about optimum control analysis of a ten-floor reinforced concrete. Trombetti & Silvestri (2004, 2007) studied analytical formulation and efficiency of MPD system in shear-type structures with viscous dampers.

With the rapid developments of new techniques, new materials and new structural systems, more and more high-rise buildings and super skyscrapers are built recently. Wind load has become key factor in structural design of high-rise buildings. So, it's much important and necessary for researchers to do studies on vibration damping control, which is one of the most safely and effective methods to reduce vibration responses. However, the studies on this aspect are limited. Authors of this paper (Wang, 2008 & Zhou, 2008) have down researches on optimal vibration control by genetic algorithms (GA) method and Rayleigh damping method and the results showed that this methods were feasible to optimize structural wind-induced responses. In this paper, improved genetic algorithms (IGA) are presented to optimize structural responses with damping control.

2 WIND-INDUCED VIBRATION RESPONSES

The governing motion equation of MDOF subjected to wind loads is expressed as:

$$M\ddot{u} + C\dot{u} + Ku = F(t) \quad (1)$$

Where M, C and K are $n \times n$ mass, damping and stiffness matrix; $\ddot{u}$, $\dot{u}$ and u are $n \times 1$ acceleration, velocity and displacement vectors; $F(t)$ is $n \times 1$ force vector of wind loads.

High-rise buildings are known as flexible systems with low damping ratio. The off-diagonal parts of the structural damping matrix are usually assumed to be neglected and then the structural damping matrix is considered to be satisfying orthogonality condition. So, N independent equations with SDOF can be obtained from equation (1):

$$\ddot{q}_i + 2\xi_i\lambda_i\dot{q}_i + \lambda_i^2 q_i = F^*_i(t) \quad (i = 1, 2, \cdots, N) \quad (2)$$

Where $F_i^*(t)$ is generalized wind load of ith mode; λ_i is base frequency of ith mode; ξ_i is damping ratio of ith mode.

Fluctuating wind is assumed to be zero-mean Gaussian stationary stochastic process (Zhao, 2006 & Ou, 1991). According to theory of stochastic vibration theory, cross correlation function of generalized fluctuating wind forces can be expressed as:

$$R_{F_i^*F_j^*}(\tau) = E\left[F_i^*(t)F_j^*(t+\tau)\right] = \frac{1}{M_i^*M_j^*}\{\varphi\}_i^T\left[R_{\{p\}}(\tau)\right]\{\varphi\}_j \tag{3}$$

Where $[R_{\{p\}}(\tau)] = E[F(t)F(t+\tau)^T]$ is cross correlation function matrix of F^*.

So, power-spectrum density function of fluctuating wind-load can be obtained:

$$S_{F_i^*F_j^*}(\omega) = \frac{1}{M_i^*M_j^*}\{\varphi\}_i^T[S_F]\{\varphi\}_j S_f(\omega) \tag{4}$$

Where $[S_F]$ and $S_f(\omega)$ are coefficient matrix of power-spectrum density function and normal fluctuating wind-speed power-spectrum separately, which can be defined as:

$$\begin{cases} [S_F] = \gamma_{ij}P_iP_j & (i, j = 1, 2, \cdots, N) \\ S_f(\omega) = \dfrac{4\pi x^2}{3\omega\left(1+x^2\right)^{4/3}} & (x = \dfrac{1200n}{\bar{v}_{10}}) \end{cases} \tag{5}$$

Where P_i is wind force of ith point; ω is frequency of fluctuating wind and $\omega = 2\pi n$; $\bar{v}_{10}$ is the mean wind speed at 10 m height; γ_{ij} is cross correlation function between i^{th} point and j^{th} point.

From equation (2) and equation (4), self power-spectrum density function of general coordinate $q_i(t)$ can be obtained:

$$S_{q_i}(\omega) = \left|H_i(\omega)\right|^2 S_{F_i^*}(\omega) \tag{6}$$

Where $S_{F_i^*}(\omega)$ is self power-spectrum density function, which can be obtained easily by equation (3) when $i = j$; $H(\omega)$ is transform function.

So, the self power-spectrum density of structural responses can then be obtained:

$$\begin{cases} S_{u_i}(\omega) = \sum_{j=1}^{N}\left(\varphi_{ij}\right)^2 S_{q_j}(\omega) = \sum_{j=1}^{N}\left(\varphi_{ij}\right)^2\left|H_j(\omega)\right|^2 S_{F_j^*}(\omega) = \sum_{j=1}^{N}\left(\varphi_{ij}\right)^2\Lambda_j\left|H_j(\omega)\right|^2 S_f(\omega) \\ S_{\dot{u}_i}(\omega) = \omega^2 S_{u_i}(\omega) \\ S_{\ddot{u}_i}(\omega) = \omega^4 S_{u_i}(\omega) \end{cases} \tag{7}$$

Where $\Lambda_j = \frac{1}{(M_j^*}\,)^2\{\varphi\}_j^T[S_F]\{\varphi\}_j$.

Structural responses can be expressed as follows:

$$\begin{cases} \sigma_{u_i}^2 = \int_{-\infty}^{\infty} S_{u_i}(\omega)d\omega = \sum_{j=1}^{N}\left(\varphi_{ij}\right)^2\Lambda_j\int_{-\infty}^{\infty}\left|H_j(\omega)\right|^2 S_f(\omega)d\omega \\ \sigma_{\dot{u}_i}^2 = \int_{-\infty}^{\infty}\omega^2 S_{u_i}(\omega)d\omega = \sum_{j=1}^{N}\left(\varphi_{ij}\right)^2\Lambda_j\int_{-\infty}^{\infty}\omega^2\left|H_j(\omega)\right|^2 S_f(\omega)d\omega \\ \sigma_{\ddot{u}_i}^2 = \int_{-\infty}^{\infty}\omega^4 S_{u_i}(\omega)d\omega = \sum_{j=1}^{N}\left(\varphi_{ij}\right)^2\Lambda_j\int_{-\infty}^{\infty}\omega^4\left|H_j(\omega)\right|^2 S_f(\omega)d\omega \end{cases} \tag{8}$$

3 IMPROVED GENETIC ALGORITHMS

3.1 *Basic principle*

Genetic algorithms (GA) presume that the potential solution of any problem is an individual and can be represented by a set of parameters. These parameters are regarded as the genes of a chromosome and can be structured by a string of values in binary form. A positive value, generally known as a fitness value, is used to reflect the degree of "goodness" of the chromosome for the problem that would be highly related to its objective value. Genetic algorithms are heuristic random search techniques based on the concept of natural selection and natural genetics of population, and so they are of "population-based" method of searching large combinatorial design spaces to find the optimum combination of design variables. Detailed discussion on the mechanisms of GA can be found in Holland (1975). GA has been proven to be a versatile and effective approach of solving optimization problems and have been used in many research fields. The basic steps of Genetic algorithm can be expressed as follows:

1. Some cells with problem information will be created by stochastic manner, and each cell is made up of one kinds of code (such as binary encoder, Gray encoder and so on);
2. The sufficiency of target function for each cell in grope will be appraised. Higher sufficiency cells can be reserved and washout lower cells;
3. New daughter lines can be created by there kinds of methods, namely ① father cells creates daughter cells by cross method, ② father cells creates daughter cells by mutation method, ③ excellent father cells are copied and form daughter cells;
4. The new created gropes perform next genetic optimum operation, and then cycle operation for step 2 and step 3 until optimization result arrives predefined target.

3.2 *Improvement of genetic algorithms*

Although there are lots of application cases of GA in many fields, such as civil engineering and mechanical engineering, some improvements should be down so as to obtain better optimal results. Cell-efficiency, cross operator and mutation operator will be improved in this paper.

Central limit theorem is adopted to design cell-efficiency function in order to avoid occurrence of deceptive phenomena of conventional cell-efficiency in GA (Wang, 2005 & Liu 2009). According to the central limit theorem, evolutional cell-efficiency function obey normal distribution (Guo, 2006). If there are n evolutional cells and corresponding cell-efficiencies are $x_1, x_2, \cdots, x_n$, then:

$$\begin{cases} \bar{x} = \dfrac{x_1 + x_2 + \cdots + x_n}{n} \\ S^2 = \dfrac{\left(x_1 - \bar{x}\right)^2 + \left(x_2 - \bar{x}\right)^2 + \cdots + \left(x_n - \bar{x}\right)^2}{n} \end{cases} \tag{9}$$

Obviously, $\bar{x}$ and S^2 are the unbiased estimates of $E(x)$ and σ^2 respectively. So, cell-efficiency function will obey $N(\bar{x}, S^2)$ distribution approximately.

Cell-efficiency function can be defined as follows:

1) Calculate average cell-efficiency $\bar{x}$ and secondary average center distance S^2;
2) Standardization: $x_i' = (x_i - \bar{x})/S(i = 1, 2, \cdots, n)$;.
3) Calculate probability cell-efficiency:

$$\begin{cases} f_i = 0.5 + \int_0^{x_i'} \frac{1}{\sqrt{2\pi}} e^{-t^2/2} dt & \text{(if calculating maximum value)} \\ f_i = 0.5 - \int_0^{x_i'} \frac{1}{\sqrt{2\pi}} e^{-t^2/2} dt & \text{(if calculating minimum value)} \end{cases} \tag{10}$$

Cross operator will be improved as follows:

1) Define group $S = \{s_1, s_2, \cdots, s_n\}$;
2) Arrange group randomly: $S' = \{s_1', s_2', \cdots, s_n'\}$;
3) Define two endpoints, s_i and s_i', and search variable, λ, for one-dimensional search. So, any point in the search line can be expressed as $s = s_i + \lambda(s_i' - s_i), \lambda \in [0, 1]$;
4) If good search variable λ' is obtained, then next individual will be generated $s = s_i + \lambda(s_i' - s_i)$.

As we know, most mutation operators in traditional GA are natural mutation, which pay more attention to group diversity but ignore the cell-efficiency of mutation individual (Li, 2006). But in fact, good cell-efficiency must be ensured in mutation process to get reasonable result. So, artificial mutation technology is introduced to deal with mutation operator to keep the good cell-efficiency of mutation individual. In mutation process from $X = \{x_1, x_2, \cdots, x_i, \cdots, x_n\}$ to $X' = \{x_1, x_2, \cdots, x_i', \cdots, x_n\}$, if genetic value at mutation point x_i lies in $[U_{\min}^i, U_{\max}^i]$, then following equation can be obtained:

$$x_i' = \begin{cases} x_i + \Delta(t, U_{\max}^i - x_i) & \text{(if random(0,1)=0)} \\ x_i + \Delta(t, x_i - U_{\min}^i) & \text{(if random(0,1)=1)} \end{cases} \tag{11}$$

Where $\Delta(t, y) = y(1 - \gamma^{(1-t/T)})$, γ is random variable obeying uniform distribution and $\gamma \in [0, 1]$. Comparing target function values between X and X', if X' is better than X', then the mutation process will be accepted.

3.3 *Target function*

Target functions(TF) are needed for IGA to decide selection, cross and mutation characteristic of creating new daughter cells. Apparently, selection of TF has great influence to optimal result. In this paper, in order to perform comparison analysis, six TFs, from *TF1* to *TF6*, are defines according to the deduction in section 2.

TF1: Top-acceleration response variance of structural first mode is designed as *TF1*:

$$TF1 = \sigma_{\ddot{u}_N} = (\varphi_{N1})\sqrt{\Lambda_1 \int_{-\infty}^{\infty} \omega^4 |H_1(\omega)|^2 S_f(\omega) d\omega} \tag{12}$$

TF2: Mean-value of drift ratio variance of the first mode is designed as *TF2*:

$$TF2 = \frac{1}{N}\sum_{i=1}^{N}\frac{\sigma_{u_i}}{h_i} = \frac{1}{N}\sum_{i=1}^{N}\frac{1}{h_i}(\varphi_{i1})\sqrt{\Lambda_1 \int_{-\infty}^{\infty} |H_{\Delta 1}(\omega)|^2 S_f(\omega) d\omega} \tag{13}$$

$$\text{Where} \begin{cases} H_\Delta(\omega) = T \cdot H(\omega) \\ T = \begin{bmatrix} 1 & 0 & & & & 0 \\ -1 & 1 & 0 & & & \\ 0 & -1 & 1 & 0\cdots & & \\ & & & \ddots & & \\ & & & & & 0 \\ 0 & & & 0 & -1 & 1 \end{bmatrix} \end{cases}$$

TF3: Top-acceleration response variance of ten structural modes is designed as *TF3*:

$$TF3 = \sigma_{\ddot{u}_N} = \sqrt{\sum_{j=1}^{10} (\varphi_{Nj})^2 \Lambda_j \int_{-\infty}^{\infty} \omega^4 |H_j(\omega)|^2 S_f(\omega) d\omega} \tag{14}$$

TF4: Mean-value of drift ratio variance of ten structural modes is designed as *TF4*:

$$TF4 = \frac{1}{N}\sum_{i=1}^{N}\frac{\sigma_{u_i}}{h_i} = \frac{1}{N}\sum_{i=1}^{N}\frac{1}{h_i}\sqrt{\sum_{j=1}^{10} (\varphi_{ij})^2 \Lambda_j \int_{-\infty}^{\infty} |H_{\Delta j}(\omega)|^2 S_f(\omega) d\omega} \tag{15}$$

TF5: Considering both *TF3* and *TF4*:

$$TF5 = TF3 \times TF4 \tag{16}$$

TF6: Base-shear variance of the first mode is designed as *TF6*:

$$TF6 = k_1 \sigma_{u,1} = k_1 \varphi_{11} \sqrt{\Lambda_1 \int_{-\infty}^{\infty} |H_1(\omega)|^2 S_f(\omega) d\omega} \tag{17}$$

4 CASE STUDY

4.1 *Computational model*

A 20-storey steel building is taken as an example in this study. The basic parameters of the building are given in Table 1 and the first ten periods and frequencies of the building are shown in Table 2.

Some first-phase preparations are here presented in order to perform further optimum analysis. Firstly, Linear viscous dampers are installed in the example building and the whole damping coefficients $\hat{c}$ of all viscous dampers in different cases are defined to be equal, $\hat{c} = 1.02 \times 10^9$ N·s/m, so as to analysis structural dynamic performance comparatively. Secondly, according to the deduction in section 2, structural optimum analysis programs are developed based on Matlab software to calculate structural wind-induced responses.

4.2 *Result and discussion*

Figure 1 shows damping coefficient distribution along structural height for different cases ("IGA" means

Table 1. Basic parameters of the building.

Height of the building (m)	72
Story-mass (kg)	$m_1 = m_2 = \cdots = m_{20} = 3.2 \times 10^4$
Story-stiffness (N/m)	$k_1 = \cdots = k_4 = 3.5 \times 10^7$
	$k_5 = \cdots = k_{10} = 3.0 \times 10^7$
	$k_{11} = \cdots = k_{16} = 2.5 \times 10^7$
	$k_{17} = \cdots = k_{20} = 2.2 \times 10^7$
Field type	D
Basic wind pressure (kN/m^2)	1.0
Basic wind speed (m/s)	40.0

Table 2. The first ten periods and frequencies of the building.

Mode	1	2	3	4	5
T_i (s)	2.664	0.935	0.564	0.409	0.319
λ_i (rad/s)	2.358	6.718	11.137	15.345	19.686
Mode	6	7	8	9	10
T_i (s)	0.263	0.226	0.198	0.178	0.162
λ_i (rad/s)	23.868	27.785	31.734	35.265	38.782

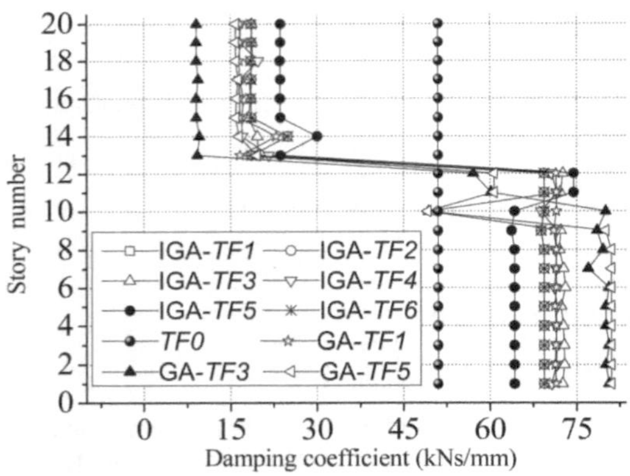

Figure 1. Damping coefficient distribution along height.

Improved Genetic Algorithms, "GA" means traditional Genetic Algorithms, "*TF0*" means damping distribution without optimization). Figure 1 shows that damping distributions along height have similar tendency for optimization cases: damping coefficients in lower stories are all bigger than that of upper stories and great mutation occurs in 13th story. Damping coefficients in lower stories for case "IGA-*TF3*" are less than case "GA-*TF3*" but bigger than case "GA-*TF3*" in upper stories, and same phenomenon can also be found in other cases. Damping distribution without optimization shows changeless along the height.

Standard deviation of story displacement and acceleration for the ten computational cases are shown in Figures 2, 3. Comparison of wind-induced dynamic

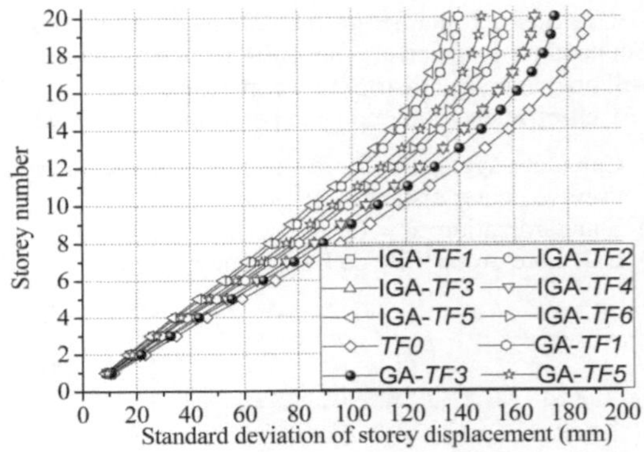

Figure 2. Standard deviation of storey displacements.

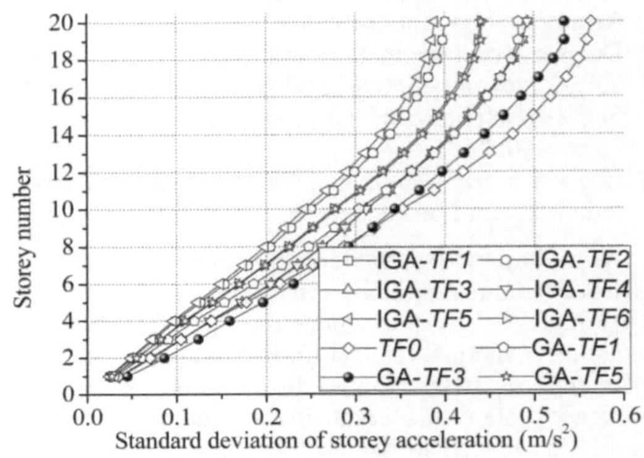

Figure 3. Standard deviation of storey accelerations.

responses, such as top displacement, maximum drift ratio and top acceleration, is given in Table 3. After examining the two figures and table, it is apparent that:

1) Standard deviations of the structural responses exist difference to some extent in different target functions, which can be verified easily from cases "IGA-*TF1*", "IGA-*TF3*", "IGA-*TF5*" in Figure 2 or Figure 3. According to the vibration-absorbing effect, the best case among the ten cases is "IGA-*TF5*", and then "IGA-*TF1*" and "IGA-*TF2*". It is suggested that both displacement and acceleration responses should be considered in target functions so as to obtain the best optimal results.
2) Standard deviations of structural responses increase in a certain extent when considering influence of higher modes (such as cases "IGA-*TF3*", "IGA-*TF4*"). Vibration-absorbing effects of cases "IGA-*TF1*" and "IGA-*TF3*" for top displacement are 25.4% and 10.27% respectively, and similar phenomenon can be found in others systems. Therefore, the influence of higher modes must be considered in structural analysis, especially for high-rise buildings.
3) Structural responses optimized by IGA are less than that by GA in same target functions. The vibration-absorbing effects of top displacement, maximum

Table 3. Comparison of wind-induced dynamic responses.

Standard deviation	*TF0* I	IGA-*TF1* II	(I-II)/I	IGA-*TF3* III	(I-III)/I	IGA-*TF5* IV	(I-IV)/I	GA-*TF5* V	(I-V)/I
Top displacement (mm)	187.32	139.81	25.4%	168.08	10.27%	135.61	27.61%	148.46	20.75%
Maximum drift ratio (/1000)	3.6	2.68	25.56%	3.23	10.28%	2.66	26.11%	2.95	18.06%
Top acceleration (m/s^2)	0.56	0.4	28.57%	0.49	12.5%	0.39	30.36%	0.44	21.43%

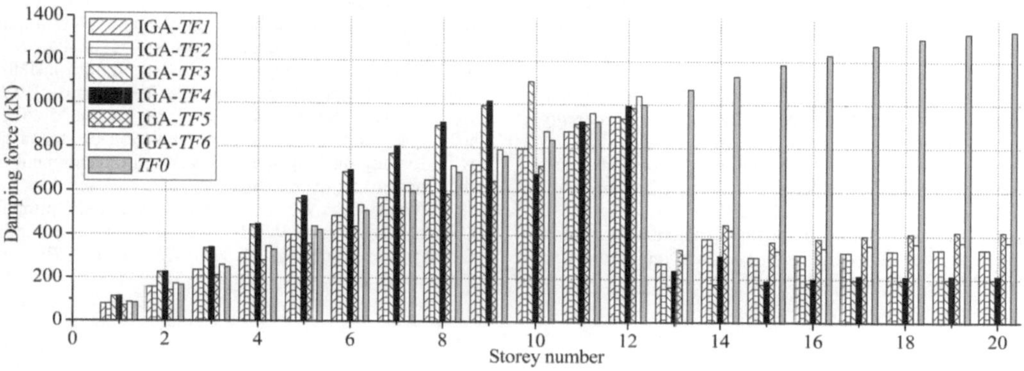

Figure 4. Storey distribution of damping forces.

drift ratio and top acceleration for case "IGA-*TF5*" are 27.61%, 26.11% and 30.36%, but 16.78%, 14.44% and 17.86% for case "GA-*TF5*". The reason is that some disadvantages in GA, such as premature convergence and fraud behavior, could be improved in a certain extent in IGA, which is helpful for the obtaining the optimal results.

Storey distributions of damping force along the height are shown in Figure 4. Similar change tendency of damping force can be found for the six optimization cases, namely damping forces increase gradually from the 1st to 12th story, and mutation occurs at the 13th. However, the damping forces for case "*TF0*" increase gradually along the height and there is not mutation occurs. So, it can be obtained that IGA method can distribute damping coefficients along the height rationally according to the specific structural properties and make full use of energy dissipation capability of dampers.

5 CONCLUSIONS

IGA are introduced and mainly studied in this paper to optimize the structural dynamic responses of high-rise buildings. Some conclusions can be summarized as follows:

1) IGA optimum method can distribute damping coefficients along structural height rationally according to the specific structural properties and make full use of energy dissipation capability of dampers.
2) IGA is better than GA in reducing wind-induced vibration responses after improvement in the cell-efficiency, cross operator and mutation operator.
3) Influence of higher modes is suggested to be considered in computational process of structural dynamic responses in order to obtain reasonable and reliable results, especially for high-rise buildings.
4) There are some differences existing for structural responses when different target functions are used. Both displacement and acceleration responses are advised to be included in target function so as to obtain the best optimal results.

ACKNOWLEDGEMENT

This project is supported by the Doctor Special Fund of Higher School (200810780001) and Natural Science Foundation of Guangdong Province (8351009101000001).

REFERENCES

De Silva, C.W. 1981. An algorithm for the optimal design of passive vibration controllers for flexible systems. *Journal of Sound Vibration* 75(4): 495–502.

Guo, G.S. Gong, D.W. & Hac, G.S. 2006. Interactive Genetic Algorithms with Fitness Adjustment. *Journal of China University of Mining and Technology* 16(4):480–484.

Holland, J.H. 1975. *Adaptation in Natural and Artificial Systems*. Ann Arbor, MI: University of Michigan Press.

Liu, D.H. Feng, Y. Yuan, S.C. & Wang, R. 2009. Optimal design of impact machine for hydraulic roofbolter based on improved genetic algorithm. *J. Xi'an Univ. of Arch. & Tech* 41(1): 121–125.

Li, J.H. Ding, X.Q. & Wang, S.A. 2006. Family genetic algorithms based on gene exchange and its application. *Journal of Systems Engineering and Electronics* 17(4): 864–869.

OU, J.P. Wu, B. & Long, X. 1998. Parameter analysis of Passive energy dissipation systems. *Earthquake Engineering and Engineering Vibration* 18(1): 60–70.

Qu, W. L. 1991. *Wind-Induced Vibration Control Design of High-Rise and Super High-Rise Buildings*. Wuhan: Pressed by Survey Science and Technology University of Wu Han.

Singh, M.P. & Moreschi, L.M. 2001. Optimal seismic response control with dampers. *Earthquake Engineering and Structural Dynamics* 30: 553–572.

Tsuji, M. & Nakamura, T. 1996. Optimum viscous dampers for stiffness design of shear buildings. *The Structural Design of Tall Buildings* 5: 217–234.

Takewaki, I. 1997. Optimal damper placement for minimum transfer functions. *Earthquake Engineering and Structural Dynamics* 26:1113–1124.

Takewaki, I. 1998. Optimal damper positioning in beams for minimum dynamic compliance. *Computer methods in applied mechanics and engineering* 156: 363–373.

Trombetti, T. & Silvestri, S. 2004. Added viscous dampers in shear-type structures: the effectiveness of mass proportional damping. *Journal of Earthquake Engineering* 8(2): 275–313.

Trombetti T. & Silvestri S. 2007. Novel schemes for inserting seismic dampers in shear-typesystems based upon the mass proportional component of the Rayleigh damping matrix. Journal of Sound and Vibration 302(3): 486–526.

Wang, W. Nie, C. & Zhao, W. 2005. Optimization Framework Supporting Integrated Design of Complex Product. *Computer Simulation* 22(11):252–255.

Wang, D.Y. & Zhou, Y. 2008. Wind-induced vibration control research of high-rise buildings based on GA optimum theory. China Civil Engineering Journal 41(1): 42–47.

Zhou, Y. Xu, Z.D. & Deng, X.S. 1998. Optimum installation of the dampers in the viscoelastic structures. *World Information on Earthquake Engineering* 14(3): 15–20.

Zhao, J.F. & Xie, B.Y. 2006. The simulation of stochastic wind loading on long-span bridges and development of the program. *Structural Engineers* 22(2):42–49.

Zhou, Y. Wang, D.Y. Wu, Y. & Deng, X.S. 2008. Optimum Control Research of Wind- Induced Vibration comfort for High-Rise Buildings. *WIND AND STRUCTURES, An International Journal* 11(6): 497–512.

Modelling and Computation in Engineering – Zhu (ed)
© 2011 Taylor & Francis Group, London, ISBN 978-0-415-61516-7

Risk analysis of tunnel shield machine driving-in and driving-out construction process in tunnelling engineering

D.L. Yang & R.J. Zheng
School of Civil Engineering, Tongji University, Shanghai, China

H.Z. Guo
Shenzhen Metro Ltd, Shenzhen, China

ABSTRACT: Based on Zhenzhou Metro M1 line project, against to the tunnelling machine driving-in and driving-out construction process, this paper provides a new risk analysis method, which evaluates risk level of working conditions by the Delphi method and evaluates the weightcoefficient of risk-causing factors in various working conditions by analytical hierarchy process (AHP). It makes the risk analysis to be distinct, focused, simple and practical. The result of risk analysis can be used to optimize shield construction.

1 INTRODUCTION

In China, by using shielded construction in the Huangpu River underwater highroad tunnel project. Since 1966 and with ever-ascending urban subway construction in recent years, the shield construction has been extensively applied in tunnel construction for its merits, such as small impact of surface environment, wide range of application, high indemnificatory of construction quality, etc. But, the characters, as the concealment, complexity and technical difficulties of tunnel engineering, make accidents happen in the shield construction from time to tome, also being with greater probability especially in the process of shield machine driving-in and driving-out construction (Zhao, et al., 2004; Chen, et al., 2006; Cui, 2009). Tunnel construction risk analysis, which is based on the risk management theory to analyze and deal with the construction risk factors, caused the attention of numerous researchers and achieved fruitful achievements (Einstein, 1996; Vik, et al., 2000; Yoo, et al., 2003; Eskesen, et al., 2004; Zhou, et al., 2006; Li, et al., 2009). Nonetheless, specific to the process of the accident-prone shield machine driving-in and driving-out construction, although relevant researchers have studied on it (Zhao, et al., 2004; Chen, et al., 2006; Cui, 2009), the risk analysis theory is not yet perfect, and applied research on new risk analysis technique is still full of theoretical and practical value.

This paper combined with the first phase of Zhengzhou metro line 1, against to the shield tunneling construction stage, and proposed to assess the risk level of operating mode by the Delphi method, to evaluate the weight of risk-causing factors in various operating modes by AHP, to conduct the risk analysis of shield tunneling in construction. It makes the risk analysis to be distinct, focused, simple and practical.

2 RISK IDENTIFICATION

The first phase of Zhengzhou metro line 1 is west Kaixuan road, east the Sports Center Station that stands 26.2 kilometers, for a total of 20 stations. It is through the bustling downtown. Risk analysis in the process of shield machine driving-in and driving-out construction is necessary for the complex underground pipeline, the levity geological conditions, the higher water tables of many sections, and lack of technological experience of subway construction in Zhengzhou. Based on practical shield tunneling construction experience, the risk in the process of shield machine driving-in and driving-out construction mainly exists in the following four working conditions: assembling and hoisting of shield machines; shield excavation; shield arrival; temporary project and equipment dismantlement.

2.1 *Risk identification in hoisting and assembling of shield machines*

As huge equipment, shield machine has complicated structure; after block hoisting of the machine, Workers needs to assemble the parts within the working shaft for quite a while. Anything would cause accidents if the construction was in rainy day and the drainage was not quite ready, or field management was in chaos, or people didn't install the machine in accordance with the drawings, even improper command in equipment installation process, etc. Risk-causing factors in such condition are shown in table 1.

2.2 *Risk identification in shield excavation*

In the condition of shield excavation, the accidents are probable because of the improper or ineffective soil reinforcement, the unsuitable arrangement of construction process, too powerful of the shield thrust, the

Table 1. Risk-causing factors in the process of shield machine driving-in and driving-out construction.

Working conditions	Risk-causing factors
Hoisting and assembling of shield machines	a. Hoisting equipment failure; b. Shield parts clash and breakage; c. Crane overturning; d. Debugging failure; e. Delay
Shield excavation	a. Workers or equipment clash and breakage in removing chamber; b. Water leakage or collapse of working face; c. Sliding instability of tailskin reaction frame; d. Shield posture error; e. Too high soil mass reinforcement intensity; f. Shield support displacement or damage
Shield arrival	a. Workers or equipment clash and breakage in removing chamber; b. Entrance collapse; c. Severe settlement around the entrance; d. Reception scaffold pedestal displacement; e. Foundation collapse
Temporary project and equipment dismantlement	a. Damage to workers caused by chamber fragment falling b. Shield cutter head destruction due to removing chamber by blasting; c. Equipment overturning; d. Damage to shield parts

Table 2. Probability interval of risk occurrence.

Grades	A	B	C	D
Description of the accident probability interval	Impossible $P < 0.01\%$	Infrequence $0.01\% \leq P < 0.1\%$	Accidental $0.1\% \leq P < 1\%$	Possibility $1\% \leq P < 10\%$

Table 3. The standard of risk losses.

Grade	1	2	3	4	5
Description	Negligible	Priority	Serious	Very serious	Disastrous

bad shield posture, the thoughtless fixed way on shield pedestal, too powerful of the reverse (back) thrust, and the amount of excavation soil is more than that of floating soil, etc. Risk-causing factors in such working mode are shown in table 1.

2.3 *Risk identification in shield arrival*

In the working condition of shield arrival, the accidents are likely due to the improper way to dismantle temporary retaining wall, the improper stratum reinforcement, the ineffective waterproof curtain, the inappropriate shield tunneling thrust and speed as the shield reached; and the improper control of shield posture and axis, etc. Risk-causing factors in such working mode are shown in table 1.

2.4 *Risk identification in temporary project and equipment dismantlement*

In the condition of temporary engineering and equipment dismantlement, accidents might be caused by the disordered field management, inappropriate construction procedure and thin safety consciousness of builders, etc. Risk-causing factors in such working mode are shown in table 1.

3 ASSESS THE RISK LEVEL OF WORKING CONDITIONS BY THE DELPHI METHOD

3.1 *Determine risk assessment standard*

In risk assessment, risk acceptance principles must be defined firstly and risk assessment standard shall be ensured to assess the risk. the risk assessment standards of this paper are based on Guidelines of risk management for tunnel engineering of the International Tunneling Association (ITA) and Guidelines of risk management for metro tunneling and underground engineering of the Ministry of Construction of China. According to the probability of risk occurrence, it can be divided into 5 grades as shown in table 2. In tunnel engineering, once risk accident happen, it will bring great damage to the project, third party or peripheral environment. Considering with various degrees of damage, risk losses standard can be set as shown in table 3. According to the different levels of risk and

Table 4. The risk rank evaluation matrix.

Risk		Accident Loss Negligible	Priority	Serious	Very serious	Disastrous
Probability of occurrence	A:P < 0.01%	1A	2A	3A	4A	5A
	B:0.01% ≤ P < 0.1%	1B	2B	3B	4B	5B
	C:0.1% ≤ P < 1%	1C	2C	3C	4C	5C
	D:1% ≤ P < 10%	1D	2D	3D	4D	5D
	E:P ≥ 10%	1E	2E	3E	4E	5E

Table 5. The acceptance criteria of different risk level and the corresponding risk control countermeasures.

Grade	Risk	Acceptance criteria	Control countermeasures
First-level	1A,2A,1B,1C	Negligible	Not nesessary to manage
Second-level	3A,2B,3B,2C,1D,1E	Admissible	Come into notice necessary to routine control
Third-level	4A,5A,4B,3C,2D,2E	Acceptable	Come into notice need to monitor and control
Fourth-level	5B,4C,5C,3D,4D,3E	Unacceptable	Need high-level decision-making Need warning and prevention measures
Fifth-level	5D,4E,5E	Refused	Stop immediately need measures of Making corrections and avoiding

Table 6. Risk evaluation in the process of shield machine driving-in and driving-out construction.

The number	Working conditions	Risk probability of occurrence	Consequential loss	Risk level
1	Shield machines hoisting and assembling	C	3	Third-level
2	Shield excavation	D	3	Fourth-level
3	Shield arrival	D	3	Fourth-level
4	Temporary project and Equipment dismantlement	C	2	Second-level

loss of accident, the risk rank evaluation matrix can be established as shown in table 4. Different risk needs different risk management and control measures. The acceptance criteria of different risk level and the corresponding risk control countermeasures can be set as shown in table 5.

3.2 *Assess the risk level of working conditions by the Delphi method*

Against to the process of shield machine driving-in and driving-out construction, the paper proposed to assess the risk level of working conditions by the Delphi method. Firstly the research team engage ten experts in shield construction to assess the risk level of working conditions in the process of shield machine driving-in and driving-out construction. The assessment of the risk of working conditions is shown in table 6 according to the risk assessment standards of this paper. Hoisting and assembling of shield machines is third-level risk; temporary project and equipment dismantlement risk is second-level; in the working condition of shield excavation and arrival, for the extensive excavation of working shaft wall, it is difficult to guarantee that the excavated surface would be absolutely stable, additionally, the great depth of working shaft and the poor geological condition cause the risk to be very high as forth-level.

4 CALCULATE THE WEIGHTCOEFFICIENT OF RISK-CAUSING FACTORS IN VARIOUS WORKING CONDITIONS BY AHP

Based on AHP, the first is to establish the judgment matrices through the investigation and study, thus realizes the quantitative analysis on the qualitative problems. The AHP can be used to analyze and calculate the weightcoefficient of risk-causing factors in various working modes. The implication of sacle in the judgment matrices is shown in table 7.

In the condition of hoisting and assembling of shield machines, the following judgment matrix A is formulated according to experts judgment on the significance of five risk-causing factors of risk identification.

$$A = \begin{bmatrix} 1 & 1/3 & 1/9 & 1/2 & 1/3 \\ 3 & 1 & 1/7 & 3 & 1 \\ 9 & 7 & 1 & 9 & 7 \\ 2 & 1/3 & 1/9 & 1 & 1/3 \\ 3 & 1 & 1/7 & 3 & 1 \end{bmatrix} \quad (1)$$

Table 7. The implication of sacle in the judgment matrices.

Scale	Implication
1	Signify the same importance weight when Comparing with two factors
3	Signify that the former is lightly more important than the latter when Comparing with two factors.
5	Signify that the former is obviously more important than the latter when Comparing with two factors.
7	Signify that the former is markedly more important than the latter when Comparing with two factors
9	Signify that the former is extremely more important than the latter when Comparing with two factors
2, 4, 6, 8	signify intermediate value between adjacent judgement aforementioned
Reciprocal	If the implication of factor i comparing with j is Aij then the implication of factor j comparing with i is 1/Aij.

The result as follows can be obtained by using colume vector normalization of matrix operation:

$$\begin{bmatrix} 0.06 & 0.03 & 0.07 & 0.03 & 0.03 \\ 0.17 & 0.10 & 0.09 & 0.18 & 0.10 \\ 0.50 & 0.72 & 0.66 & 0.55 & 0.72 \\ 0.11 & 0.03 & 0.07 & 0.06 & 0.03 \\ 0.17 & 0.10 & 0.09 & 0.18 & 0.10 \end{bmatrix}$$

The characteristic vector as follows can also be obtained by summing of row vector:

$$\begin{bmatrix} 0.23 \\ 0.65 \\ 3.16 \\ 0.31 \\ 0.65 \end{bmatrix}$$

The result ω as follows can be obtained by using colume vector normalization of matrix operation:

$$\omega = \begin{bmatrix} 0.04 \\ 0.16 \\ 0.57 \\ 0.08 \\ 0.16 \end{bmatrix} \quad (2)$$

Therefore

$$A\omega = \begin{bmatrix} 0.23 \\ 0.65 \\ 3.16 \\ 0.31 \\ 0.65 \end{bmatrix} \quad (3)$$

The characteristic root can be approximately obtained as follows:

$$\lambda_{\max} = \frac{1}{n}\sum_{j=1}^{m}\frac{\omega_i}{A\omega_i} = 4.66 \quad (4)$$

Then the coincident indicator coincidence indicator CI can be obtained by calculating as follows:

$$CI = \frac{\lambda_{\max} - n}{n-1} = \frac{4.66-5}{5-1} = -0.08 \quad (5)$$

Then the coincidence proportion CR can be obtained by calculating as follows:

$$CR = \frac{CI}{RI} = \frac{-0.08}{1.12} = -0.071 \quad (6)$$

where 1.12 is the mean random consistency index (RI)corresponding the five risk-causing factors based on the method of AHP.

Because $|CR| = 0.071 < 0.1$,The coincidence of judgment matrix A is acceptable.

Therefore

$$\omega = \begin{bmatrix} 0.04 \\ 0.16 \\ 0.57 \\ 0.08 \\ 0.16 \end{bmatrix} \quad (7)$$

It is acceptable to regard ω as the weightcoefficient of risk-causing factors in hoisting and assembling of shield machines.

Similarly, the weightcoefficient of risk-causing factors in other working conditions can be calculated. risk assessment summary combined with the Delphi method is shown in table 8.

According to the table 8, it is effective to grasp the risk level of working conditions in the process of shield machine driving-in and driving-out construction and the weightcoefficient of risk-causing factors not only on the whole, but also taking the mutual relationship of them into account. It makes the risk analysis to be distinct, focused, simple and practical. Construction technicians should have intuitive knowledge of risk level and risk-causing factors in specific construction stage, so it is helpful to master the risks further, to improve the operational of risk prevention, which can effectively reduce blind construction, to accomplish know fairly well and to make the maximum possible to reduce or avoid engineering accidents.

5 CONCLUSION

On the basis of risk Analysis of tunnel shield machine driving-in and driving-out construction process in tunnelling engineering in this paper, shield machinery hoisting and assembling is third-level risk and need to be monitored and controlled; shield excavation and shield arrival are fourth-level risk, and need high-level decision-making and preventive action; temporary project and equipment dismantlement is second-level risk and need routine control.

According to risk analysis in this paper, it is effective to grasp the risk level of working conditions not only

Table 8. Risk evaluation and weight coefficient of risk-causing factors in the process of shield machine driving-in and driving-out construction.

Working conditions	Risk level	The number of risk-causing factors	The weight Coefficient of risk-causing factors
hoisting and assembling of shield machines	third-level	a	0.04
		b	0.16
		c	0.57
		d	0.08
		e	0.16
shield excavation	fourth-level	a	0.04
		b	0.1
		c	0.56
		d	0.05
		e	0.14
		f	0.12
shield arrival	fourth-level	a	0.09
		b	0.6
		c	0.04
		d	0.14
		e	0.12
temporary project and equipment dismantlement	second-level	a	0.23
		b	0.09
		c	0.38
		d	0.3

on the whole, but also taking the mutual relationship of risk-causing factors into account. It makes the risk analysis to be distinct, focused, simple and practical. Construction technicians should have intuitive knowledge of risk level and risk-causing factors in specific construction stage, so it is helpful to master the risks further, to improve the operational of risk prevention and to make the maximum possible to reduce or avoid engineering accidents.

Risk Analysis method in this paper is more perfect, which has the advantage of combining the qualitative analysis of the Delphi method with the quantitative analysis of AHP.

REFERENCES

Chen, L., Huang, H.W., 2006. Risk loss analysis during construction of soft soil shield tunnel.Chinese Journal of Underground Space and Engineering, 2(1):74~78.

Cui, J.J., 2009. Risks and Countermeasures for Construction of Shield Tunnel. Tunnel Construction, 29(4):376~396

Einstein, H.H., 1996. Risk and risk analysis in rock engineering. Tunnelling and Underground Space Technology, 11(2):141~155.

Eskesen, S.D., Tengborg, P., Kampann, J., etal. 2004. Guidelines for tunnelling risk management: International Tunnelling Association. Tunnelling and Underground Space Technology, 19:217~237.

Li, G., Huang. H.W.,2009. Risk analysis on Arriving into Shaft of Super Large Diameter Shield Machine under Water. Chinese Journal of Underground Space and Engineering, 5:1421~1426.

Vik, E.A., Sverdrup, L., Kelley, A. et al., 2000. Experiences from environmental risk management of chemical grouting agents used during construction of the Romeriksporten tunnel. Tunnelling and Underground Space Technology, 15(4):369~378.

Yoo, C., Kim, J. H., 2003. A web-based tunneling-induced building/utility damages assessment system. Tunnelling and Underground Space Technology, 18: 497~511.

Zhao, J., Dai, H.J., 2004. Shield break-in and break-out techniques in soft soil for shield driven tunnel. Chinese Journel of Rock Mechanics Engineering, 7(3):5147~5152.

Zhou, H.B., He, X.X., et al. 2005. Risk identification and handling in the shield tunneling of metro in soft soil. Modern Tunnelling Technology, 43(2):10~14.

Modelling and Computation in Engineering – Zhu (ed)
© 2011 Taylor & Francis Group, London, ISBN 978-0-415-61516-7

Research on CPR parameters impact on hemodynamic effects based on mathematics model

Lin Xu
Department of Math & Information Technology, Hanshan Normal University, Chaozhou, Guangdong, P.R. China

Xiaoming Wu
School of Bioscience and Bioengineering, South China University of Technology, Guangzhou, Guangdong, P.R. China

ABSTRACT: This paper describes quantitative analysis and research for cardiopulmonary resuscitation (CPR) parameters, which are chest compression, chest decompression, lower limbs compression and frequency, in order to determine how various parameters impact on hemodynamic effects. At first, a mathematics model of human circulatory system used to research CPR has been established. Changing one parameter or several parameters of CPR on the model respectively, the hemodynamic effects of CPR to the blood circulatory system model will be performed on the model. Two technical indicators and three hemodynamic parameters are used to analyze, which two technical indicators are systolic maximum and diastolic maximum, and three hemodynamic parameters are aortic pressure, right atrial pressure and coronary perfusion pressure. According to five groups of experiment, an optimal hemodynamic effects will be obtained under the proper parameter values are set.

1 INTRODUCTION

Sudden cardiac arrest is one of the most critical clinical syndromes in all of emergency disease. A cardiopulmonary resuscitation (CPR) is a simple and effective cure means for those sudden cardiac arrest patients, which is to help patients have a relatively effective blood circulation and respiration through compression and ventilation. The CPR can help to restore patients' heartbeat and breathing, and to protect cardiac, brain and other vital organs. Whether hemodynamic effects are good or bad will have a direct impact on the survival rate and prognosis for patients.

Recently, most CPR technologies claim that their methods can up to optimal hemodynamic effects, such as active compression-decompression CPR (ACD-CPR), active compression-decompression CPR with enhanced external counter-pulsation and inspiratory impedance threshold valve (AEI-CPR) and so on (Hengxin 1995). However, because of difficulty of obtaining clinical data and many uncertain factors for patients, it is difficult to analyze quantitatively.

A study of CPR results showed that the quality of CPR was partially determined by frequency, magnitude and time sequence of external force, confirming the belief of Peter Safar (Babbs 2006). Babbs & Eunok Jung (2005) have built a CPR mathematics model, and also researched the optimal waveforms for external chest and abdomen compression and decompression during CPR, on the model.

In this paper, we perform five experiments on an established CPR mathematics model created by group members to study hemodynamic effect when change one CPR control parameter value while others unchanged. There are two technical indicators and three hemodynamic parameters used to analyze. Two technical indicators are systolic maximum and diastolic maximum. Three hemodynamic parameters are aortic pressure, right atrial pressure and coronary perfusion pressure.

According to results, different control parameters cause different trend to the same indicators, as well as changing the same CPR parameter value also causes different trend to various indicators. Therefore, there is always a set of CPR control parameter values can be made to reflect hemodynamic effects achieve optimal status.

2 MATHEMATICAL MODELING

2.1 *Cardiovascular system model for CPR*

To research the hemodynamic effects of CPR, it is necessary to not only model human blood circulatory system but also consider mechanical coupling of the external force on the chest and lower limbs. Figure 1 shows a diagram of mathematical model for CPR. Applying various external pressures on the model, hemodynamic effect of CPR is displayed.

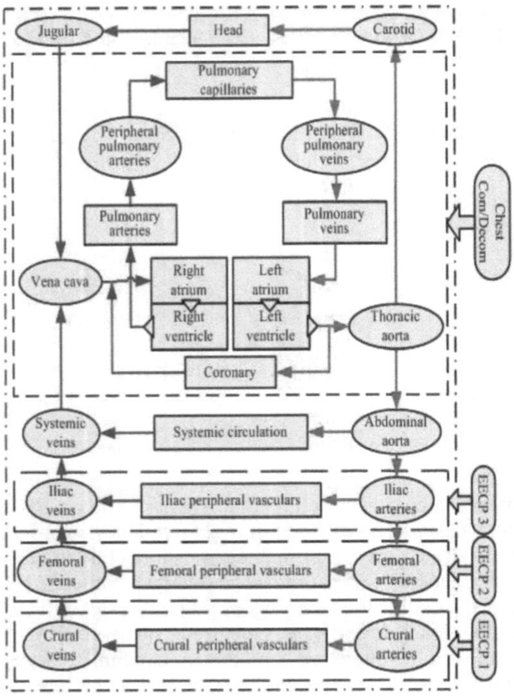

Figure 1. Diagram of cardiovascular system model for CPR.

There are two parts in the model. One part is the blood circulatory system which is contained in the point dotted line rectangle. The blood circulatory system model is based upon human anatomy and physiology. However, a blood circulatory system is very complex and the ideal hypothesis is necessary. Therefore, the model is a lumped model. Another part is external force acting on the chest and lower limbs which are represented by rounded rectangle on the right of diagram. EECP, that is enhanced external counter-pulsation, is a technology which external force imposes on the lower limbs during recoil phase of chest compression to enhance blood regurgitation. EECP can be divided into three parts. EECP1, EECP2 and EECP3 are the compression pressure imposed on the crus, thigh and hip, respectively. Four dashed rectangles indicated control range of external force. This model has been established on the Matlab/Simulink by Yanru (2008).

2.2 *Modeling principle*

The mathematics model is established denoting the heart as a power pump, the ventricles as Frank-Starling time-varying elastic models, the valves as ideal diodes, the vascular system as elastic tubes, the blood flow as an incompressible Newtonian fluid, and the micro-circulation as a lumped system model (Xiaoxiang 2003).

According to Navier-Stokes equation and Ohm's Law, the model is established by an analog circuit.

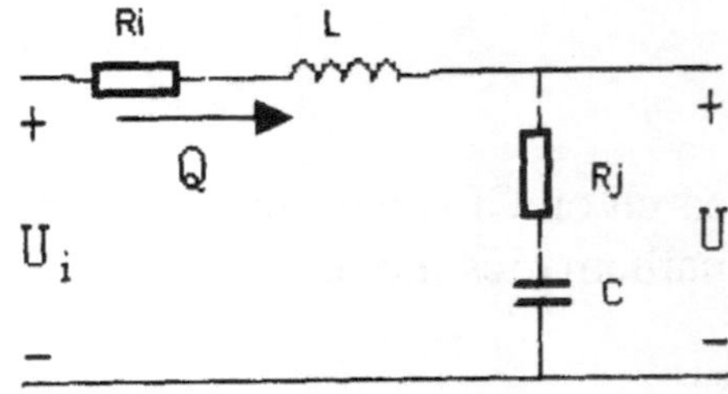

Figure 2. Analog circuit of a vascular compartment.

Table 1. The transform between physiological parameter and engineering parameter.

Fluid parameter	Physiologic variable (Units)	Electrical analog (Units)
Pressure	Pressure difference (mmHg)	Voltage difference *P* (Volts)
Flux	Blood flow (L/min)	Current flow *I* (Amps)
Viscosity	Blood resistance (mmHg/L/min)	Resistance *R* (Ohms)
Elastic Coefficient	Compliance (mL/mmHg)	Capacitance C (Farads)
Inertia	Intertance (g/cm^4)	Inductance L (Henrys)

Figure 2 shows that a vascular compartment would be simulated to a resistor paralleled with a capacitor. Table 1 shows the relationship between the electrical parameters and the physiologic variables.

Corresponding to Figure 1, the mathematics model is built with rules as follow: 1) great vessels, such as thoracic aorta, pulmonary arteries, has more compliance than their resistance, so they were only simulated to capacitors; 2) some small vessels, such as capillary net, has more resistance than their compliance, so they were only simulated to resistances. The Actual model is described by Yanru (2008).

3 EXPERIMENTS

3.1 *Experimental program*

Five experiments are performed on the mathematics model, and each experiment corresponding to a CPR technology. In addition, in each experiment, only one CPR control parameter value is changed at allowed range, and others unchanged. And then, according to the result, we can analyze each parameter quantitatively impacting on hemodynamic effects.

The CPR control parameters include chest compression, chest decompression, lower limb compression, frequency.

3.1.1 *S-CPR*

The first experiment is performing standard CPR (S-CPR) on the model. S-CPR is a CPR technology which compression imposes only on the chest and the ratio compression and relaxation is 1:1. In the experiment, unchanged control parameters are set as follows: 1) compression frequency is $100\,\text{min}^{-1}$; 2) chest decompression and lower limbs compression is 0Nt. The changed parameter is chest compression. The value of chest compression changes from 100 Nt to 500 Nt, step is 50 Nt.

3.1.2 *ACD-CPR*

The second experiment is performing active compression-decompression CPR (ACD-CPR) on the model. ACD-CPR is a CPR technology which compression and decompression impose on the chest and the ratio compression and decompression is 1:1. In the experiment, unchanged control parameters are set as follows: 1) compression frequency is $100\,\text{min}^{-1}$; 2) the value of chest compression is 400 Nt. The changed parameter is chest decompression. The value of chest decompression changes from 0 Nt to 200 Nt, step is 100 Nt.

3.1.3 *Standard CPR coupled with EECP*

The third experiment is performing S-CPR coupled with EECP on the model. In the experiment, unchanged control parameters are set as follows: 1) compression frequency is $100\,\text{min}^{-1}$; 2) the value of chest compression is 400 Nt; 3) the interval of time of EECP is 0.05 s. The changed parameter is lower limbs compression. The value of lower limbs compression changes from 0 mmHg to 500 mmHg, step is 200 mmHg.

3.1.4 *AEI-CPR*

The fourth experiment is performing active compression-decompression CPR with EECP and inspiratory impedance threshold valve (AEI-CPR) on the model. AEI-CPR is a CPR technology, proposed in 1994, to improve the aortic pressure and myocardial perfusion pressure. In the experiment, unchanged control parameters are set as follows: 1) compression frequency is $100\,\text{min}^{-1}$; 2) the value of chest compression is 400 Nt; 3) the value of EECP is 300 mmHg. The changed parameter is chest decompression. The value of chest decompression changes from 0 Nt to 200 Nt, step is 100 Nt.

3.1.5 *Frequency*

The fifth experiment is performing AEI-CPR on the model. In the experiment, unchanged control parameters are set as follows: 1) the value of chest compression is 400 Nt; 2) the value of chest decompression is 200 Nt; 3) the value of lower limbs compression is 300 mmHg. The changed parameter is compression frequency. The value of frequency changes from $60\,\text{min}^{-1}$ to $140\,\text{min}^{-1}$, step is 20.

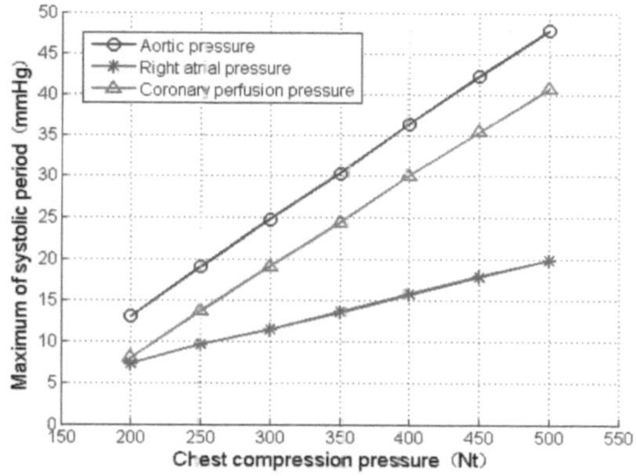

Figure 3. The systolic maximum of the first experiment.

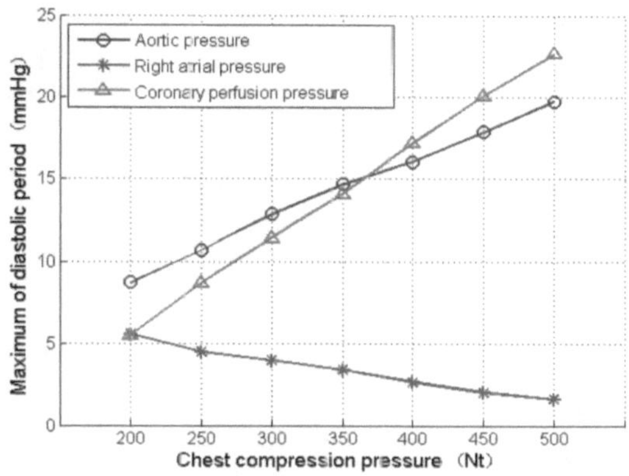

Figure 4. The diastolic maximum of the first experiment.

3.2 *Results*

Figure 3–10 display the trends of systolic and diastolic maximum about aortic pressure, right atrial pressure and coronary perfusion pressure (CPP), respectively, after performing five experiments descript above on the model. In these figures, the curve "–⊖–" expresses thoracic aortic pressure, the curve "–*–" expresses right artial pressure and the curve "–△–" expresses CPP. According to CPR Criterion, the maximal CPP (CPP = thoracic aortic pressure minus right atrial pressure) during diastole is an important indicator to evaluate the quality of CPR techniques. The results of five experiments are described as follows.

Figure 3–4 show the results of the first experiment. From Figure 3–4, it can be seen that the maximal aortic pressure and CPP increase about 5 mmHg and the maximal right atrial pressure increase about 2 mmHg in the systolic per chest compression pressure increased by 50 Nt. This shows that increasing chest compression pressure can effectively improve the systolic aortic pressure and CPP. Although the diastolic aortic pressure's change is not significant (Figure 4), right atrial pressure displays a significantly downward trend. The more compression pressure imposed on the thorax,

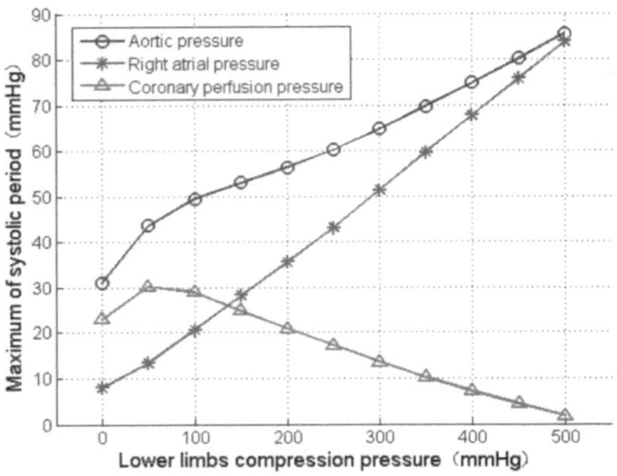

Figure 5. The systolic maximum of the third experiment.

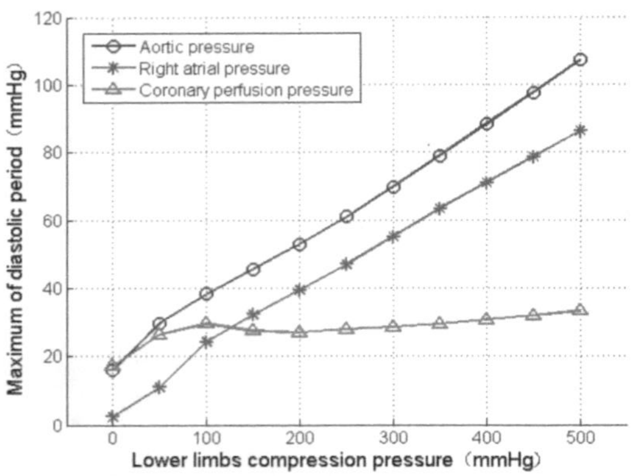

Figure 6. The diastolic maximum of the third experiment.

the greater the downward displacement of thorax, and then the greater the automatic retracted displacement of thorax. Thus, it can make intrathoracic pressure decreased rapidly. Therefore, CPP can be increased by increasing chest compression pressure.

In the second experiment, hemodynamic effects have no significant improvement when change chest decompression pressure.

Figure 5–6 show the results of the third experiment. From Figure 5–6, it can be seen that the maximal aortic pressure and right atrial pressure have significant growth with lower limbs compression pressure increasing. The CPP has small amplitude growth when the EECP force less than 100 mmHg. While the CPP drops sharply when EECP force greater than 100 mmHg. Causing such a result is because addition of artery blood returning back to the heart leads to an increase of thoracic aortic pressure. This shows that appropriate increase in EECP can improve the hemodynamic effects, but too much will be counterproductive. In the other hand, it is necessary to find a method to decrease right artial pressure.

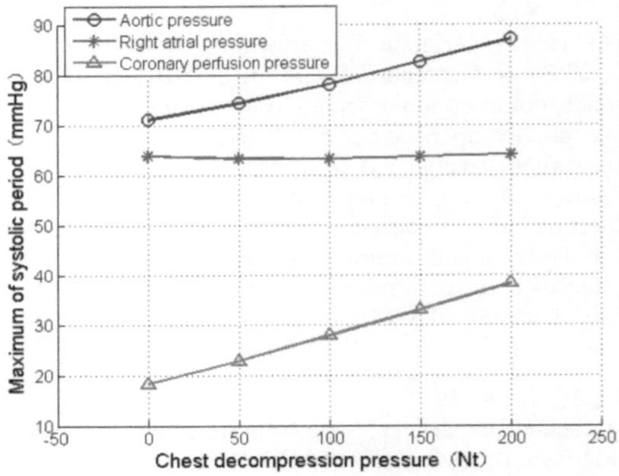

Figure 7. The systolic maximum of the fourth experiment.

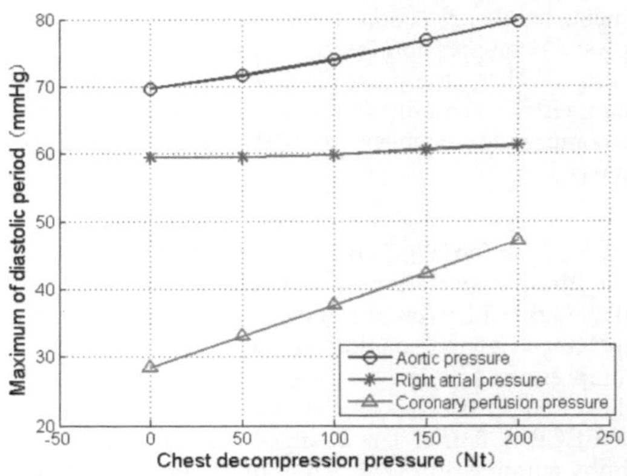

Figure 8. The diastolic maximum of the fourth experiment.

Figure 7–8 show the results of the fourth experiment. From Figure 7–8, it can be seen that AEI-CPR can improve hemodynamic effects effectively. It is because that chest compression can increase aortic pressure and chest decompression can ensure right atrial pressure has no dramatic growth when the lower limbs compression pressure increasing.

Figure 9–10 show the results of the fifth experiment. From Figure 9–10, it can be seen that when frequency is 100/min, the systolic aortic pressure maximum reach maximum, that is 88.139 mmHg. In the systolic, with frequency increases, the thoracic aortic pressure appears downward trend, while frequency impact on right atrial pressure is unobvious. Thus the trend of CPP has the same as aortic pressure. In the diastolic, with the frequency increases, although aortic pressure and right atrial pressure are both decrease slightly, because right atrial pressure decreases much more than aortic pressure, therefore CPP displays increase trend.

According to results of five experiments, the influence of CPR control parameters on hemodynamic effects is listed in Table 2. In the Table 2, S_{Pao}, D_{Pao}, S_{Pra}, D_{Pra}, S_{CPP} and D_{CPP} are indicated systolic aortic

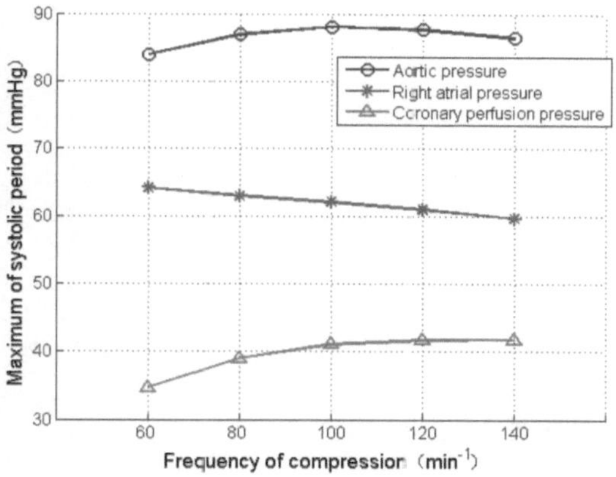

Figure 9. The systolic maximum of the fifth experiment.

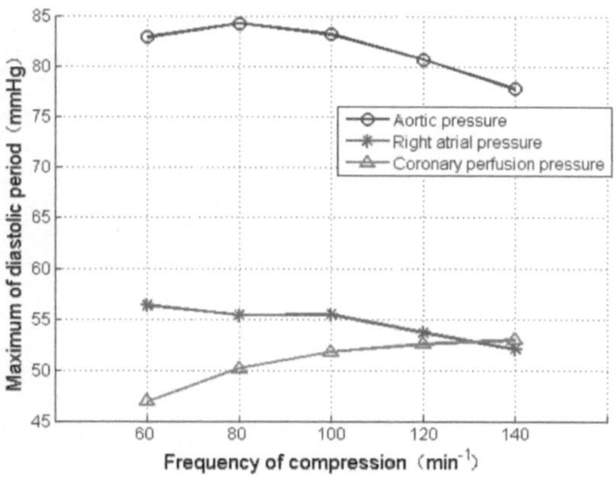

Figure 10. The diastolic maximum of the fifth experiment.

Table 2. The comparative analysis of CPR control parameters impacting on hemodynamic effects.

Parameter	S_{Pao}	D_{Pao}	S_{Pra}	D_{Pra}	S_{CPP}	D_{CPP}
Chest comp	↑	↑	↑	↓	↑	↑
Chest dec	↑	↑	—	↓	↑	↑
EECP	↑	↑	↑	↑	↓	↑↓
Frequency	↑↓	↓	—	↓	↑↓	↑

pressure, diastolic aortic pressure, systolic right atrial pressure, diastolic right atrial pressure, systolic CPP and diastolic CPP, respectively. The tag "↑", "↓" and "-" represents the change trend rise, decline and smooth, respectively.

4 CONCLUSION

From Table 2, two conclusions can be obtained. At first, different control parameters cause different trend to the same indicators. For example, for the diastolic phase of right atrial pressure, an increasing chest compression pressure, chest decompression pressure or frequency can cause it drop down, while an increasing EECP can cause it rise. Secondly, changing each CPR parameter value can also cause different trend to various indicators. For instance, increasing the EECP can cause addition both thoracic aortic pressure and right atrial pressure in the diastolic phase. Summary, there is always a set of CPR control parameter values can be made to reflect hemodynamic effects achieve optimal status.

ACKNOWLEDGEMENT

This paper is supported by projects Hanshan normal university youth science foundation (LQ200907).

REFERENCES

Babbs, C.F. 2006, Design of Near-Optimal Waveforms for Chest and Abdominal Compression and Decompression in CPR Using Computer-Simulated Evolution. *Resuscitation* 68: 277–293.

Hengxin, Yuan & Zhensheng, Zheng 1995, The Hemodynamic Study of the Chest Compression Assisted with Enhanced External Counter-pulsation. *Emergency Medicine*, 4(3): 140–144.

Jung, E. & Babbs, C.F. 2005, Optimal Control Theory Applied to a Difference Equation Model for Cardiopulmonary Resuscitation. *Mathematical Models and Methods in Applied Sciences* 15(10): 1519–1531.

Xiaoming, Wu & Yanru, Zhang 2008, The Computer Model in View of Hemodynamic Effects of Electro Ventilation Double Pump CPR. *APCMBE 2008*: 562–566.

Xiaoxiang, Zheng (ed.) 2003, *Physiological System Simulating and Modeling*. Beijing: Beijing Institute of Technology Press.

Yanru, Zhang 2008, The Hemodynamic Effects Analysis of the New CPR Technique-Electro Ventilation Double Pump CPR£ºStudies in the Computer Model. *ICBBE* 2008: 1761–1764.

Modelling and Computation in Engineering – Zhu (ed)
© 2011 Taylor & Francis Group, London, ISBN 978-0-415-61516-7

Infinite element method for solving open boundary field problem and its application in resitivity well-logging

Zhanghong Tang
College of Material Science and Engineering, Beijing University of Technology, Beijing, China

Jiansheng Yuan
Dept. of Electrical Engineering, Tsinghua University, Beijing, China

Jun Zhu & Weibing Yan
China Petroleum Logging (CPL) Co., Ltd., Xi'an, China

ABSTRACT: An infinite element method (IEM) is introduced in this paper to solve open boundary field problems. The method converts the original infinite field domain into bounded field domain, which can save the memory and CPU time greatly. This algorithm is successfully applied to the 3-D numerical mode matching method (NMM) for resistivity well-logging. Comparing with the original finite element method used in 3-D NMM for solving the well-logging problem, this new approach decreases the total number of nodal points to form the 2-D Finite Element Method (FEM) matrices, which are the matrices of generalized matrix eigenvalue equation used in 3-D NMM. Results show that the new method is almost 7 times faster than the original one.

1 INTRODUCTION

The FEM is widely used in the electromagnetic calculation. For the open boundary problems, the conventional FEM must place an artificial boundary either far away or close to the near field domain. When the boundary is far, a Dirichlet or Neumann boundary condition is applied to the artificial boundary (Fan & Liu 2000). This method is simple, but need a large number of FEM nodes for a given accuracy. Some hybrid methods have been presented when the boundary is close, such as the FEM are used in the interior domain and all the nodes on the surface of the boundary are coupled with the boundary integral method (Pearson et al. 1989). This method can be more accurate with fewer unknowns, however, the coupled matrix of FEM and the boundary integral method is no longer symmetric or sparse and as a result, more solution time is need.

Another efficient method to process open boundary problem is the IEM (Towers et al. 1993), which has two types: one is to define a kind of infinite element for outer unbounded domain and then apply new shape functions for the infinite elements. This method is direct and easy to be merged into current FEM format. However, it can't process exterior domain which contains more than one medium, especially when the interfaces of different mediums extend to infinite. The other one is the method described in this paper which doesn't have such limitation. It divides the whole domain into bounded interior domain and unbounded exterior domain at first, and then transforms the exterior domain to a bounded domain and applies similar FEM to the transformed domain, at last, the total field is the sum of these two parts: the interior domain which is close to and includes the source and the exterior domain is the rest of domain and the final equations are formed by coupling the nodes of interior and exterior domains which are lies in the common boundary of these two domains. This method is especially suitable to the 2-D open boundary field problem. Furthermore, it is easy to be integrated into current FEM package and gives more accurate results.

The NMM method is an efficient algorithm which is widely used in well-logging calculation. When the 2-D distribution of medium is arbitrary in every layer, a 3-D NMM method is introduced in Fan & Liu (2000). This method established and solved the generalized eigenvalue matrices by the 2-D FEM problem of every layer at first, then a series analytical formula are formed in *z* direction to calculate the field of the whole domain. This method is much faster than 3-D FEM since it decomposes the 3-D solution to a series of 2-D eigenvalue solutions. However, the eigenvalue matrix of each layer is formed by general 2-D FEM and a large calculation domain is selected, as a result, a truncate error is not avoidable and large number of FEM nodes are generated.

In this paper, a FEM/IEM based method to solve the 2-D open boundary problem is developed. Compared to the general 2-D FEM, this method can solve the above problem with less unknowns but without truncate error. It is successfully applied to the 3-D NMM method used in resistivity well-logging.

2 2-D FEM/IEM FORMULATIONS

In 3-D NMM method, the correspondent 2-D eigenvalue problem of a layer can be solved by a generalized matrix eigenvalue equation (Fan & Liu 2000)

$$\overline{\mathbf{A}}\cdot\mathbf{v}_j = k_{jz}^2\overline{\mathbf{B}}\cdot\mathbf{v}_j \tag{1}$$

The variables of equation (1) are described in Fan & Liu (2000), and the matrix $\boldsymbol{A}$ and $\boldsymbol{B}$ of (1) is obtained as follows

$$\begin{aligned}\overline{\mathbf{A}} &= \sum_e \iint_e \sigma^e\left(\partial\mathbf{N}/\partial x\cdot(\partial\mathbf{N}/\partial x)^T + \partial\mathbf{N}/\partial y\cdot(\partial\mathbf{N}/\partial y)^T\right)d\Omega \\ \overline{\mathbf{B}} &= \sum_e \iint_e \sigma^e\mathbf{N}\cdot(\mathbf{N})^T d\Omega\end{aligned} \tag{2}$$

where $\mathbf{N} = [N_1, N_2, \ldots]^T$ are the shape function vector of every node which are numbered globally and σ^e is the conductivity of every element.

To calculate the matrices $\boldsymbol{A}$ and $\boldsymbol{B}$, the following formula is considered at first:

$$\begin{aligned}F(\mathbf{N}) = \sum_e \iint_e \Big(\sigma(x,y)\big(\partial\mathbf{N}/\partial x\cdot(\partial\mathbf{N}/\partial x)^T \\ +\partial\mathbf{N}/\partial y\cdot(\partial\mathbf{N}/\partial y)^T\big) + \sigma(x,y)\mathbf{N}\cdot\mathbf{N}^T\Big)d\Omega\end{aligned} \tag{3}$$

Open boundary problems can be solved without boundary truncation by the infinite element method. Firstly, the open domain is divided into interior and exterior parts so the interior part contains the source and fields of principal interest and the remained part is exterior one. Then, an artificial boundary Γ is placed around the inner region Ω_i, which also separates Ω_i from the outer region Ω_o.

Figure. 1 shows an illustration of the transformation. As shown in Fig. 1, after transformation, the outer region Ω_o would be transformed to the transformed region Ω_t which is inside the inner-outer interface Γ of the original domain. After such a transformation, the principal continuity conditions must be satisfied on the interface Γ.

Consider the two regions Ω_i and Ω_o, integral equation (3) can be calculated by the sum of Ω_i and Ω_o. Furthermore, since the inner region can be handled by the standard FEM, only the exterior domain is considered.

For the outer region Ω_o, a coordinate transformation $\Re$ is applied to map it into a finite transformed region Ω_t. Fig. 1 shows the transformation. Because of continuity conditions at the interface Γ, the transformation

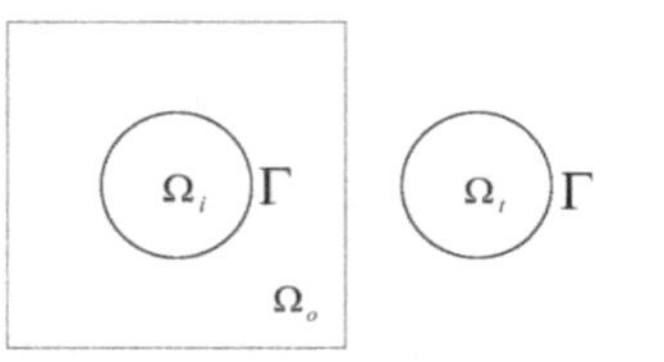

Figure 1. The Kelvin transformation maps the exterior of a circle to the interior of another circle.

requires that the boundary Γ map into itself: $\Re\Gamma = \Gamma$. Under such transformation, the exterior domain of integral equation (3) is transformed to an integral over Ω_t from the coordinate system x, y (original) into x', y' (new). The gradient of every node's shape function N_i should be written in the new coordinate system as follows (Silvester & Ferrari 1996):

$$\begin{bmatrix}\partial N_i/\partial x \\ \partial N_i/\partial y\end{bmatrix} = \begin{bmatrix}\partial x'/\partial x & \partial y'/\partial x \\ \partial x'/\partial y & \partial y'/\partial y\end{bmatrix}\begin{bmatrix}\partial N_i/\partial x' \\ \partial N_i/\partial y'\end{bmatrix} = \mathbf{J}\begin{bmatrix}\partial N_i/\partial x' \\ \partial N_i/\partial y'\end{bmatrix} \tag{4}$$

where N_i is the ith node's shape function when nodes are numbered globally and $\mathbf{J}$ denotes the Jacobian matrix of the transformation. The area integral transforms by the usual rule:

$$d\Omega = dxdy = |\mathbf{J}|^{-1}dx'dy' = |\mathbf{J}|^{-1}d\Omega' \tag{5}$$

The gradient vector of shape functions in (3) can be written as follows according to (4):

$$\begin{aligned}\sigma(x,y)\nabla\mathbf{N}_i\cdot(\nabla\mathbf{N}_j)^T &= \sigma(x',y')(\mathbf{J}\cdot\nabla'\mathbf{N}_i)\cdot(\mathbf{J}\cdot\nabla'\mathbf{N}_j)^T \\ &= \sigma(x',y')(\nabla'\mathbf{N}_i)\cdot\mathbf{J}\cdot\mathbf{J}^T\cdot(\nabla'\mathbf{N}_j)^T\end{aligned} \tag{6}$$

where $\sigma(x',y') = \sigma(\Re(x), \Re(y))$, $\nabla'\mathbf{N}_i = [\partial N_i/\partial x' \partial N_i/\partial y']^T$ are gradient vector of shape functions and i, j are the global numbers of nodes.

According to (5) and (6), the exterior domain of integral equation (3) can be written in the new coordinate system as follows after the transformed domain discretized:

$$\begin{aligned}F_o(\mathbf{N}) = \sum_e^{e\in\Omega_t} \iint_e \Big(\sigma(x',y')\nabla'\mathbf{N}\cdot\mathbf{J}\cdot\mathbf{J}^T\cdot(\nabla'\mathbf{N})^T/|\mathbf{J}| \\ +\sigma(x',y')\mathbf{N}\cdot\mathbf{N}^T/|\mathbf{J}|\Big)d\Omega'\end{aligned} \tag{7}$$

where $\nabla'\mathbf{N} = [(\nabla'\mathbf{N}_1)^T, (\nabla'\mathbf{N}_2)^T, \cdots]^T$ are the gradient matrix of all nodes' shape functions.

By now, the matrices $\mathbf{A}$ and $\mathbf{B}$ in (2) can be calculated by equation (7).

3 KELVIN TRANSFORMATION

The simplest coordinate transformation $\Re$ that satisfies the needs of the finite element analysis is the Kelvin transformation. Silvester & Ferrari (1996) gives the Kelvin transformation under the polar coordinate system. However, it is not easy to model with triangular or quadrilateral elements. To overcome this shortcoming, the equivalent transformation under Cartesian coordinate system is proposed as follows:

$$x' = R^2y/r^2, \quad y' = R^2x/r^2, \qquad r^2 = x^2 + y^2 \ge R^2 \tag{8}$$

(8) shows that by Kelvin transformation, the region $r^2 \geq R^2$ is transformed into the region $(r')^2 = (x')^2 + (y')^2 \leq R^2$.

Under Kelvin transformation, the Jacobian matrix **J** of (4) is as follows:

$$\mathbf{J} = (r'/R)^2 \mathbf{J}^0 \tag{9}$$

where $\mathbf{J}^0 = \frac{1}{(r')^2} \begin{bmatrix} -2x'y' & (x')^2 - (y')^2 \\ (y')^2 - (x')^2 & -2x'y' \end{bmatrix}$.

Since matrix $\mathbf{J}^0$ is orthogonal, equation (7) can be simplified as follows under Kelvin transformation:

$$\begin{aligned} F_o(\mathbf{N}) = \sum_{e}^{e\in\Omega_i} \iint_e \Big(& \sigma(x',y')\nabla'\mathbf{N}\cdot\mathbf{J}\cdot\mathbf{J}^T\cdot(\nabla'\mathbf{N})^T / |\mathbf{J}| \\ & + \sigma(x',y')(R/r')^4\,\mathbf{N}\cdot\mathbf{N}^T / |\mathbf{J}| \Big) d\Omega' \end{aligned} \tag{10}$$

The element stiff matrix of (10) can be calculated by the Gauss integral.

4 APPLICATION OF IEM TO 3-D NMM FOR RESISTIVITY WELL-LOGGING

A typical resistivity well-logging graph is shown in Figure 2. In the figure, the well is vertical and the pole is placed off-axis of the well, which is filled with mud. The ground can be regarded as a horizontally-layered multiregional medium and each layer is modeled as homogeneous in the vertical direction.

Under such medium distribution, a 3-D NMM method is proposed (Fan & Liu 2000), which reduces the original 3-D problem into a series of 2-D eigenvalue problems in x-y plane and an analytical solution in z direction. In x-y plane, the 2-D eigenvalue problem is solved by the 2-D FEM equations in every ground layer, in which the medium is distributed arbitrarily in the whole x-y plane. For such an open boundary 2-D FEM problem, a general method is to place an artificial boundary far away enough to the well region and then set the boundary value to zero (Fan & Liu 2000). This method is simple but need large memory and CPU time for enough precision, especially when the thickness of ground layers and cylindrical poles are very small. In this case more eigenvalues and associated eigenmodes are necessary to give a convergent field solution comparing to the model which only considers point source and large layer thickness (Fan & Liu 2000).

To overcome the above problems, the IEM is applied to create the matrices 2-D eigenvalue equation of every layer used by 3-D NMM. In this application, the whole x-y plane is divided into two regions, which is separated by the boundary of well, because the poles only move inside the well and the interested region is the poles.

In previous work (Chew 1990), only point sources are considered to analyze such electromagnetic field. It did not agree to the working conditions of logging tools which have cylindrical electrodes. Therefore, the Green's function for potential at a point excited by a thin circle electrode is established and the apparent resistivity is obtained by the numerical solution of the integral equations (Nie et al. 1997).

In this paper, a simple and direct method is introduced to process cylindrical electrodes. Since the surface of a conductor can satisfy equipotential condition naturally when it is placed in the field, an alterative method is to act electrodes as a kind of medium with large conductivity (for example, 1.0e + 7 S/m). Under such consideration, the cylindrical current sources can be acted as point sources inside the cylindrical electrodes. Similarly, for the cylindrical insulated material between two electrodes, a very small conductivity (for example, 1.0e−7 S/m) is filled inside it. Basing on these two preprocesses, the 3-D NMM is applied to solve a well logging problem without changing the source to analyze well-logging tool's logging response in a medium with an arbitrary number of horizontal layers when the tool is off-axis in borehole.

For a vertical well with rotational symmetric ground layers, the electrical field distribution can be acted as symmetric to the plane $y = 0$ since the tool can be acted as off-axis along the x axis. As a result, only half of the region needs to be considered in the calculation. Figure 3 shows the meshes of the interior domain (inside the borehole) and transformed domain of exterior domain.

For a horizontal well, the matrices of 2-D eigenvalue equation are established in the whole y-z plane. In other words, the 2-D numerical result is obtained in the y-z plane and the analytical result is obtained in the x direction when applying 3-D NMM, since the medium in every layer is homogenous along x direction. Figure 4 shows the geometry of the well and the correspondent vertical ground layers (in the figure a 3-layer ground model is shown).

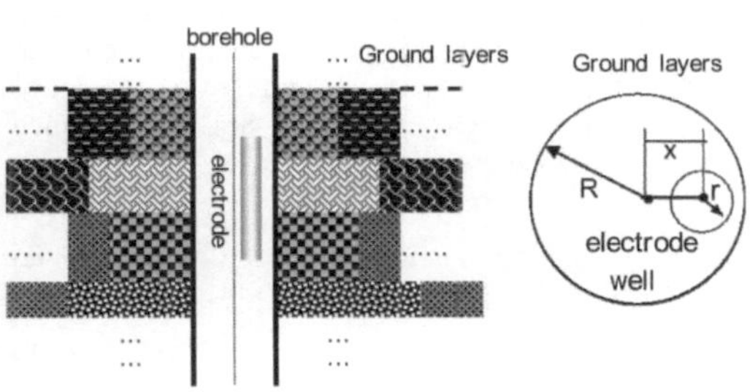

Figure 2. A geometry structure of resistivity well-logging (vertical well), (a) view from x-z plane, (b) view from x-y plane.

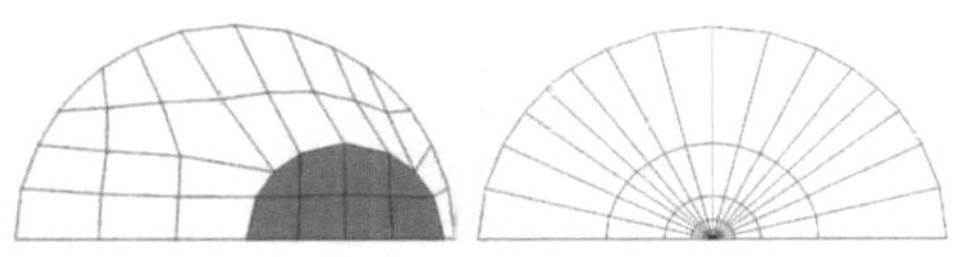

Figure 3. 2-D Meshes of the whole domain when the well is vertical, (a) interior domain together with electrode filled with red color, (b) transformed domain of exterior domain.

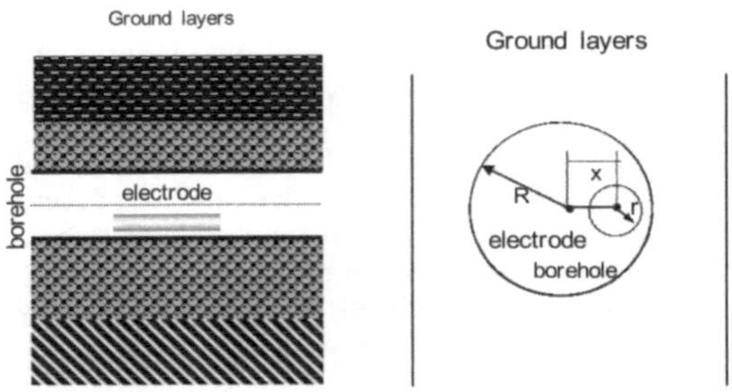

Figure 4. A geometry structure of resistivity well-logging (horizontal well), (a) view from *x-z* plane, (b) view from *y-z* plane.

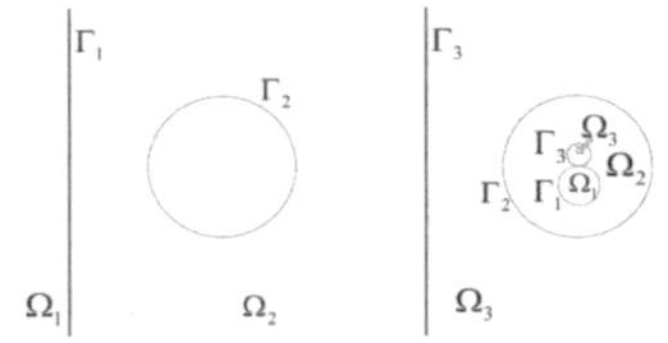

Figure 5. Transform the exterior infinite domain to another finite domain when considering a horizontal well, (a) regions before transformation, (b) regions after transformation.

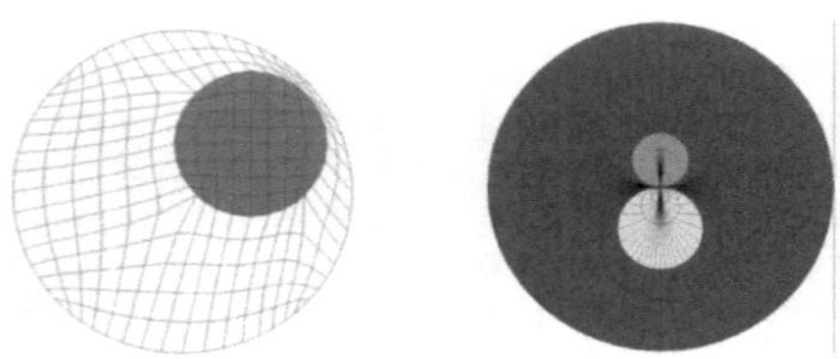

Figure 6. 2-D Meshes of the whole domain when the well is vertical, (a) mesh of interior domain together with electrode filled with red color, (b) mesh of transformed domain of exterior domain as shown in Figure 5.

As shown in Figure 4, when the well is horizontal, the electrical field distribution is no longer symmetric since the location of well and electrodes is arbitrary. In Figure 4(b) or Figure 5(a), three different mediums are separated by two lines, according to the equations (11) or (12) the transformed domain should be separated by two different circles. Figure 5(b) shows the transformed domain.

Figure 6 shows the 2-D mesh of the whole domain (Thompson et al. 1998). Since the electrical field distribution is no longer symmetric when the well is horizontal, the whole region should be considered in mesh generation and numerical calculation.

Note that in Figure 3(b) and Figure 6(b), the mesh near original point should be denser than the mesh near the common boundary of interior and exterior domains because the original point in transformed domain is mapped from the infinite point in original exterior domain.

Based on the meshes shown in Figure 3 or Figure 6, the matrices $\boldsymbol{A}$ and $\boldsymbol{B}$ of the 2-D eigenvalue equation

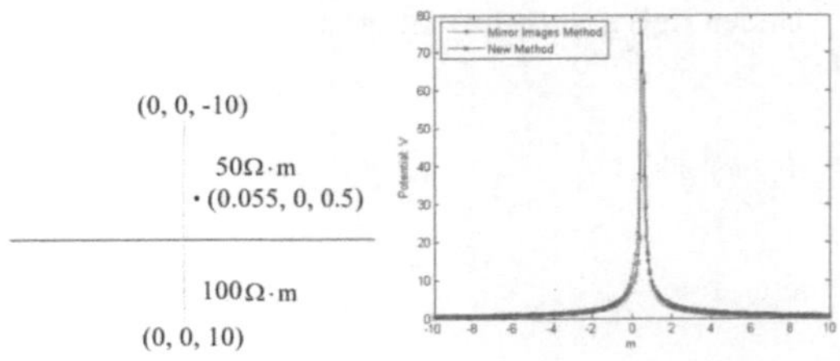

Figure 7. Results comparison, (a) the structure of simple model, (b) comparing of new method results and analytical results.

in every layer can be established according to (12) and the standard FEM formula.

5 NUMERICAL EXAMPLES

The method of FEM/IEM for open boundary problem in this paper is applied to resistivity well-logging calculation successfully. In this paper results of several examples are shown to verify the correct and speed of this method. Without special clarification, when the models are used in the 3-D FEM and general method of 3-D NMM, the truncation range for the calculation is always 100 m far away from the well-logging tools (both radial and portrait).

To verify the correct of this method, the electrical field excited by a point source in a simple two-layer homogenous ground model as shown in Figure 7(a) is calculated, since the field can be solved by the method of mirror images.

Figure 7(b) shows the electrical potential changes from $(0, 0, -10)$ to $(0, 0, 10)$ along z axis when the point source was placed at point (0.055, 0, 0.5). Two results are shown in Fig. 8(b): the method of mirror images (analytical result) and the new method presented in this paper. Results show that these two methods have good agreement.

Figure 8(a) shows a typical group of electrodes' arrange of an array laterlog tool (symmetric about *x-y* plane). By controlling electrical potential balance of electrodes in different working frequencies, the tool can work in different logging modes with different detecting depths. In these modes, electrode A_0 is the main current supply electrode. Figure 8(a) also shows an example of shallow logging mode which means that electrode A_1 is the shielding electrode of A_0 and the current returns to A_2, A_3, A_4, A_5 and A_6.

For the mode shown in Figure 8(a), the balance equations is as follows

$$I_0 = 0.5, \quad \sum_{i=0}^{6} I_i = 0 \tag{11a}$$

$$u_0 = u_1, \quad u_2 = u_3 = u_4 = u_5 = u_6 \tag{11b}$$

Other modes have similar balance equations.

As an example, a layered ground model which is shown in Figure 8(b) is considered in this paper. In the

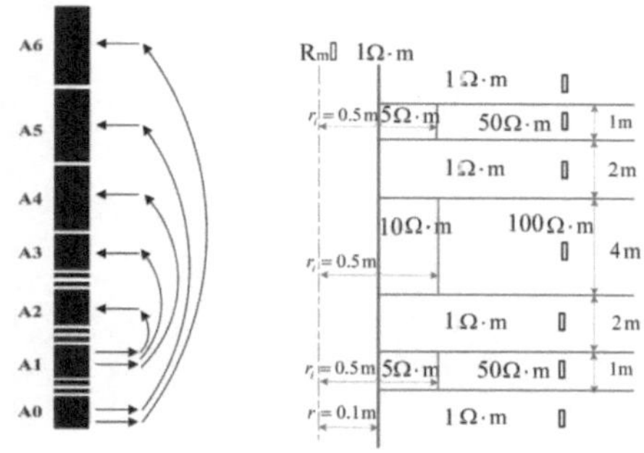

Figure 8. An example application in well-logging, (a) electrodes of array laterlog tool, (b) ground layer model structure.

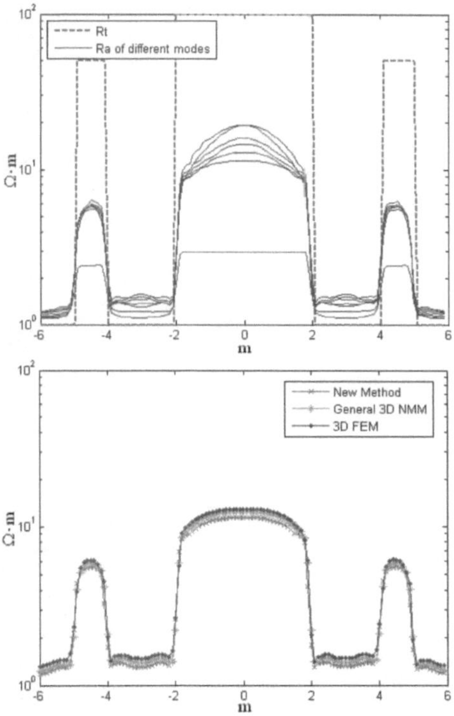

Figure 9. Results comparison (a) aperant resistivity of different logging modes by the method presented in this paper, (b) apperant resistivity of different methods for the same logging mode.

example, the borehole is vertical and its radius is 0.1 m. The tool is off-axis in the borehole and the radius of electrode is 0.045 m.

Figure 9(a) shows the apparent resistivity curves of different logging modes when the tool moves along the z direction (the original point of z is set to the middle of layer 4). In the figure, the blow curve is the result of shallow logging mode and the above curve is the result of deep logging mode. As a comparison, Figure 9(b) shows the results of a mode calculated by general 3-D FEM, general 3-D NMM and the method presented in

Table 1. Speed comparison of different methods.

	3-D FEM	3-D NMM	Method in this paper
Time (s)	475.5	16.5	2.4
Memory (MB)	3,862	236	85

this paper. Results show that these three methods have good agreement.

The comparisons of the speed and memory usage are shown in table 1. Results show that the method presented in this paper is almost 7 times faster than the general 3-D NMM and the usage of memory is only about one third of that used by the general 3-D NMM with almost the same accuracy.

6 CONCLUSIONS

By introducing the Kelvin transformation, the open boundary electromagnetic problem in 2-D can be solved efficiently without adding artificial boundary. In this paper, the Kelvin transformation is also successfully applied to the regions which are inhomogeneous in infinite point. As an application of this method, a 3-D well-logging problem is solved based on the 3-D NMM which was created by the general 2-D FEM previously. Results show that this method is much faster than the original method and saves much memory. This great speed up makes it possible to do real-time inverse work for the 2.5-D well-logging problem.

ACKNOWLEDGMENT

This work was supported by the Doctor Research Fund of School (No. X0009011200905).

REFERENCES

G.-X. Fan and Liu Q. H. 2000. 3-D Numerical Mode-Matching (NMM) Method for Resistivity Well-logging tools. *IEEE Trans. Antennas Propagat.* 48: 1544–1553.

Joe F. Thompson, Bharat K. Soni and Nigel P. Weatherill 1998. *Handbook of Grid Generation*. New York: CRC Press.

L. W. Pearson, R. A. Whitaker, and L. J. Bahrmascl 1989. An exact radiation boundary condition for the finite element solution of electromagnetic scattering on an open domain. *IEEE* Trans. *Map*. 25: 3046–3048.

M. S. Towers, A. McCowen, and J. A. R. Macnab 1993. Electromagnetic Scattering from an Arbitrary, Inhomogeneous 2-D Object-A Finite and Infinite Element Solution. *IEEE Trans. Antennas Propagat.* 41: 770–777.

Nie Zaiping, Yang Fong, Nie Xiaochuen, Yuan Ying and Cheng Shiyuan 1997. An efficient numerical modeling of the DC resistivity logging: theory and applications. *Journal of Electronics*. 14: 169–179.

Peter P. Silvester and Ronald L. Ferrari 1996. *Finite Elements for Electrical Engineers*. Cambridge: Cambridge University Press.

W.C. Chew 1990. *Waves and Fields in Inhomogeneous Media*. New York: Van Nostrand Reinhold.

Modelling and Computation in Engineering – Zhu (ed)
© 2011 Taylor & Francis Group, London, ISBN 978-0-415-61516-7

Author index